Functional Analysis and Economic Theory

Springer
Berlin
Heidelberg
New York
Barcelona
Budapest
Hong Kong
London
Milan
Paris
Santa Clara
Singapore
Tokyo

Y. Abramovich · E. Avgerinos
N. C. Yannelis (Eds.)

Functional Analysis and Economic Theory

With 14 Figures
and 6 Tables

Springer

Professor Yuri Abramovich
IUPUI
Department of Mathematical Sciences
Indianapolis, IN 46202
USA

Dr. Evgenios Avgerinos
University of the Aegean
Department of Education
1 Demokratias Ave.
85100 Rhodes
Greece

Professor Nicholas C. Yannelis
University of Illinois at Urbana Champaign
Department of Economics
Champaign, IL 61801
USA

Library of Congress Cataloging-in-Publication Data
Die Deutsche Bibliothek - CIP-Einheitsaufnahme
Functional analysis and economic theory: with 6 tables / Y. Abramovich ... (ed.). - Berlin; Heidelberg; New York; Barcelona; Budapest; Hong Kong; London; Milan; Paris; Santa Clara; Singapore; Tokyo: Springer, 1998
ISBN-13: 978-3-642-72224-0 e-ISBN-13: 978-3-642-72222-6
DOI: 10.1007/ 978-3-642-72222-6

Softcover reprint of the hardcover 1st edition 1998

Hardcover-Design: Erich Kirchner, Heidelberg

SPIN 10679966 42/2202-5 4 3 2 1 0 - Printed on acid-free paper

Preface

In July of 1996, the conference *Nonlinear Analysis and its Applications in Engineering and Economics* took place on the Greek island of Samos, the birthplace of Pythagoras. During this conference, a special session was held on the occasion of the 50th birthday of the well known mathematician and mathematical economist Professor Charalambos Aliprantis, who, by his numerous friends, is usually called *Roko.* The story behind this nickname is not quite clear yet; it will be investigated further and will be made public prior to his 60th birthday. (At this moment we have already found out that it has nothing to do with the famous movie *Rocco and his Brothers* even though Roko does have two brothers.)

Roko was born on the Greek island of Cephalonia on May 12, 1946, and his elementary and secondary school education took place there. At 18 he entered the Mathematics Department at the University of Athens. Upon graduation from the University of Athens he proceeded with his graduate studies at Caltech, where in 1973 he completed his Ph.D. degree in Mathematics under the supervision of Professor W. A. J. Luxemburg.

His research career can be divided into two periods. The first one, till 1981, was devoted entirely to pure mathematics. The other one, after 1981, has been subdivided between *pure mathematics* and *mathematical economics.*

The main objects of Roko's work in pure mathematics are spaces with order structure (Riesz spaces) and operators acting on them. His books (co-authored with O. Burkinshaw) *Locally Solid Riesz Spaces* and *Positive Operators*, both published by Academic Press, are highly recognized reference books used by researchers all over the world. The knowledge of Riesz spaces turned out to be rather crucial for Roko's further development in mathematical economics.

In 1981 Professor Donald Brown, an economist from Yale University, was visiting Caltech. Brown needed to speak to a mathematician who had a good knowledge of Banach lattices, and Wim Luxemburg suggested to him to talk to Roko, who was also visiting Caltech that year. That is how Roko got involved in Mathematical Economics and started his collaboration with Don. The paper of Aliprantis and Brown, *Equilibria in Markets with a Riesz Space of Commodities*, made it clear for the first time that the theory of Riesz spaces and Banach lattices is the natural setting for general equilibrium and infinite horizon economic models. Moreover, this was the first paper which utilized the idea of the *order ideal* generated by the social endowment, a concept which played a fundamental role subsequently in the work of many economists. By now Roko has written four books and over twenty papers in Economics. In

1990 Roko's presence in this field grew even further when he launched, with the support and encouragement of Springer-Verlag, a new Journal called *Economic Theory*, which has become a leading journal in the rigorous treatment of economic thought.

This volume is coming out at a special moment for Roko. He has just moved from the Department of Mathematical Sciences at IUPUI to Purdue University, where he will hold a joint appointment in the Departments of Economics and Mathematics. Together with his numerous friends we wish him all the best for this new phase of his life.

All those who are familiar with Roko know that he is a dependable, dynamic, decisive, enthusiastic, energetic, generous, honest, hard working, and extremely fair-minded individual. To make this litany of nearly perfect qualities bearable, Roko luckily also has a great sense of humor. We were together with him on a tour when one of the guides at a historic city of Perga (Turkey) mentioned in his speech that *Alexander the Great was not Greek, but rather he was Macedonian.* Roko immediately retorted: *You know, Ronald Reagan is not American; he is Californian*! Everybody laughed and a historical crisis was avoided. We are fortunate to have known Roko for a long time, and we are happy to dedicate this volume to him.

This volume contains contributions from many colleagues and friends who participated in the conference as well as from many who did not manage to attend. The papers are divided into two parts. The first part contains papers in mathematics and the second one contains papers in economics.

We conclude by expressing our thanks to all those who helped organize the conference. For financial support we are very thankful to Professor David Stocum, Dean of the School of Science at IUPUI, to Professor Bart Ng, former Chair of the Department of Mathematical Sciences at IUPUI and to Dr. Werner Müller of Springer-Verlag. Last but not least, we thank A. Zaslavsky for his help in tackling endless TEX problems during our preparation of this volume.

Y. Abramovich
Indianapolis
USA

E. Avgerinos
Rodos
Greece

N. Yannelis
Champaign
USA

Table of Contents

BIJECTIVE DISJOINTNESS PRESERVING OPERATORS

YURI ABRAMOVICH and ARKADY KITOVER

Abstract. A linear operator $T : X \to Y$ between vector lattices is said to be disjointness preserving if T sends disjoint elements in X to disjoint elements in Y. Let $T : X \to Y$ be a bijective disjointness preserving operator, and so the inverse operator T^{-1} exists. In this paper we discuss the most recent results regarding the following problem: when is T^{-1} disjointness preserving? Apart from presenting several counterexamples to this problem we also formulate many sufficient conditions for the affirmative answer to it.

Two elements x_1, x_2 of a vector lattice are called *disjoint* (in notation: $x_1 \perp x_2$) if $|x_1| \wedge |x_2| = 0$. A (linear) operator $T : X \to Y$ between vector lattices is said to be *disjointness preserving* if T sends disjoint elements in X to disjoint elements in Y, that is, $x_1 \perp x_2$ in X implies that $Tx_1 \perp Tx_2$ in Y. In some form or the other disjointness preserving operators appeared in the literature for the first time in early 30-s, but only during the last 15–20 years have they become the object of a systematic study. We mention here only several monographs [2, 8, 15, 16, 18], and a survey [10] in which these operators occupy a prominent role. (We do not even touch here any literature on the spectral properties of disjointness preserving operators.)

One of the reasons for the recent interest in the disjointness preserving operators lies in the fact that it is precisely these operators that allow a multiplicative representation as weighted composition operators; thus the disjointness preserving operators provide an abstract framework for a very important class of operators in analysis. We refer to [1] and [2] for the results in this direction. All our vector lattices are assumed to be Archimedean and considered either over $\mathbb{R}$ or over $\mathbb{C}$. We refer to [6, 17, 18] for all necessary terminology regarding vector and Banach lattices, and operators on them.

1. Main problem and related background

In this paper, we will concern ourselves with the internal problems regarding the disjointness preserving operators. The following question was posed by the first named author several years ago.

Problem A. *Let X, Y be two arbitrary vector lattices and $T : X \to Y$ be a disjointness preserving bijection. Is it true that the inverse operator T^{-1} is also disjointness preserving?*

On the one hand, the question is very simple as it addresses the very basic structure of the operators preserving disjointness but, on the other hand, it has turned out to be rather stimulating and fruitful, having generated a number of results devoted to it.

To some extent, Problem A was motivated by the following result which, implicitly, is in [4, 5] and which solves this problem for a special class of disjointness preserving operators on Banach lattices. Recall that an operator $T : X \to X$ on a vector lattice is *band preserving* if $x_1 \perp x_2$ implies that $Tx_1 \perp x_2$.

1.1. Theorem. *Let X be a Banach lattice and T be an invertible band preserving operator from X onto X. Then the inverse operator T^{-1} is also band preserving.*

The first considerable progress regarding Problem A was made in 1990 by K. Jarosz [13] who answered it in the affirmative in the case when X and Y are the classical Banach lattices of continuous functions on compact Hausdorff spaces.

A few years later a much stronger result was proved by C. Huijsmans–B. de Pagter [11] and independently by A. Koldunov [14]. To formulate their theorem, recall that a vector lattice X is said to be relatively uniformly complete ((r_u)–complete, in short) if each principal ideal in X is order isomorphic to the $C(K)$ space for some compact Hausdorff space K. The most important examples of (r_u)–complete vector lattices are Banach lattices and Dedekind complete vector lattices.

1.2. Theorem. *If X is an (r_u)–complete vector lattice and Y is a normed vector lattice, and $T : X \to Y$ is a bijective disjointness preserving operator, then T^{-1} is also disjointness preserving. Moreover, T is a regular operator.*

The list of the known facts regarding Problem A can be easily continued. We will confine ourselves to three more results. Each of them describes some special case with an affirmative solution to Problem A.

In [11] an affirmative answer is obtained in the case when X is a discrete vector lattice (and Y is an arbitrary vector lattice).

C. Huijsmans and A. Wickstead [12] proved that the conclusion of Theorem 1.1 remains true if the vector lattice X either is (r_u)–complete or has the principal projection property.

Very recently, J. Araujo, E. Beckenstein, and L. Narici [7] proved that the answer to Problem A is affirmative if $X = C(S_1)$ and $Y = C(S_2)$, where S_1, S_2 are Tykhonoff spaces and additionally either S_1 is zero–dimensional, or S_2 is connected, or S_2 is pseudocompact. The last case is, of course, an improvement of Jarosz's result but, on the other hand, it is just a special case of Theorem 1.2.

2. General answer

In spite of all the affirmative results mentioned in the previous section, Problem A has a negative solution.

2.1. Theorem. *There exists a normed vector lattice X and an injective disjointness preserving operator T from X onto itself such that the operator T^{-1} is not disjointness preserving.*

The proof of this theorem can be found in [3]. From Theorem 2.1 it follows that the condition of the (r_u)–completeness in the domain vector lattice X in Theorem 1.2 is essential. The vector lattice constructed in Theorem 2.1 is not Dedekind complete and, therefore, this leaves Problem A open for the most important class of vector lattices, namely the Dedekind complete vector lattices. However, Theorem 3.1 below provides a counterexample for this case as well. This shows that the condition of the (r_u)–completeness alone is not enough either.

In view of Theorem 2.1, Problem A should be slightly modified. Specifically, we are interested now in finding either the weakest possible conditions or various sufficient conditions under which Problem A has an affirmative solution. Some of the new sufficient conditions will be given later.

3. Further counterexamples

Here we will continue discussing Problem A and present additional counterexamples that describe rather clearly the "boundaries" beyond which Problem A has a negative solution. Recall that $L_0(0,1)$ denotes the Dedekind complete vector lattice of all (equivalence classes of) Lebesgue measurable functions on the interval $(0,1)$.

3.1. Theorem. *There exists a bijective disjointness preserving operator T on $L_0(0,1)$ such that the operator T^{-1} is not disjointness preserving.*

The conclusion of Theorem 3.1 remains true for a large class of universally complete vector lattices including the spaces $L_0(\mu)$ for any non-pathological atomless measure μ. The proof is based on a delicate work with d–bases, a concept originally used in this context in [4, 5].

It follows from Theorems 2.1 and 3.1 that both hypotheses on X and Y in Theorem 1.2 are essential. Moreover, even if X (respectively Y) belongs to the class of the very well–behaved Dedekind complete Banach lattices $L_p(0,1)$, $1 \leq p \leq \infty$, this is not enough to force an affirmative solution of Problem A for an arbitrary Y (respectively X).

An element u in a vector lattice X is a *component* of another element $x \in X$ if $u \perp x - u$. A component u of x is nontrivial if $0 < u < x$.

3.2. Definition. *A vector lattice X is said to be essentially one–dimensional if for any two non-disjoint elements x_1 and x_2 in X there are nonzero components u_i of x_i $(i = 1, 2)$ that are linearly dependent.*

The condition of the previous definition, though rather restrictive, is satisfied in a number of interesting cases.

1. X is a discrete vector lattice.

2. $X = C(\beta\mathbb{N}\setminus\mathbb{N})$ and we assume the continuum hypothesis, (CH). Indeed, then the set of all P–points is dense in $\beta\mathbb{N} \setminus \mathbb{N}$.

3. Quite remarkably, A. Gutman showed in [9] that there are examples of atomless essentially one–dimensional Dedekind complete vector lattices. Clearly, such vector lattices are very rare objects, and the majority of nonatomic vector lattices are not essentially one–dimensional. In particular, the classical spaces $L_p(0,1)$ are not essentially one–dimensional.

3.3. Theorem. *Let W be a nonatomic universally complete vector lattice which is not essentially one–dimensional. Let X be an arbitrary nonzero ideal in W. Then there is a vector sublattice Y of W and an injective disjointness preserving operator T from X onto Y such that the operator T^{-1} is not disjointness preserving.*

The conditions of Theorem 3.3 are, of course, satisfied in the case of spaces $L_p(0,1)$ that are ideals in their universal completion $W = L_0(0,1)$. Our final example shows that no matter how good an atomless Dedekind complete range space Y is, we can always find a domain vector lattice and a disjointness preserving operator on it for which Problem A has a negative solution.

3.4. Theorem. *Let Y be a nonzero ideal in $L_0(0,1)$. Then there is a vector sublattice X of $L_0(0,1)$, and an injective disjointness preserving operator T from X onto Y, such that the operator T^{-1} is not disjointness preserving.*

4. An order isomorphism problem

In all instances of having had an affirmative solution to Problem A we have been able to prove additionally a very surprising fact that the vector lattices in question are order isomorphic. This has lead us to the following problem.

Problem B. *Let X and Y be vector lattices and $T : X \to Y$ be a disjointness preserving bijection such that T^{-1} is also disjointness preserving. Is it true that X and Y are order isomorphic?*

This problem sounds as simple as Problem A does at first glance, but it is just as difficult and, in fact, in this generality, has a negative solution.

4.1. Theorem. *Let Y be a nonzero ideal in $L_0(0,1)$. Then there exists a vector sublattice X of $L_0(0,1)$, and an injective disjointness preserving operator T from X onto Y, such that the operator T^{-1} is also disjointness preserving but the vector lattices X and Y are not order isomorphic.*

The vector lattice X produced in the process of proving Theorem 4.1 is not Dedekind complete. For Dedekind complete vector lattices, Problem B is still open.

Now we are ready to present several sufficient conditions under which both Problems A and B can be solved in the affirmative. In the first two results the restrictions on one of the vector lattices are so strong that there is no need to impose any condition on the other.

4.2. Theorem. *Let X be a vector lattice that is essentially one–dimensional. Let T be an injective disjointness preserving operator from X onto a vector lattice Y. Then the operator T^{-1} is also disjointness preserving, both T and T^{-1} are regular, and the vector lattices X and Y are order isomorphic.*

Theorem 4.2 contains as a special case the case of discrete vector lattices mentioned earlier. Our next result shows that an analogue of this result holds if the range Y is a discrete vector lattice (and X is arbitrary).

4.3. Theorem. *Let X be an arbitrary vector lattice, Y be a discrete vector lattice, and T be an injective disjointness preserving operator from X onto Y. Then the operator T^{-1} is also disjointness preserving, both T and T^{-1} are regular, and the vector lattices X and Y are order isomorphic.*

We have a rather meager supply of vector lattices for which the conditions on a single vector lattice are enough to have affirmative solutions to Problems A and B (for example, essentially one–dimensional vector lattices for X and discrete vector lattices for Y). And certainly we do not claim that Theorems 4.2 and 4.3 characterize such vector lattices; there is a remote possibility that the other examples may exist among vector lattices $C(K)$.

5. New information about the inverse

We have not had yet a chance to say something special about the class of Banach lattices, which is, by far, the most important class. But our next theorem does indicate some very special property of this class. Even though the inverse of a disjointness preserving operator on such spaces may fail to preserve disjointness, nevertheless, there appears a "mysterious" constant showing that this inverse is very "close" to preserving disjointness. The corresponding result, Theorem 5.3, together with Theorem 5.4 are two main results in this paper. To state the former we need one more definition, which is meant to introduce some quantitative gauge to measure how far T^{-1} is from preserving disjointness.

5.1. Definition. *Let $T : X \to Y$ be a disjointness preserving operator between vector lattices. We will write that $d(T) = d(T, X, Y) \leq n$ for some $n \in \mathbf{N}$ if from the conditions*

$$\bigwedge_{i=1}^{m} |x_i| > 0, \ x_i \in X \quad \text{and} \quad Tx_i \perp Tx_j, \ i \neq j$$

it follows that $m \leq n$.

The meaning of the expressions $d(T) = n$, $d(T) \geq n$ and $d(T) = \infty$ is self-evident. Clearly, the inverse operator T^{-1} is disjointness preserving if and only if $d(T) = 1$. So, the smaller is the integer $d(T)$ the "closer" is the operator T to having a disjointness preserving inverse. We need one more definition.

5.2. Definition. *A vector lattice X has a* ***cofinal family of band–projections*** *if for each nonzero band B in X there is a nonzero band–projection P with its range in B.*

Typical examples of vector lattices with a cofinal family of band–projections are vector lattices with the principal projection property (in particular, the Dedekind σ–complete vector lattices).

5.3. Theorem. *Let X be a Banach lattice with a cofinal family of band-projections and let Y be an arbitrary vector lattice. If $T : X \to Y$ is a bijective disjointness preserving operator, then $d(T) \leq 2$.*

An interesting and very difficult question still left open in connection with Theorem 1.2 is as follows. *Does the conclusion of that theorem remain true if we exchange the conditions imposed upon X and Y between these vector lattices, that is, if we assume that X is a normed vector lattice and that Y is (r_u)–complete?* The following theorem is the strongest result we have proved in this direction.

5.4. Theorem. *Let X be a normed vector lattice with a cofinal family of band–projections and let Y be an (r_u)–complete vector lattice. If $T : X \to Y$ is a bijective disjointness preserving operator, then the operator T^{-1} is also disjointness preserving.*

Under the conditions of Theorem 5.4, Problem B also has an affirmative solution. Indeed, by Theorem 5.4, $T^{-1} : Y \to X$ is a disjointness preserving bijection from an (r_u)–complete vector lattice onto a normed vector lattice, and so by Theorem 1.2 the operator T^{-1} is regular. Therefore, in view of [1], this operator T^{-1} allows a multiplicative representation. This easily implies that X and Y are order isomorphic.

Though we believe that the "cofinal" condition in Theorem 5.4 is nonessential, our proof depends on it. It has taken a completely different technique to prove the following supplement to Theorem 5.4, which is free of this extra assumption.

5.5. Theorem. *Let $X = C[0,1]$ and let T be a bijective disjointness preserving operator from X onto an (r_u)–complete vector lattice Y. Then the operator T^{-1} is also disjointness preserving, the operator T is regular, and the vector lattices X and Y are order isomorphic.*

An extended version of this work with the detailed proofs will be published elsewhere.

References

1. Y. A. Abramovich, Multiplicative representation of operators preserving disjointness, *Netherl. Acad. Wetensch. Proc. Ser. A* 86 (1983), 265–279.
2. Y. A. Abramovich, E. L. Arenson, and A. K. Kitover, *Banach $C(K)$-modules and operators preserving disjointness*, Pitman Research Notes in Mathematical Series #277, Longman Scientific & Technical, 1992.
3. Y. A. Abramovich and A. K. Kitover, A solution to a problem on invertible disjointness preserving operators, *Proc. Amer. Math. Soc.* 126 (1998), 1501–1505.
4. Y. A. Abramovich, A. I. Veksler, and A. V. Koldunov, Operators preserving disjointness, *Dokl. Akad. Nauk USSR* 248 (1979), 1033–1036.
5. Y. A. Abramovich, A. I. Veksler, and A. V. Koldunov, Operators preserving disjointness, their continuity and multiplicative representation, *Linear operators and their appl. Sbornik Nauchn. Trudov, Leningrad* (1981), 13–34.
6. C. D. Aliprantis and O. Burkinshaw, *Positive Operators*, Academic Press, New York & London, 1985.
7. J. Araujo, E. Beckenstein, and L. Narici, When is a separating map biseparating?, *Archiv der Math.* 67 (1996), 395–407.

8. A. V. Bukhvalov, A. E. Gutman, V. B. Korotkov, A. G. Kusraev, S. S. Kutateladze, and B. M. Makarov, *Vector lattices and integral operators*, Mathematics and its Applications, #358, Kluwer, Dordrecht, 1996 (Translated from the 1992 Russian original).

9. A. Gutman, Locally one–dimensional K–spaces and σ–distributive Boolean algebras, *Sibirian Adv. Math.* 5 (1995), 99–121.

10. C. B. Huijsmans, Disjointness preserving operators on Banach lattices, *Operator Theory in Function Spaces and Banach Lattices*, Operator Theory Advances and Applications 75 (1995), 173–189.

11. C. B. Huijsmans and B. de Pagter, Invertible disjointness preserving operators, *Proc. Edinburgh. Math. Soc.* (2) 37 (1993), 125–132.

12. C. B. Huijsmans and A. W. Wickstead, The inverse of band preserving and disjointness preserving operators, *Indag. Math.* 3 (1992), 179–183.

13. K. Jarosz, Automatic continuity of separating linear isomorphisms, *Canad. Math. Bull.* 33 (1990), 139–144.

14. A. V. Koldunov, Hammerstein operators preserving disjointness, *Proc. Amer. Math. Soc.* 123 (1995), 1083–1095.

15. P. Meyer-Nieberg, *Banach Lattices*, Springer, Berlin Heidelberg New York, 1991.

16. H. H. Schaefer, *Banach Lattices and Positive Operators*, Springer, Berlin Heidelberg New York, 1974.

17. B. Z. Vulikh, *Introduction to the theory of partially ordered spaces*, Wolters–Noordhoff Sci. Publication, Groningen, 1967.

18. A. C. Zaanen, *Riesz Spaces* II, North–Holland, Amsterdam, 1983.

Yuri Abramovich
Department of Mathematical Sciences
IUPUI
Indianapolis, IN 46260
USA
email: yabramovich@math.iupui.edu

Arkady Kitover
Department of Mathematics
CCP
Philadelphia, PA 19130
USA
email: akitover@ccp.cc.pa.us

A REMARK ON POSITIVE ISOTROPIC RANDOM VECTORS*

ALVARO ARIAS and ALEXANDER KOLDOBSKY

Abstract. A random vector $X = (X_1, \ldots, X_n)$ is positive isotropic if $P(X_1 > 0) > 0$, its coordinates are nonnegative and identically distributed random variables, and there exists a function $c : \mathbb{R}^n_+ \to \mathbb{R}_+$ so that for every $a \in \mathbb{R}^n_+ \setminus \{0\}$, the random variables $\sum a_i X_i$ and $c(a)X_1$ are identically distributed. We study the properties of the function $c(\cdot)$, and prove that $c(\cdot)$ cannot be a norm unless the coordinates of X are equal with probability 1.

1. Introduction

It is well known that for every $q \in (0,1)$ there exist q-stable random vectors with nonnegative coordinates. The classical example is given by the measure μ_q on $\mathbb{R}^n_+ = \{x \in \mathbb{R}^n : x_i \geq 0, \; i = 1, \ldots, n\}$, whose Laplace transform is equal to $\exp(|x_1|^q + \cdots + |x_n|^q)$ (see [3, 16]). Any random vector $(X_1, \ldots, X_n)$ with joint distribution equal to μ_q has the property that for every $a = (a_1, \ldots, a_n) \in \mathbb{R}^n_+$, the random variable $a_1X_1 + \cdots + a_nX_n$ has the same distribution as $\|a\|_q X_1$, where $\|a\|_q = (|a_1|^q + \cdots + |a_n|^q)^{\frac{1}{q}}$. Geometrically, this means that one–dimensional projections of μ_q are equal up to a constant parameter.

In this note, we consider a more general class of random vectors:

1.1. Definition. *Let $X_1, \ldots, X_n$ be identically distributed nonnegative (with probability 1) random variables with $P(X_1 \neq 0) > 0$. A random vector $X = (X_1, \ldots, X_n)$ is positive isotropic if there exists a function $c : \mathbb{R}^n_+ \to \mathbb{R}_+$ so that for every $a \in \mathbb{R}^n_+$, the random variables $(a, X) = \sum a_i X_i$ and $c(a)X_1$ are identically distributed. We call $c(\cdot)$ the norming functional of X.*

There are several examples of norming functionals $c(\cdot)$. In addition to the example above, it is known that if $f_1, \ldots, f_n \in L_q$ $(0 < q < 1)$ are nonnegative functions of norm one, then the functional $c(a) = \|a_1 f_1 + \cdots + a_n f_n\|_{L_q}$ is the norming functional of a positive isotropic random vector in $\mathbb{R}^n$ (see Proposition 2.7 below). Let us mention here that positive isotropic random vectors were used to construct an isometric embedding of the space L_r in $L_p(L_q)$, where $p \leq r \leq q$ (see [12, 11]).

If we drop the condition that the X_i's are nonnegative and allow a to be an arbitrary vector from $\mathbb{R}^n$, we get the definition of isotropic random vectors introduced by Eaton [2]. The study of isotropic random vectors has a long history and is closely connected with different problems of probability theory

*Research supported in part by the NSF Grants DMS 9321369 and DMS 9531594.

and functional analysis (see surveys [5, 9]). One of the main problems in this area is to characterize those functions $c(\cdot)$ that can appear as norming functionals of isotropic vectors. This problem is related to the study of positive definite functions, because $c(\cdot)$ is the norming functional of an isotropic random vector if and only if there exists a function f on $\mathbb{R}$ so that $f(c(x))$ is a continuous positive definite function on $\mathbb{R}^n$ (i.e., it is the Fourier transform of a finite measure on $\mathbb{R}^n$). It appears to be quite difficult to check positive definiteness of functions of this type. For example, the question of whether the function $\exp(-\|x\|_q^\beta)$ is positive definite for any $q > 2,\ \beta > 0,\ n > 2$ was posed by Schoenberg in 1938 [13] and remained open for more than fifty years. The final answer was given in [8] (for $q = \infty$) and in [4] (for $2 < q < \infty$), where it was shown that these functions were not positive definite. In fact, [8, 7, 15, 14] also show that the norms of the spaces $l_q^n,\ q > 2,\ n > 2$, cannot appear as norming functionals of any isotropic vectors. Note that classical results of P. Levy [6] and Schoenberg [13] imply that the functions $\exp(-\|x\|_q^\beta)$ are positive definite for every $0 < q \leq 2$ and $0 < \beta \leq q$. Furthermore, these functions are norming functionals of stable vectors.

We show below that the "positive" versions of the latter questions are much easier. Namely, we prove that the only norm which can appear as the function $c(\cdot)$ in Definition 1.1 is the l_1^n–norm, and only when the coordinates of X are equal to each other with probability 1.

2. Properties of norming functionals

Let us first prove that norming functionals are continuous.

2.1. Proposition. *The function* $c : \mathbb{R}_+^n \to \mathbb{R}_+$ *appearing in Definition* 1.1 *has the following properties*:

(i) *for every* $a \in \mathbb{R}_+^n$ *and* $k > 0$, $c(ka) = kc(a)$;

(ii) $c(a) = 0$ *if and only if* $a = 0$;

(iii) $c(e_i) = 1$ *for each* $i \leq n$, *where the* e_i*'s are the standard basis in* $\mathbb{R}^n$;

(iv) $c(\cdot)$ *is continuous on* $\mathbb{R}_+^n$.

Proof. All properties follow from the condition that $P(X_1 \neq 0) > 0$. The proofs for (i), (ii), and (iii) are immediate. We will check (iv). Let $a^k = (a_1^k, \ldots, a_n^k) \in \mathbb{R}_+^n$ be a sequence converging to $a = (a_1, \ldots, a_n) \in \mathbb{R}_+^n$. Since the Laplace transform of the joint distribution of $(X_1, X_2, \ldots, X_n)$ is continuous, we have

$$\mathbb{E}e^{-(a^k,X)} = \mathbb{E}e^{-a_1^k X_1 - \cdots - a_n^k X_n} \longrightarrow \mathbb{E}e^{-a_1 X_1 - \cdots - a_n X_n} = \mathbb{E}e^{-(a,X)}.$$

By the assumption, we get that

$$\mathbb{E}e^{-c(a^k)X_1} \longrightarrow \mathbb{E}e^{-c(a)X_1}.$$

Since the Laplace transform of the distribution of X_1 is strictly decreasing, we conclude that $c(a^k) \to c(a)$. ■

The next fact establishes the connection between positive isotropic vectors and the Laplace transform. Compare Proposition 2.2 with Proposition 1 from [5] to see that the Laplace transform replaces the Fourier transform in the corresponding property of isotropic random vectors.

2.2. Proposition. *The vector X is positive isotropic with norming functional $c(\cdot)$ if and only if the Laplace transform of the measure μ has the form $\mathcal{L}\mu(x) = f(c(x))$, where $f : \mathbb{R}_+ \to \mathbb{R}_+$ is a non-constant function.*

Proof. For every $a \in \mathbb{R}^n_+$, $a \neq 0$, denote by ν_a the probability measure on $\mathbb{R}_+$ which is the image of the measure μ under the mapping $x \mapsto (a,x)/c(a)$ from $\mathbb{R}^n_+$ to $\mathbb{R}_+$.

For every positive number k and every $a \in \mathbb{R}^n_+$, $a \neq 0$, we have

$$\begin{aligned}\mathcal{L}\mu(ka) &= \int_{\mathbb{R}^n_+} \exp(-k(a,x))\, d\mu(x) = \int_{\mathbb{R}^n_+} \exp(-kc(a)(a,x)/c(a))\, d\mu(x) \\ &= \int_{\mathbb{R}_+} \exp(-kc(a)y)\, d\nu_a(y) = \mathcal{L}\nu_a(c(ka)). \end{aligned} \tag{2.1}$$

Suppose that X is positive isotropic with norming functional $c(\cdot)$. Then, the measures ν_a's are equal to the distribution of the vector X_1. Denote $f = \mathcal{L}\nu_a$. Then, the representation for $\mathcal{L}\mu$ immediately follows from (2.1). The function f is non-constant, because $P(X_1 > 0) > 0$ and ν_a cannot be supported in the origin.

Conversely, suppose that for every $k > 0$ and $a \in \mathbb{R}^n_+$, $\mathcal{L}\mu(ka) = f(c(ka))$. Since k is arbitrary, it follows from (2.1) that $\mathcal{L}\nu_a \equiv f$, for every $a \in \mathbb{R}^n_+$, $a \neq 0$. By the uniqueness theorem for the Laplace transform, we see that all the measures ν_a are equal. Hence, the random vectors $(a,X)/c(a)$ are identically distributed, and the result follows from the fact that $c(e_i) = 1$ for every i. ■

A function $\phi : \mathbb{R}^n_+ \to \mathbb{R}_+$ is called completely monotone if ϕ is infinitely differentiable in the interior of $\mathbb{R}^n_+$, and for every choice of nonnegative integers $\alpha_1, \ldots, \alpha_n$

$$(-1)^{\alpha_1+\cdots+\alpha_n} \frac{\partial^{\alpha_1+\cdots+\alpha_n}\phi}{\partial x_1^{\alpha_1}\ldots\partial x_n^{\alpha_n}} \geq 0$$

at every point from the interior of $\mathbb{R}^n_+$. The celebrated Bernstein theorem states that a function ϕ on $\mathbb{R}^n_+$ is the Laplace transform of a probability measure on $\mathbb{R}^n$ if and only if ϕ is continuous, completely monotone, and $\phi(0) = 1$ (see [1]).

Remark. It follows from the proof of Proposition 2.2 that the function f corresponding to a positive isotropic vector is the Laplace transform of a measure supported on $[0, \infty)$. Hence, f is a continuous and completely monotone on $\mathbb{R}_+$. On the other hand, Proposition 2.2 implies that a functional $c(\cdot)$, satisfying the conditions of Proposition 2.2, is a norming functional of a positive isotropic vector if and only if there exists a non-constant completely monotone function f on $\mathbb{R}_+$ so that $f(c(x))$ is completely monotone on $\mathbb{R}^n_+$.

We will show that a norming functional cannot satisfy the triangle inequality, except for the trivial case when $P(X_1 = X_2 = \cdots = X_n) = 1$.

2.3. Proposition. *Let $c(\cdot)$ be the norming functional of a positive isotropic vector X. Then, for every $\alpha, \beta > 0$, $c(\alpha e_1 + \beta e_2) \geq \alpha + \beta$. Equality holds only if $P(X_1 = X_2) = 1$.*

Proof. Suppose that $c(\alpha e_1 + \beta e_2) \leq \alpha + \beta$. Since X_1 and X_2 are nonnegative, and X_1 has the same distribution as X_2 and $(\alpha X_1 + \beta X_2)/c(\alpha e_1 + \beta e_2)$, we can apply Hölder's inequality:

$$\begin{aligned}
\int \exp(-X_1)\,dP &= \int \exp\Big(-\frac{\alpha X_1 + \beta X_2}{c(\alpha e_1 + \beta e_2)}\Big)\,dP \\
&\leq \int \exp\Big(-\frac{\alpha X_1 + \beta X_2}{\alpha + \beta}\Big)\,dP \\
&= \int \exp\Big(-\frac{\alpha}{\alpha + \beta} X_1\Big) \exp\Big(-\frac{\beta}{\alpha + \beta} X_2\Big)\,dP \\
&\leq \Big(\int \exp(-X_1)\,dP\Big)^{\frac{\alpha}{\alpha+\beta}} \Big(\int \exp(-X_2)\,dP\Big)^{\frac{\beta}{\alpha+\beta}} = \int \exp(-X_1)\,dP\,.
\end{aligned}$$

By the equality condition in Hölder's inequality, $X_1 = X_2$ with probability 1. But this means that $c(\alpha e_1 + \beta e_2) = \alpha + \beta$. ∎

2.4. Corollary. *The only norm that can appear as the norming functional of a positive isotropic random vector in $\mathbb{R}^n$ is the norm of the space l_1^n.*

Proof. Suppose that $c(a)$ is a norm. Use Proposition 2.3 to conclude that all the random variables $X_1, \ldots, X_n$ are equal to each other with probability 1, and notice that this implies that $c(a) = a_1 + \cdots + a_n$. ∎

Another consequence of Proposition 2.3 is that the coordinates of positive isotropic vectors cannot have finite moments of order greater than 1.

2.5. Corollary. *Let X be a positive isotropic vector in $\mathbb{R}^n$, $n \geq 2$, so that $P(X_1 = X_2 = \cdots = X_n) < 1$. Then, for every $q \geq 1$, the q–th moment of X_1 does not exist; i.e.* $\int X_1^q \, dP = \infty$.

Proof. Let $c(\cdot)$ be the norming functional of X. Suppose that $\int X_1^q \, dP < \infty$. For every vector $a \in \mathbb{R}^n_+$, $a \neq 0$, we have

$$\Big(\int (a, X)^q \, dP \Big)^{1/q} = c(a) \Big(\int X_1^q \, dP \Big)^{1/q}. \tag{2.2}$$

The left–hand side is the norm of a subspace of L_q, where $q \geq 1$, so $c(a)$ is a norm. By Corollary 2.4, $c(a) = a_1 + \cdots + a_n$, and the coordinates are equal with probability 1. This contradicts our assumptions. ∎

Note that the coordinates of the standard positive q–stable $(0 < q < 1)$ random vector, which were mentioned at the beginning of this note, have finite moments only for orders less than q (see [16]).

If equality (2.2) holds with $c(a) = a_1 + \cdots + a_n$, the coordinates of X are equal for any random vector X with nonnegative coordinates (not necessarily positive isotropic).

2.6. Proposition. *Suppose that $q > 1$ and X is any random vector in $\mathbb{R}^n$ with positive coordinates satisfying the equality*

$$(a_1 + \cdots + a_n)^q = \int (a, X)^q \, dP$$

for every $a \in \mathbb{R}^n_+$. Then, there exist constants $k_1, \ldots, k_n$ so that $P(k_1 X_1 = k_2 X_2 = \cdots = k_n X_n) = 1$.

Proof. Take the derivatives by a_1 of both sides of (2.2) and use Hölder's inequality (recall that $q > 1$):

$$q(a_1 + \cdots + a_n)^{q-1} = q \int (a, X)^{q-1} X_1 \, dP \leq$$

$$\leq q \Big(\int (a, X)^q \, dP \Big)^{(q-1)/q} \Big(\int X_1^q \, dP \Big)^{1/q} = q(a_1 + \cdots + a_n)^{q-1}.$$

By the equality condition in Hölder's inequality, we conclude that there exists a constant k_1 so that $P\big((a, X) = k_1 X_1\big) = 1$. Repeating the same argument for each X_i, $i = 2, \ldots, n$, we get the desired result. ∎

To complete this article, we show a large class of positive isotropic vectors whose norming functionals are quasi-norms. This fact is known to specialists.

2.7. Proposition. *Let $f_1, \ldots, f_n \in L_q$ $(0 < q < 1)$ be the nonnegative functions with norm 1. Then the functional $c(a) = \|a_1 f_1 + \cdots + a_n f_n\|_{L_q}$ is the norming functional of a positive isotropic random vector in $\mathbb{R}^n$.*

Sketch of the proof. Consider the function $g(a) = \exp(-(c(a))^q)$. Looking at the derivatives of $\log(g(a))$ and using logarithmic differentiation one can see that g is completely monotone. Then the result follows from Bernstein's theorem and Proposition 2.2. ■

References

1. C. Berg, J. P. R. Christensen, and P. Ressel, *Harmonic analysis on semigroups*, Springer–Verlag, 1984.
2. M. Eaton, On the projections of isotropic distributions, *Ann. Stat.* 9 (1981), 391–400.
3. W. Feller, *An introduction to probability theory and its applications*, Vol. 2, Wiley & Sons, 1971.
4. A. Koldobsky, Schoenberg's problem on positive definite functions, *Algebra and Analysis* 3, no. 3 (1991), 78–85 (in Russian); English translation in *St. Petersburg Math. J.* 3 (1992), 563–570.
5. A. Koldobsky, Positive definite functions, stable measures and isometries on Banach spaces, *Lect. Notes in Pure Appl. Math.* **175** (1996), 275–290.
6. P. Levy, *Théory de l'addition de variable aléatoires*, Gauthier–Villars, Paris, 1937.
7. A. Lisitsky, One more proof of Schoenberg's conjecture, preprint.
8. J. Misiewicz, Positive definite functions on ℓ_∞, *Statist. Probab. Lett.* 8 (1989), 255–260.
9. J. Misiewicz, Sub-stable and pseudo-isotropic processes, preprint.
10. J. Misiewicz and Cz. Ryll–Nardzewski, Norm dependent positive definite functions, *Lecture Notes in Math.* **1391** (1987), 284–292.
11. Y. Raynaud, Almost isometric methods in some isomorphic embedding problems, *Contemporary Mathematics* **85** (1989), 427–445.
12. Y. Raynaud, Sous espaces l^r et geometrie des espaces $L^p(L^q)$, *C. R. Acad. Sci. Paris*, Ser. I, 301 (1985), 299–302.
13. I. J. Schoenberg, Metric spaces and positive definite functions, *Trans. Amer. Math. Soc.* **44** (1938), 522–536.
14. V. Zastavny, Positive definite functions depending on the norm, *Russian J. Math. Phys.* **1** (1993), 511–522.

15. V. Zastavny, Positive definite norm dependent functions, *Dokl. Ross. Acad. Nauk* 325 (1992), 901–903.

16. V. M. Zolotarev, *One–dimensional stable distributions*, Amer. Math. Soc., Providence, 1986.

Alvaro Arias
Division of Mathematics and Statistics
University of Texas at San Antonio
San Antonio, TX 78249
USA
email: arias@math.utsa.edu

Alexander Koldobsky
Division of Mathematics and Statistics
University of Texas at San Antonio
San Antonio, TX 78249
USA
email: koldobsk@math.utsa.edu

A TOPOLOGICAL APPROACH TO THE STUDY OF FUZZY MEASURES

GIUSEPPINA BARBIERI and HANS WEBER

Abstract. A topological approach to the study of fuzzy measures is developed. To do so we need (instead of a clan of fuzzy sets) a more general structure of the domain of the fuzzy measures. This structure is defined by means of some equations. Our general setting allows us to treat simultaneously fuzzy measures, group–valued measures on Boolean rings, and linear operators on Riesz spaces. We deal with extension and decomposition theorems. Also we study connected, totally disconnected, and compact MV–algebras.

Introduction

During the last thirty years, topological methods introduced by Fréchet–Nikodým have been systematically applied to the examination of measures on Boolean rings. The main aim of this paper is to develop a topological approach to the study of "fuzzy measures" (= T_∞–valuations in the sense of [4]). A T_∞–valuation is a function μ defined on a clan $\mathcal{C}$ of fuzzy sets such that $\mu(f+g) = \mu(f) + \mu(g)$ if $f, g \in \mathcal{C}$ and $f + g \leq 1$; a clan of fuzzy sets is a family of $[0,1]$–valued functions on a set X such that $1 \in \mathcal{C}$ and $(f-g) \vee 0 \in \mathcal{C}$ whenever $f, g \in \mathcal{C}$.

Here we transfer the method of [20], used to study FN–topologies and measures on Boolean rings, to the theory of fuzzy measures. So we need, as a domain for the fuzzy measures, instead of a clan of fuzzy sets a more general structure, which is equationally defined and therefore closed with respect to quotients and uniform completions. A structure which satisfies these requirements is introduced in Definition 1.3 under the name of a "Δ-ℓ–semigroup." Examples of Δ-ℓ–semigroups are MV–algebras (see Definition 1.5), in particular, clans of fuzzy sets, Boolean rings, and positive cones of commutative ℓ–groups. Therefore, taking Δ-ℓ–semigroups as a domain for measures allows us — and this is the second aim of this paper — to treat at the same time (real–valued) T_∞–valuations in the sense of [4], group–valued (finitely additive) measures on Boolean rings and homomorphisms on ℓ–groups, in particular, linear operators on Riesz spaces. For example, Decomposition Theorem 5.3 contains new decomposition theorems for T_∞–valuations, Traynor's decomposition theorem [16, Theorem 1.2′] — for group–valued measures on rings, and the decomposition theorem [2, Theorem 3.3] for homomorphisms on commutative ℓ–groups.

This paper is organized as follows. In Section 1, we recall some basic facts about Δ-ℓ–semigroups and MV–algebras that will be needed later. In Section 2, we present the basic properties of those uniformities on Δ-ℓ–semigroups or MV–algebras that make their operations uniformly continuous. In Section 3, these uniformities are used for proving the extension theorem for measures on Δ-ℓ–semigroups, which contains as special cases the extension theorem for T_∞–measures (= σ–order continuous T_∞–valuations), the known extension theorem for group–valued measures on Boolean rings, and the extension theorem for homomorphisms on commutative ℓ–groups, in particular, for linear operators on Riesz spaces. The main result of Section 4 says that the space of all exhaustive uniformities on an MV–algebra that make its operations uniformly continuous is a complete Boolean algebra. This fact is the basic tool in the proof of the decomposition theorems of Section 5. Section 6 contains the Hammer–Sobczyk decomposition theorem for measures on Δ-ℓ–semigroups. Moreover, we study connected MV–algebras and we give a complete description of all totally disconnected and all compact MV–algebras.

1. Preliminaries: Δ-ℓ–semigroups and MV–algebras

As a domain of the measures studied in this paper we take partially ordered (p.o.) Δ–semigroups introduced in [19] under the name "$(\leq, +, \Delta)$–Bereiche."

1.1. Definition ([19], [24, Definition 1.1]). *We call $(P, \leq, +, \Delta)$ a* ***p.o. Δ–semigroup*** *if*

(P0) *$(P, +)$ is a commutative semigroup, $\leq$ is a partial order on P, and $\Delta : (y, x) \mapsto y\Delta x$ is a map from $\{(y, x) \in P^2 : x \leq y\}$ into P such that for all $x, y, z \in P$ the following conditions hold:*

(P1) $x \leq x + y$,

(P2) $x \leq y \leq z + x \;\Rightarrow\; y\Delta x \leq z$,

(P3) $y = y\Delta x + x$ *whenever* $x \leq y$.

1.2. Proposition. *Let $(P, \leq, +, \Delta)$ be a p.o. Δ–semigroup. Then*

(a) *P has a smallest element 0. For all $x \in P$ we have*

$$x\Delta x = 0 \quad \text{and} \quad x + 0 = x\Delta 0 = x.$$

(b) *For $x, y, z, t \in P$, we have:*

(i) *If $x \leq y$, then $x + z \leq y + z$.*

(ii) *If $x \leq y \leq z$, then $z\Delta x = (z\Delta y) + (y\Delta x)$.*

(iii) *If $x \leq y \leq z \leq t$, then $z\Delta y \leq t\Delta x$.*

(c) *If* $x, y \in P$ *and* y *is a lower bound of a net* (x_α) *in* P *with* $x_\alpha \uparrow x$, *then* $x_\alpha \Delta y \uparrow x \Delta y$.

Proof. (a), (b) are contained in [24, 1.3]. We prove (c). By (b) (iii), the net $(x_\alpha \Delta y)$ is increasing and $x \Delta y$ is an upper bound of $(x_\alpha \Delta y)$. If t is an upper bound of $(x_\alpha \Delta y)$, then $x_\alpha = (x_\alpha \Delta y) + y \leq t + y$, hence $x \leq t + y$ and $x \Delta y \leq t$. ∎

1.3. Definition. *We call a p.o.* Δ*–semigroup* $(P, \leq, +, \Delta)$, *which is a lattice with respect to* $\leq$, *a* **Δ-ℓ–semigroup**. *In a* Δ-ℓ*–semigroup* P, *the partially defined operation* Δ *is extended to* P^2 *by the formula* $x \Delta y := (x \vee y) \Delta (x \wedge y)$.

A motivation for the axioms of a Δ-ℓ–semigroup is the one that x, y should be close if $x \Delta y$ is near zero with respect to the uniformities we are interested in; see 2.1.

Throughout this paper, let $(P, \leq, +, \Delta)$ *be a* Δ-ℓ*–semigroup.*

1.4. Proposition. *For* $x, y, z, x_i, y_i \in P$, *we have*:

(a) $(x \wedge y) + (x \vee y) = x + y$.

(b) $(x + y) \wedge (x + z) = x + (y \wedge z)$.

(c) $(x + y) \vee (x + z) = x + (y \vee z)$.

(d) P *is a distributive lattice.*

(e) (i) $x \leq x \Delta y + y$,

(ii) $x \Delta y \leq z$ *iff* $x \leq z + y$ *and* $y \leq z + x$,

(iii) $x \Delta y = 0$ *iff* $x = y$,

(iv) $x \Delta z \leq (x \Delta y) + (y \Delta z)$.

(f) *If* $\circ$ *is one of the operations* $+, \Delta, \vee, \wedge$, *then*

(i) $(x \circ z) \Delta (y \circ z) \leq x \Delta y$,

(ii) $(x_1 \circ y_1) \Delta (x_2 \circ y_2) \leq (x_1 \Delta x_2) + (y_1 \Delta y_2)$.

(g) *If* $z \leq x_1 + x_2$, *then* $z_1 := z \wedge x_1 \leq x_1$, $z_2 := z \Delta z_1 \leq x_2$, *and* $z = z_1 + z_2$.

(h) *If* $z \leq x_1 + \cdots + x_n$, *then there are elements* $z_i \in P$ *such that* $z_i \leq x_i$ *and* $z = z_1 + \cdots + z_n$.

(k) *If* y *is an upper bound of a net* (x_α) *in* P *with* $x_\alpha \downarrow x$, *then* $y \Delta x_\alpha \uparrow y \Delta x$.

Proof. For (a) to (h), see [24, 1.5]. We prove (k). By 1.2 (b) (iii), $(y \Delta x_\alpha)$ is an increasing net and $y \Delta x$ is an upper bound of $(y \Delta x_\alpha)$. If t is an upper bound of $(y \Delta x_\alpha)$, then $y \leq t + x_\alpha$, hence $y \leq t + x$ by [19, 1.2.1] and $y \Delta x \leq t$. ∎

Examples of Δ-ℓ–semigroups are given in [24, 1.9]. In particular, positive cones of ℓ–groups, Boolean rings (with $x + y := x \vee y$), clans of fuzzy sets or, more generally, MV–algebras are Δ-ℓ–semigroups.

1.5. Definition. *An* ***MV–algebra*** $(L, +, ' ; 0, 1)$ *is a commutative semigroup* $(L, +)$ *with zero element* 0, *a constant* $1 \in L$, *and a unary operation* $' : L \to L$ *satisfying the following axioms:*

(L1) $x + 1 = 1$,

(L2) $x'' = x$,

(L3) $0' = 1$,

(L4) $x + x' = 1$,

(L5) $(x' + y)' + y = (x + y')' + x$,

for all $x, y \in L$.

By [12], any MV–algebra is a distributive lattice with greatest element 1, where the order relation is defined by

$$x \leq y \quad \text{iff} \quad x' + y = 1,$$

moreover, the supremum and infimum are given by the following formulas:

$$x \vee y = (x + y')' \quad \text{and} \quad x \wedge y = (x' \wedge y')'.$$

1.6. Proposition ([24, 4.3 (a)]). *Let* $(L, +, ' ; 0, 1)$ *be an MV–algebra. Define for* $x, y \in L$

$$x \leq y \quad \textit{iff} \quad x' + y = 1 \quad \textit{and}$$

$$y \Delta x := (x + y')' \quad \textit{whenever} \quad x \leq y.$$

Then $(L, \leq, +, \Delta)$ *is a* Δ-ℓ*–semigroup and* $1 = \max L$.

As we will see in Section 4 and Section 6, a special role is played by the set $I(P) := \{x \in P : 2x = x\}$ of idempotent elements of P and the center (in the lattice theoretical sense, see [3, p. 67]) $C(P)$ of P if P is bounded.

1.7. Theorem ([24, 1.5 (1), 2.2, 2.3, 2.6]). *Assume that* P *has a greatest element* 1 *and put* $x' := 1 \Delta x$ *for* $x \in P$. *Then*

(a) $C(P)$ *is the largest Boolean subalgebra of* $(P, \vee, \wedge, 0, 1)$.

(b) $C(P) = \{x \in P : \; x \wedge x' = 0\} = \{x \in P : \; x$ *has a complement in* $P\}$
$= \{x \in P : \; x'$ *is the complement of* $x\}$.

(c) *For* $a \in C(P)$, *the intervals* $[0,a]$ *and* $[0,a']$ *are subalgebras of* $(P,+,\Delta,\vee,\wedge)$ *and* $x \mapsto (x\wedge a, x\wedge a')$ *defines an isomorphism (compatible with* $+,\Delta,\wedge,\vee$*) from* P *onto* $[0,a]\times[0,a']$.

(d) $I(P)$ *is a sublattice of* P.

(e) $C(P) \subset I(P)$. *If* $x'' = x$ *for every* $x \in P$, *then* $C(P) = I(P)$.

A subset N of P is called *solid* (in P) if $P \ni x \le y \in N$ implies $x \in N$. An *ideal* in P is a non-empty, additively closed, solid subset of P.

There is a natural correspondence between ideals and congruences in P, in the special case of MV–algebras already observed by Chang [5]. If N is an ideal in P, then

$$x \sim y \quad \text{iff} \quad x\Delta y \in N$$

defines a congruence relation with respect to $\vee,\wedge,+,\Delta$; if $\sim$ is a congruence relation on P with respect to $\vee,\wedge,+,\Delta$, then $N := \{x \in P : x \sim 0\}$ is an ideal in P and $x \sim y$ iff $x\Delta y \in N$. This correspondence can be obtained as special case of 2.1.

For any ideal N in P, the quotient $P/N := P/{\sim}$ is then also a Δ-ℓ–semigroup, since Δ-ℓ–semigroups are equationally defined [24, 1.6].

We call a group–valued function μ on P a *measure* if $\mu(y\Delta x) = \mu(y) - \mu(x)$ whenever $x, y \in P$ and $x \le y$. Examples of measures in this sense are T_∞–valuations (mentioned at the beginning of the introduction).

For any measure μ on P, $N(\mu) := \{x \in P : \mu(z) = 0 \text{ for all } z \in [0,x]\}$ is an ideal in P.

1.8. Theorem ([24, 4.6]). *If* P *has a greatest element* 1 *and* μ *is a measure on* P *with values in a commutative group, then the quotient* $\hat{P} := P/N(\mu)$ *becomes an MV–algebra if one defines* $x' := 1\Delta x$; *moreover*, $I(\hat{P}) = C(\hat{P})$.

1.9. Proposition. *Assume that* P *has a greatest element* $a \in C(P)$, a' *is its unique complement, and* μ *is a measure on* P *with values in an Abelian group. Then* $\mu_a(x) := \mu(a \wedge x)$ *and* $\mu_{a'}(x) := \mu(a' \wedge x)$ *define measures on* P *and* $\mu = \mu_a + \mu_{a'}$.

1.9 immediately follows from 1.7 (c).

2. Topological Δ-ℓ–semigroups and topological MV–algebras

A *lattice uniformity* is a uniformity on a lattice which makes the lattice operations $\wedge$ and $\vee$ uniformly continuous. The set $\mathcal{LU}(P)$ of all lattice uniformities on P is a complete lattice with the discrete uniformity as greatest element and the trivial uniformity as smallest element. The uniformities, which are here of interest, are the lattice uniformities on P, which also make the operations $+$ and Δ uniformly continuous; the set of these uniformities we denote by $\mathcal{LUA}(P)$. It follows from [25] that for any family $(u_\alpha)_{\alpha\in A}$ in $\mathcal{LUA}(P)$, $\inf u_\alpha$ and $\sup u_\alpha$ built in the lattice of all uniformities on P belong to $\mathcal{LUA}(P)$; in particular, $\mathcal{LUA}(P)$ is a sublattice of $\mathcal{LU}(P)$. For our purpose it is a fundamental fact that the uniformities of $\mathcal{LUA}(P)$ are uniquely determined by their 0–neighborhood filters. Denote by $\mathcal{FNA}(P)$ the set of all filters $\mathcal{F}$ on P with the following property:

$$\forall F\in\mathcal{F}\quad \exists G\in\mathcal{F}\quad \forall x\in P\quad \forall y,z\in G\quad x\le y+z\Rightarrow x\in F. \qquad (*)$$

Equivalent to $(*)$ are the two conditions that $F+F\in\mathcal{F}$ for any $F\in\mathcal{F}$ and that $\mathcal{F}$ has a base of solid sets.
The symbol $\mathcal{FNA}(P)$ is to remind one of FN–topologies defined on Boolean rings.

2.1. Theorem.

(a) $\psi: u\mapsto u(0)$ *defines an order isomorphism from* $\mathcal{LUA}(P)$ *onto* $\mathcal{FNA}(P)$. *Here,* $u(0)$ *denotes the* 0*–neighborhood system in* (P,u).

(b) *For* $u\in\mathcal{LUA}(P)$ *and* $F\in u(0)$, *the sets* $F^\Delta := \{(x,y)\in P^2 : x\Delta y\in F\}$ *form a base of* u.

Proof. (i) Obviously, ψ is well–defined, i.e., $u(0)\in\mathcal{FNA}(P)$ for $u\in\mathcal{LUA}(P)$ and ψ is monotone.

(ii) $\psi(u)\subset\psi(v)$ implies $u\subset v$ for $u,v\in\mathcal{LUA}(P)$: Let $U\in u$. By [21, 1.1.3] we may assume that $(a,b)\in U$ implies $[a\wedge b, a\vee b]^2\subset U$. Choose $U'\in u$ with $U'+U'\subset U$ and $V,W\in v$ with $V(0)\subset U'(0)$ and $W\Delta W\subset V$. We show that $W\subset U$; this implies $U\in v$. Let $(x,y)\in W$. Then $(0,x\Delta y) = (x\Delta x, x\Delta y)\in W\Delta W\subset V$, hence $x\Delta y\in V(0)\subset U'(0)$ and $(0,x\Delta y)\in U'$. Therefore $(x\wedge y, x\vee y) = (x\wedge y, x\wedge y) + (0, x\Delta y)\in U'+U'\subset U$. It follows that $(x,y)\in U$ since $(x,y)\in[x\wedge y, x\vee y]^2$.

(iii) Let $\mathcal{F}\in\mathcal{FNA}(P)$. Then the sets F^Δ, $F\in\mathcal{F}$, form a base of a filter v. It follows immediately from 1.4 (f) that the operations $\vee,\wedge,+,\Delta$ are uniformly continuous. Hence $v\in\mathcal{LUA}(P)$. Obviously, $\psi(v)=\mathcal{F}$. ∎

We say that two uniformities u,v on a set X *permute* if for any $U\in u$ and $V\in v$ there are $U'\in u$ and $V'\in v$ with $U'\circ V'\subset V\circ U$.

2.2. Proposition. *Any two uniformities of* $\mathcal{LUA}(P)$ *permute.*

Proof. Let $u, v \in \mathcal{LUA}(P)$, U and V be solid 0–neighborhoods of (P, u) and (P, v), respectively. By 2.1, it is enough to show that $V^\Delta \circ U^\Delta \subset U^\Delta \circ V^\Delta$. Let $(x, y) \in V^\Delta \circ U^\Delta$ and $p \in P$ with $(x, p) \in V^\Delta$ and $(p, y) \in U^\Delta$. Then $a := p\Delta y \in U$ and $b := x\Delta p \in V$. Moreover, $q := (x+a) \wedge (y+b) \leq x+a$ and, by 1.4 (b), $x \leq (x+a+a) \wedge (y+a+b) = q+a$; hence, by 1.4 (e) (ii), $x\Delta q \leq a \in U$ and therefore $x\Delta q \in U$. Similarly it can be proved that $q\Delta y \in V$. Therefore $(x, q) \in U^\Delta$, $(q, y) \in V^\Delta$, and $(x, y) \in U^\Delta \circ V^\Delta$. ■

It is proved in [25] that if L is a lattice and $\mathcal{L}$ is a sublattice of $\mathcal{LU}(L)$ such that any two uniformities of $\mathcal{L}$ permute, then $\mathcal{L}$ is distributive. Therefore we have the following.

2.3. Corollary. *$\mathcal{LUA}(P)$ is a distributive complete lattice.*

2.3 can be also deduced from 2.1 and the following proposition.

2.4. Proposition. *Let $\mathcal{F}_1, \mathcal{F}_2 \in \mathcal{FNA}(P)$. Then the sets $F_1 \cap F_2$ and $F_1 + F_2$ ($F_i \in \mathcal{F}_i$) form a base of $\mathcal{F}_1 \vee \mathcal{F}_2$ and $\mathcal{F}_1 \wedge \mathcal{F}_2$, respectively.*

Proof. We only prove the statement about $\mathcal{F}_1 \wedge \mathcal{F}_2$. Let $F_i \in \mathcal{F}_i$, and then $G_i \in \mathcal{F}_i$ chosen for F_i according to $(*)$. Using the decomposition of 1.4 (h), it can easily be seen that $y, z \in G_1 + G_2$ and $P \ni x \leq y + z$ imply $x \in F_1 + F_2$. Therefore the filter $\mathcal{F}$ generated by the sets $F_1 + F_2$ $(F_i \in \mathcal{F}_i)$ belongs to $\mathcal{FNA}(P)$. Since any filter of $\mathcal{FNA}(P)$ coarser than $\mathcal{F}_1$, and $\mathcal{F}_2$ is also coarser than $\mathcal{F}$, one obtains $\mathcal{F} = \mathcal{F}_1 \wedge \mathcal{F}_2$. ■

A function $\eta : P \to [0, \infty]$ is called *subadditive* if $\eta(x+y) \leq \eta(x)+\eta(y)$, and *monotone* if $x \leq y$ implies $\eta(x) \leq \eta(y)$ for $x, y \in P$. $\eta : P \to [0, \infty]$ is called a *submeasure* if it is monotone, subadditive, and $\eta(0) = 0$. If $\eta : P \to [0, \infty]$ is a submeasure, then $d(x, y) := \eta(x\Delta y)$ defines a pseudometric generating a uniformity of $\mathcal{LUA}(P)$. Conversely, we have the following.

2.5. Theorem ([19, Satz 5.2.4]). *Any uniformity $u \in \mathcal{LUA}(P)$ is generated by a system $(\eta_\alpha)_{\alpha \in A}$ of submeasures on P, i.e., the sets $\{(x, y) \in P^2 : \eta_\alpha(x\Delta y) \leq \varepsilon\}$ $(\alpha \in A,\ \varepsilon > 0)$ form a subbase for u.*

2.5 can also be easily deduced from [22, 1.2] and 2.1.

$\eta : P \to [0, \infty]$ is called *σ–subadditive* if $x, x_n \in P$ $(n \in \mathbb{N})$ and $x = \sup_n \sum_{i=1}^n x_i =: \sum_{n=1}^\infty x_n$ imply $\eta(x) \leq \sum_{n=1}^\infty \eta(x_n)$. $\eta : P \to [0, \infty]$ is called a *σ–submeasure* if it is monotone, σ–subadditive, and $\eta(0) = 0$.

2.6. Proposition. *A submeasure is σ–subadditive iff for every 0–neighborhood U in (P, η) there is a sequence (U_n) of 0–neighborhoods in (P, η) such that*

$$x_n \in U_n \ (n \in \mathbb{N}), \quad x \in P, \quad x = \sum_{n=1}^\infty x_n \quad \textit{imply} \quad x \in U.$$

For the proof, see [19, 5.2.3]. Observe here that the completeness assumption is superfluous since, for $x = \sum_{n=1}^{\infty} x_n$ and $y_n := (\sum_{i=1}^{n} x_i)\Delta(\sum_{i=1}^{n-1} x_i)$, the supremum $\sup_{m\in\mathbb{N}} \sum_{i=n}^{n+m} y_i =: \sum_{i=n}^{\infty} y_i$ exists, namely $x\Delta(\sum_{i=1}^{n-1} x_i) = \sum_{i=n}^{\infty} y_i$ by 1.2 (b) (ii) and (c), and, moreover, $y_n \leq x_n \quad (n \in \mathbb{N})$.

2.7. Corollary ([19, 5.2.4]). *A uniformity $u \in \mathcal{LUA}(P)$ is generated by a system of σ–submeasures iff for every 0–neighborhood U in (P, u) there is a sequence (U_n) of 0–neighborhoods in (P, u) such that*

$$x_n \in U_n \quad (n \in \mathbb{N}), \quad x \in P, \quad x = \sum_{n=1}^{\infty} x_n \quad \textit{imply} \quad x \in U.$$

With the aid of 1.2 (c) and 1.4 (k) it is easy to see that the condition stated in 2.7 for $u \in \mathcal{LUA}(P)$ is equivalent to the condition (σ) studied in [21, Section 3] for arbitrary lattice uniformities.

A uniformity $u \in \mathcal{LUA}(P)$ is called (*locally*) *exhaustive* if every (bounded) monotone sequence in (P, u) is Cauchy, and *σ–order continuous* (*order continuous*) if every order convergent monotone sequence (net) converges in (P, u) to its order limit. Exhaustivity and (σ) imply σ–order continuity; σ–order continuity implies (σ) and also, in the σ–complete case, exhaustivity (see [21, 8.1.2]).

2.8. Proposition. *Let $(L, +, ' ; 0, 1)$ be an MV–algebra. As in* 1.6, *we consider it also as a Δ-ℓ–semigroup.*

(a) *A uniformity on L belongs to $\mathcal{LUA}(L)$ iff it makes $+$ and $'$ uniformly continuous.*

(b) *$u \in \mathcal{LUA}(L)$ is σ–order continuous (order continuous) iff every decreasing sequence (net) with order limit* 0 *converges to* 0 *in (L, u) iff every increasing sequence (net) with order limit* 1 *converges to* 1 *in (L, u).*

(c) *$u \in \mathcal{LUA}(L)$ is exhaustive iff every decreasing sequence is Cauchy iff every increasing sequence is Cauchy.*

Proof. (a) Use the formulas $x \vee y = (x' + y)' + y$, $(x \vee y)' = x' \wedge y'$ (see [12]), $y\Delta x = (x + y')'$ for $x \leq y$, in particular, $1\Delta x = x'$ (see 1.6).

(b), (c) Use that $x \mapsto x'$ defines a dual isomorphism and that $x_n \downarrow x$ implies $x_n \Delta x \downarrow 0$. ∎

A statement analogous to 2.8 (b), (c) is not true for Δ-ℓ–semigroups:

2.9. Example. Let u be the discrete uniformity on an infinite chain $C = \{a_n : n \in \mathbb{N}\} \cup \{0\}$ with $a_n \downarrow 0$. By defining $x + y := x \vee y$ and $y\Delta x := y$ if

$x < y$, C becomes a complete Δ-ℓ–semigroup. Since $x_\alpha \uparrow x$ implies $x_\alpha \to x(u)$, any increasing sequence in (C, u) is Cauchy. But (C, u) is not σ–order continuous and not exhaustive.

Analogously, if u is the discrete uniformity on an infinite chain $C = \{a_n : n \in \mathbb{N}\} \cup \{1\}$ with $a_n \uparrow 1$, then u is not σ–order continuous and not exhaustive although $x_\alpha \downarrow x$ implies $x_\alpha \to x(u)$. ■

3. Extension of measures on Δ-ℓ–semigroups

3.1. The μ–uniformity

In this section, let $(G, +)$ be a Hausdorff topological Abelian group and $\mu : P \to G$ a measure.

3.1.1. Proposition ([24, 3.2 and 3.3]).

(a) *μ is modular, i.e., $\mu(x \wedge y) + \mu(x \vee y) = \mu(x) + \mu(y)$ for $x, y \in P$.*

(b) *If $x, y, t \in P$, $x \le y$, and $t \le y\Delta x$, then $\mu(x + t) = \mu(x) + \mu(t)$.*

3.1.2. Proposition. *Let $|\cdot|_\alpha$, $\alpha \in A$, be a system of group seminorms on G, which generates the topology of G. Put*

$$\tilde{\mu}_\alpha(x) := \sup\left\{\left|\mu(t)\right|_\alpha : t \in [0, x]\right\} \quad \textit{for} \quad \alpha \in A \quad \textit{and} \quad x \in P.$$

Then

(a) *$\tilde{\mu}_\alpha : P \to [0, \infty]$ is a submeasure,*

(b) *$\tilde{\mu}_\alpha(x\Delta y) = \sup\left\{\left|\mu(t) - \mu(s)\right|_\alpha : \quad s, t \in P,\ x \wedge y \le s \le t \le x \vee y\right\}$
$=: d_{\mu,\alpha}(x, y)$ for $x, y \in P$ and $\alpha \in A$.*

Proof. (a) Let $x, y, z \in P$ and $x \le y + z$. For any $t \in [0, x]$ we have $\mu(t) = \mu(t \wedge y) + \mu(t\Delta(t \wedge y))$. Since $t\Delta(t \wedge y) \le z$ by 1.4 (g), we have

$$\left|\mu(t)\right|_\alpha \le \left|\mu(t \wedge y)\right|_\alpha + \left|\mu(t\Delta(t \wedge y))\right|_\alpha \le \tilde{\mu}_\alpha(y) + \tilde{\mu}_\alpha(z).$$

It follows that $\tilde{\mu}_\alpha(x) \le \tilde{\mu}_\alpha(y) + \tilde{\mu}_\alpha(z)$.

(b)($\ge$) Let $s, t \in P$ with $x \wedge y \le s \le t \le x \vee y$. Then $\left|\mu(t) - \mu(s)\right|_\alpha = \left|\mu(t\Delta s)\right|_\alpha \le \tilde{\mu}_\alpha(x\Delta y)$ since $t\Delta s \le x\Delta y$ by 1.2 (b) (iii). Therefore $d_{\mu,\alpha}(x, y) \le \tilde{\mu}_\alpha(x\Delta y)$.

($\le$) Let $t \in P$, $t \le x\Delta y$. By 3.1.1 (b), we have $\mu((x \wedge y) + t) = \mu(x \wedge y) + \mu(t)$, hence $\left|\mu(t)\right|_\alpha = \left|\mu((x \wedge y) + t) - \mu(x \wedge y)\right|_\alpha \le d_{\mu,\alpha}(x, y)$. ■

With the notation of 3.1.2, the pseudometrics $d_{\mu,\alpha}$, $\alpha \in A$, generate the *μ–uniformity*, this is the weakest lattice uniformity that makes μ uniformly continuous; see [23, Section 3]. By 3.1.2, the μ–uniformity is generated by the system $(\tilde{\mu}_\alpha)_{\alpha\in A}$ of submeasures. Therefore the μ–uniformity belongs to $\mathcal{LUA}(P)$.

The μ–uniformity can also be easily described without using group seminorms on G. The sets $U^* := \{x \in P : \mu([0,x]) \subset U\}$, where U is a 0–neighborhood in G, form a base of a filter of $\mathcal{FNA}(P)$. The μ–uniformity is the uniformity which corresponds to this filter according to 2.1.

3.1.3. Proposition. *For $u \in \mathcal{LUA}(P)$, the following conditions are equivalent:*

(1) *$\mu : (P,u) \to G$ is continuous at 0.*

(2) *$\mu : (P,u) \to G$ is uniformly continuous.*

(3) *The μ–uniformity is weaker than u.*

Proof. Denote by v the μ–uniformity. Obviously, (2) is equivalent to (3), i.e. to $v \subset u$; and (1) is equivalent to $v(0) \subset u(0)$. But $v(0) \subset u(0)$ iff $v \subset u$ by 2.1. ■

μ is called (*locally*) *exhaustive* if $(\mu(x_n))$ is Cauchy for any (bounded) monotone sequence $(x_n)_{n\in\mathbb{N}}$ in P. μ is called *σ–order continuous* if $(\mu(x_n))$ converges to $\mu(x)$ whenever $(x_n)_{n\in\mathbb{N}}$ is a monotone sequence order–converging to x in P; replacing here sequences by nets, one gets the definition of *order continuous* measures.

3.1.4. Proposition ([23, 3.5, 3.6]). *μ is exhaustive (locally exhaustive, σ–order continuous, order continuous) iff the μ–uniformity is exhaustive (locally exhaustive, σ–order continuous, order continuous, respectively).*

3.1.5. Proposition.

(a) *μ is σ–order continuous iff $x_n \uparrow x$ implies $\mu(x_n) \to \mu(x)$ for any sequence (x_n) in P and $x \in P$.*

(b) *μ is order continuous iff $x_\alpha \uparrow x$ implies $\mu(x_\alpha) \to \mu(x)$ for any net (x_α) in P and $x \in P$.*

(c) *μ is exhaustive iff $(\mu(x_n))_{n\in\mathbb{N}}$ is Cauchy for any increasing sequence (x_n) in P.*

(d) *μ is locally exhaustive iff $(\mu(x_n))_{n\in\mathbb{N}}$ is Cauchy for any decreasing sequence (x_n) in P.*

Proof. The implication $(\Rightarrow)$ obviously holds in (a) to (d).

(a)$(\Leftarrow)$ If $x_n \downarrow x$, then $x_1\Delta x_n \uparrow x_1\Delta x$ by 1.4 (k), hence $\mu(x_1) - \mu(x_n) = \mu(x_1\Delta x_n) \to \mu(x_1\Delta x) = \mu(x_1) - \mu(x)$ and therefore $\mu(x_n) \to \mu(x)$.

(d)($\Leftarrow$) If (x_n) is an increasing sequence in P with a supremum x, then $(x\Delta x_n)$ is decreasing, and hence $(\mu(x\Delta x_n))_{n\in\mathbb{N}}$ is a Cauchy sequence. Since $\mu(x\Delta x_n) = \mu(x) - \mu(x_n)$, the sequence $(\mu(x_n))$ is Cauchy, too.

The proof of (b), (c) is similar to that of (a), (d). ∎

In 3.1.5 (a), (b) one cannot replace increasing sequences (nets) by decreasing ones:

3.1.6. Example. Define $\mu : [0,\infty] \to \mathbb{R}$ by $\mu(x) = 0$ if $x < +\infty$ and $\mu(+\infty) = 1$. Then $x_\alpha \downarrow x$ implies $\mu(x_\alpha) \to \mu(x)$, but μ is not σ–order continuous. (Here, $x + y$ in $[0,+\infty]$ is defined as usual, $x\Delta x = 0$ and $y\Delta x = y - x$ for $x < y$).

If P is an MV–algebra, then μ is σ–order continuous (order continuous) iff $x_n \downarrow 0$ implies $\mu(x_n) \to 0$ for any sequence (net) in P, cf. 2.8. ∎

3.1.7. Proposition. *For $n \in \mathbb{N}$, an $\mathbb{R}^n$–valued measure on P is exhaustive iff it is bounded.*

Proof. Passing to the components, we may assume that the measure is real–valued. For $n = 1$, Proposition 3.1.7 is a special case of [23, 2.7] (and it can also be verified as in the Boolean case). This will follow from the next lemma, where $t_i \in \{0,\ x_i\Delta x_{i-1}\}$. ∎

3.1.8. Lemma. *Let $x_0 \le \cdots \le x_n$ be a finite chain in P and $P \ni t_i \le x_i\Delta x_{i-1}$ $(i = 1,\ldots,n)$. Then $\mu(\sum_{i=1}^n t_i) = \sum_{i=1}^n \mu(t_i)$.*

Proof. By induction. For $n = 1$, there is nothing to prove.

$(n-1)\to(n)$: Applying 3.1.1 (b) with $x = \sum_{i=1}^{n-1} t_i$, $y = x_n$, $t = t_n$, we obtain

$$\mu\left(\sum_{i=1}^{n} t_i\right) = \mu(x+t) = \mu(x) + \mu(t) =$$

$$= \mu\left(\sum_{i=1}^{n-1} t_i\right) + \mu(t_n) = \sum_{i=1}^{n-1} \mu(t_i) + \mu(t_n) = \sum_{i=1}^{n} \mu(t_i).$$

∎

3.2. An extension theorem

In this section, $(G,+)$ will denote a complete Hausdorff topological Abelian group.

We will present an extension theorem for G–valued measures on P, see 3.2.2. For that we transfer a proof well known in the Boolean case to our more general context.

3.2.1. Lemma. *Assume that P satisfies the following two conditions:*

(1) *$x_n \uparrow x$ implies $y \wedge x_n \uparrow y \wedge x$,*

(2) *$x_n \uparrow x$ implies $x_n + y \uparrow x + y$.*

Let $0 \in S \subset P$ and $\eta : S \to [0, \infty]$ be a function with $\eta(0) = 0$. Then

$$\eta^*(a) := \inf \left\{ \sum_{n=1}^{\infty} \eta(x_n) : \quad x_n \in S \ (n \in \mathbb{N}), \quad a \wedge \sum_{i=1}^{n} x_i \uparrow a \right\}$$

defines a σ–submeasure on P. Moreover, $\eta = \eta^*|_S$ if and only if $x, x_n \in S$ and $x \wedge \sum_{i=1}^n x_i \uparrow x$ implies $\eta(x) \le \sum_{n=1}^\infty \eta(x_n)$.

Proof. Since $\eta(0) = 0$, we have $\eta \ge \eta^*|_S$, in particular, $\eta^*(0) = 0$.

Let $a, b, a_n \in P$ with $b \le a = \sum_{n=1}^\infty a_n$. We show that $\eta^*(b) \le \sum_{n=1}^\infty \eta^*(a_n)$. We may assume that $\eta^*(a_n) < \infty$ for each $n \in \mathbb{N}$. Let $\varepsilon > 0$ and $x_{mn} \in S$ with $\sup_{m \in \mathbb{N}} a_n \wedge \sum_{i=1}^m x_{in} = a_n$ and $\sum_{m=1}^\infty \eta(x_{mn}) \le \eta^*(a_n) + \varepsilon 2^{-n} \quad (n \in \mathbb{N})$.

Then we have $b \wedge \sum_{i,j \le m} x_{ij} \uparrow b$: If $z \in P$ is an upper bound of the elements $b \wedge \sum_{i,j \le m} x_{ij}$, $m \in \mathbb{N}$, then $b \wedge \sum_{j=1}^n (a_j \wedge \sum_{i=1}^m x_{ij}) \le b \wedge \sum_{i,j \le m} x_{ij} \le z$ for $n \le m$ and by (1), (2) $b \wedge \sum_{j=1}^n a_j = \sup_m b \wedge \sum_{j=1}^n (a_j \wedge \sum_{i=1}^m x_{ij}) \le z$ $(n \in \mathbb{N})$, hence $b = \sup_n b \wedge \sum_{j=1}^n a_j \le z$. It follows that $\eta^*(b) \le \sum_{n,m=1}^\infty \eta(x_{nm})$ $\le \sum_{n=1}^\infty \eta^*(a_n) + \varepsilon$.

The second statement, which gives a criterion for $\eta = \eta^*|_S$, obviously holds. ■

It can be shown by examples that in 3.2.1 neither of the assumptions (1) and (2) is superfluous. On the other hand, both of these assumptions are satisfied if P is an MV–algebra or the positive cone of an ℓ–group.

3.2.2. Theorem. *Assume that P satisfies conditions (1) and (2) of 3.2.1. Let S be a subalgebra of $(P, +, \Delta, \vee, \wedge)$ and $\mu : S \to G$ a σ–order continuous measure. Then*

(a) *There is a uniformity $u^* \in \mathcal{LUA}(P)$, which satisfies (σ) and induces on S the μ–uniformity. The closure $\bar{S}$ of S in (P, u^*) is a subalgebra of P. μ has a unique continuous measure extension $\bar{\mu} : \bar{S} \to G$.*

(b) *If P is Dedekind σ–complete and μ is locally exhaustive, then $\bar{\mu}$ is σ–order continuous and $\bar{S}$ is Dedekind σ–complete. If P is σ–complete and μ is exhaustive, then $\bar{S}$ is σ–complete. (Here,* sup *and* inf *are always taken in P).*

Proof. (a) Let $|\cdot|_\alpha$, $\alpha \in A$, be a system of group seminorms generating the topology of G. Define for $\alpha \in A$

$$\tilde{\mu}_\alpha(x) := \sup\left\{\left|\mu(t)\right|_\alpha : \quad t \in S,\ t \leq x\right\} \quad (x \in S),$$

$$\mu_\alpha^*(a) := \inf\left\{\sum_{n=1}^{\infty} \tilde{\mu}_\alpha(x_n) : \quad x_n \in S,\ a \wedge \sum_{i=1}^{n} x_i \uparrow a\right\} \quad (a \in P).$$

Let us show that $x \wedge \sum_{i=1}^n x_i \uparrow x$ implies $\tilde{\mu}_\alpha(x) \leq \sum_{n=1}^\infty \tilde{\mu}_\alpha(x)$ for $x, x_n \in S$. Take $t \in S$, $t \leq x$. Then

$$\left|\mu\left(t \wedge \sum_{i=1}^{n} x_i\right)\right|_\alpha \leq \tilde{\mu}_\alpha\left(\sum_{i=1}^{n} x_i\right) \leq \sum_{i=1}^{n} \tilde{\mu}_\alpha(x_i) \leq \sum_{i=1}^{\infty} \tilde{\mu}_\alpha(x_i)$$

by 3.1.2, and $\mu(t) = \lim_n \mu(t \wedge \sum_{i=1}^n x_i)$ since $t \wedge \sum_{i=1}^n x_i \uparrow t$. Therefore, $\left|\mu(t)\right|_\alpha = \lim_n \left|\mu(t \wedge \sum_{i=1}^n x_i)\right|_\alpha \leq \sum_{i=1}^\infty \tilde{\mu}_\alpha(x_i)$.

By 3.2.1, μ_α^* is a σ–submeasure extending $\tilde{\mu}_\alpha$ for $\alpha \in A$. Therefore the uniformity $u^* \in \mathcal{LUA}(P)$ generated by $(\mu_\alpha^*)_{\alpha \in A}$ satisfies (σ) by 2.7 and extends the μ–uniformity. Since μ is uniformly continuous with respect to u^*, μ has a unique continuous extension $\bar{\mu} : \bar{S} \to G$ to $\bar{S}$. Since the operations $+, \Delta, \vee, \wedge$ are uniformly continuous, $\bar{S}$ is a subalgebra of P and $\bar{\mu}$ is a measure.

(b) Since u^* satisfies (σ) and, under the assumption of (b), $u^*|_S$ is (locally) exhaustive, $\bar{S}$ is (Dedekind) σ–complete, and $u^*|_{\bar{S}}$ is σ–order continuous by [21, 8.2]. Since $\bar{\mu}$ is continuous with respect to u^*, $\bar{\mu}$ is also σ–order continuous. ∎

3.2.3. Corollary. *Assume that P is (Dedekind) σ–complete and satisfies conditions* (1) *and* (2) *of* 3.2.1. *Let S be a subalgebra of $(P, +, \Delta, \vee, \wedge)$ and $\mu : S \to G$ a σ–order continuous (locally) exhaustive measure. Then μ has a unique G–valued σ–order continuous measure extension to the (Dedekind) σ–complete subalgebra T of P generated by S. (Here,* sup *and* inf *are always taken in P).*

Proof. The existence statement follows from 3.2.2.

Uniqueness: For $i = 1, 2$ let $\nu_i : T \to G$ be a σ–order continuous measure extension of μ, $\nu := \nu_1 - \nu_2$, and u the ν–uniformity. Since u is σ–order continuous, the closure $\bar{S}$ of S in (T, u) is a (Dedekind) σ–complete subalgebra of T, hence $\bar{S} = T$. Since ν is continuous with respect to u and ν vanishes on S, we get $\nu = 0$, i.e. $\nu_1 = \nu_2$. ∎

3.2.4. Corollary. *Let L be a σ–complete MV–algebra, S a subalgebra of L, and $\mu : S \to G$ a σ–order continuous exhaustive measure. Then μ has a unique G–valued σ–order continuous measure extension to the σ–complete subalgebra generated by S.*

3.2.5. Corollary. *Let E be a Dedekind σ-complete ℓ-group, F be an ℓ-subgroup of E, and $T : F \to G$ a σ-order continuous locally exhaustive homomorphism. If E is the only subgroup of E containing F and being closed with respect to countable* sup *and* inf, *then there is a unique σ-order continuous homomorphism $\bar{T} : E \to G$ which extends T.*

Proof. Since any Dedekind σ–complete ℓ–group is commutative, the positive cone P of E is a Δ-ℓ–semigroup. Let S be the positive cone of F. Then, by 3.2.3, $\mu := T|_S$ has a unique σ–order continuous measure extension $\nu : P \to G$. The additive extension $\bar{T} : E \to G$ of ν is the unique σ–order continuous homomorphism on E extending T. ∎

The extension Theorem 3.2.4 can be used to obtain the following integral representation of bounded real–valued T_∞–measures on clans of fuzzy sets — known for T_∞–measures on tribes of fuzzy sets (= clans closed under countable supremum). Let $\mu : \mathcal{C} \to \mathbb{R}$ be a bounded real–valued T_∞–measure on a clan $\mathcal{C}$ of fuzzy sets, $\mathcal{T}$ the tribe generated by $\mathcal{C}$, $\bar{\mu} : \mathcal{T} \to \mathbb{R}$ the T_∞–measure extension of μ (see 3.2.4), and $\check{\mu}$ the restriction of μ to the σ–algebra $\check{\mathcal{T}}$ of the crisp sets of $\mathcal{T}$. Then, by [4], $\bar{\mu}(f) = \int f d\check{\mu}$ for all $f \in \mathcal{T}$, hence $\mu(f) = \int f d\check{\mu}$ for $f \in \mathcal{C}$.

3.2.5 contains as special case an extension theorem of Pellaumail [14] for group–valued Daniell integral, which is proved in the Banach space–valued case by Kluvanek [10].

4. Exhaustive topological Δ-ℓ–semigroups and MV–algebras

For $w \in \mathcal{LUA}(P)$, the space $\mathcal{LUA}(P, w) := \{u \in \mathcal{LUA}(P) : u \subset w\}$ is a complete sublattice of $\mathcal{LUA}(P)$. We look for conditions under which $\mathcal{LUA}(P, w)$ is a Boolean algebra. This is important to get decomposition theorems for measures on P, see Section 5. To prove the main results of this section we will pass to the completion $(\tilde{P}, \tilde{w})$ of (P, w).

For exhaustive w, $\tilde{P}$ is complete as a lattice and $\tilde{w}$ is order continuous, hence (understandably) $(\tilde{P}, \tilde{w})$ is easier to handle than (P, w).

4.1. Proposition. *Let w be a Hausdorff uniformity of $\mathcal{LUA}(P)$. Then*

(a) *(P, w) is a dense subalgebra of a Δ-ℓ–semigroup $\tilde{P}$ endowed with a Hausdorff complete uniformity $\tilde{w} \in \mathcal{LUA}(\tilde{P})$.*

(b) *If w is (locally) exhaustive, then $(\tilde{P}, \leq)$ is (Dedekind) complete as a lattice and $\tilde{w}$ is order continuous.*

(c) $\tilde{u} \mapsto \tilde{u}|_P$ *defines an order isomorphism from* $\mathcal{LUA}(\tilde{P}, \tilde{w})$ *onto* $\mathcal{LUA}(P, w)$. *Here,* $\tilde{u}$ *is (locally) exhaustive iff* $\tilde{u}|_P$ *is.*

Proof. (a) Let $(\tilde{P}, \tilde{w})$ be the Hausdorff uniform completion of (P, w). Then $+, \Delta, \vee, \wedge$ defined on P^2 have a uniformly continuous extension to $\tilde{P}^2$. So $\tilde{P}$ becomes a Δ-ℓ–semigroup since Δ-ℓ–semigroups are equationally defined. Obviously, P is a subalgebra of $(\tilde{P}, +, \Delta, \vee, \wedge)$.

(b) is a special case of [21, 6.15].

(c) The analogous statement for lattice uniformities is given in [23, 3.8 (a)]. Therefore it is enough to observe that for any lattice uniformity $\tilde{u}$ on $\tilde{P}$ coarser than $\tilde{w}$ we have

$$\tilde{u} \in \mathcal{LUA}(\tilde{P}) \qquad \text{iff} \qquad \tilde{u}|_P \in \mathcal{LUA}(P).$$

Here, the implication ($\Rightarrow$) is obvious. Now let $\tilde{u}|_P \in \mathcal{LUA}(P)$. Then the closures in $(\tilde{P}, \tilde{w})$ of the 0–neighborhoods in $(P, \tilde{u}|_P)$ form a filter of $\mathcal{FNA}(\tilde{P})$. The induced uniformity $\tilde{v} \in \mathcal{LUA}(\tilde{P})$ is coarser than $\tilde{w}$ and $\tilde{v}|_P = \tilde{u}|_P$ (see 2.1), hence $\tilde{v} = \tilde{u}$ and $\tilde{u} \in \mathcal{LUA}(\tilde{P}, \tilde{w})$. ■

For any $w \in \mathcal{LUA}(P)$, let $N(w) := \overline{\{0\}}^w$ be the closure of $\{0\}$ in (P, w). If $\mu : P \to G$ is a measure with values in a Hausdorff topological Abelian group and w is the μ–uniformity, then $N(w) = N(\mu)$.

4.2. Proposition. *Let* $w \in \mathcal{LUA}(P)$. *Then* $N(w)$ *is an ideal in* P. *Denote by* $\hat{a}$ *the equivalence class of* $\hat{P} := P/N(w)$ *containing* a *and let* $\hat{U} := \{\hat{a} : a \in U\}$ *for* $U \subset P$. *Then, for* $u \in \mathcal{LUA}(P, w)$, *the sets* $\hat{U}$ $(U \in u(0))$ *form a base of a filter of* $\mathcal{FNA}(\hat{P})$.

Denote by $\hat{u}$ *the uniformity corresponding to this filter according to* 2.1. $u \mapsto \hat{u}$ *defines an order isomorphism from* $\mathcal{LUA}(P, w)$ *onto* $\mathcal{LUA}(\hat{P}, \hat{w})$. *Hereby* u *is (locally) exhaustive iff* $\hat{u}$ *is.*

The proof of 4.2 is obvious.

4.3. Theorem. *Let* w *be an exhaustive uniformity of* $\mathcal{LUA}(P)$ *and* $(\tilde{P}, \tilde{w})$ *the completion of* $(\hat{P}, \hat{w}) := (P, w)/N(w)$ *(see* 4.1, 4.2*). Then* $q : u \mapsto \hat{u}$ *and* $r : \tilde{u} \mapsto \tilde{u}|_{\hat{P}}$ *define order isomorphisms from* $\mathcal{LUA}(P, w)$ *onto* $\mathcal{LUA}(\hat{P}, \hat{w})$ *and from* $\mathcal{LUA}(\tilde{P}, \tilde{w})$ *onto* $\mathcal{LUA}(\hat{P}, \hat{w})$, *respectively; and* $s : \tilde{u} \mapsto \sup N(\tilde{u})$ *defines a dual order isomorphism from* $\mathcal{LUA}(\tilde{P}, \tilde{w})$ *onto* $I(\tilde{P})$.

$s \circ r^{-1} \circ q : \mathcal{LUA}(P, w) \to I(\tilde{P})$ *is therefore a dual order isomorphism.*

Proof. By 4.1 and 4.2, it remains to prove that $s : \mathcal{LUA}(\tilde{P}, \tilde{w}) \to I(\tilde{P})$ is a dual isomorphism.

s is well–defined: Let $\tilde{u} \in \mathcal{LUA}(\tilde{P}, \tilde{w})$. Then $N(\tilde{u}) = \overline{\{0\}}^{\tilde{u}}$ is a closed ideal in $(\tilde{P}, \tilde{u})$, hence it is also closed with respect to $\tilde{w}$. Since $\tilde{w}$ is Hausdorff and order continuous, $N(\tilde{u})$ is a principal ideal. Therefore its maximal element $s(\tilde{u})$ is idempotent, i.e. $s(\tilde{u}) \in I(\tilde{P})$.

s is surjective: Let $a \in I(\tilde{P})$. Then the sets $[0, a] + W$, $W \in \tilde{w}(0)$, form a base of a filter $\mathcal{F} \in \mathcal{FNA}(\tilde{P})$, cf. 2.4. By 2.1, $\mathcal{F}$ generates a uniformity $\tilde{u} \in \mathcal{LUA}(\tilde{P}, \tilde{w})$. We show that $s(\tilde{u}) = a$, i.e. that $\cap_{W\in\tilde{w}(0)}([0, a] + W) = [0, a]$. Let $x \in \cap_{W\in\tilde{w}(0)}([0, a] + W)$. Then there are a net $(x_\gamma)_{\gamma\in\Gamma}$ in $[0, a]$ and a net $(y_\gamma)_{\gamma\in\Gamma}$ converging to 0 in $(\tilde{P}, \tilde{w})$ such that $x = x_\gamma + y_\gamma$ $(\gamma \in \Gamma)$. Therefore $x \le a + y_\gamma \to a$ $(\tilde{w})$, hence $x \le a$ since $(\tilde{P}, \tilde{w})$ is a Hausdorff topological lattice.

$\tilde{u} \subset \tilde{v}$ iff $s(\tilde{u}) \ge s(\tilde{v})$ for $\tilde{u}, \tilde{v} \in \mathcal{LUA}(\tilde{P}, \tilde{w})$. The implication $(\Rightarrow)$ is obvious. Now suppose that $s(\tilde{u}) \ge s(\tilde{v})$. Then the congruence corresponding to the ideal $[0, s(\tilde{u})]$ contains the congruence corresponding to $[0, s(\tilde{v})]$, i.e. $\cap_{U\in\tilde{u}}U \supset \cap_{V\in\tilde{v}}V$. It follows now from [21, 6.7] that the $\tilde{u}$–topology is weaker than the $\tilde{v}$–topology. Hence $\tilde{u} \subset \tilde{v}$ by 2.1. ■

Of particular interest are the cases in which $I(\tilde{P}) = C(\tilde{P})$; $\mathcal{LUA}(P, w)$ is then, by 4.3, a Boolean algebra. If P is an MV–algebra, then so are $\hat{P}$ and $\tilde{P}$, hence $I(\tilde{P}) = C(\tilde{P})$. It follows therefore from 4.3 that:

4.4. Corollary. *If P is an MV–algebra and w an exhaustive uniformity of $\mathcal{LUA}(P)$, then $\mathcal{LUA}(P, w)$ is a complete Boolean algebra.*

4.5. Corollary. *For $\alpha \in A$, let $\mu_\alpha : P \to G_\alpha$ be an exhaustive measure with values in a Hausdorff topological Abelian group. If w is the supremum of the μ_α–uniformities, $\alpha \in A$, then $\mathcal{LUA}(P, w)$ is a complete Boolean algebra.*

Proof. Let G be the completion of $\Pi_{\alpha\in A}G_\alpha$, where $\Pi_{\alpha\in A}G_\alpha$ is endowed with the product topology and $\mu : P \to G$ the measure defined by $\mu = (\mu_\alpha)_{\alpha\in A}$. Let $(\tilde{P}, \tilde{w})$ be the completion of $(\hat{P}, \hat{w}) = (P, w)/N(w)$ and $\tilde{\mu}$ the continuous extension of the measure $\hat{\mu} : \hat{P} = P/N(w) \to G$ defined by $\hat{\mu}(\hat{x}) = \mu(x)$. By [23, 3.8 (b)] (see 5.2) $\tilde{w}$ is the $\tilde{\mu}$–uniformity, hence $N(\tilde{\mu}) = N(\tilde{w}) = \{0\}$. Moreover, $\tilde{P}$ has a greatest element, since $\tilde{P}$ is complete by 4.1 (b). It follows by 1.8 that $\tilde{P}$ is an MV–algebra, hence $C(\tilde{P}) = I(\tilde{P})$, and $\mathcal{LUA}(P, w)$ is a Boolean algebra by 4.3. ■

5. Decomposition of measures on Δ-ℓ–semigroups

In the following, let $(G, +)$ be a complete Hausdorff topological Abelian group.

The essential tool to get decomposition theorems for measures here is Theorem 4.3.

First, we fix some notation. Let $u, v \in \mathcal{LUA}(P)$ and μ, ν be measures on P. We write $u \perp v$ iff the infimum of u and v is the trivial uniformity. If u is the μ–uniformity, we write also $\mu \perp v$ instead of $u \perp v$; if, moreover, v is the ν–uniformity, we write $\mu \perp \nu$ instead of $u \perp v$. $\mu \ll v$ means that the μ–uniformity is weaker than v, i.e. that μ is continuous with respect to v. If v is the ν–uniformity, we write $\mu \ll \nu$ instead of $\mu \ll v$.

5.1. Proposition. *Let $\lambda, \nu : P \to G$ be measures satisfying $\lambda \perp \nu$ and $\mu = \lambda + \nu$. Then the μ–uniformity is the supremum of the λ–uniformity and the ν–uniformity.*

Proof. We denote by u, v, and w, respectively, the λ–uniformity, the ν–uniformity and the μ–uniformity. Since μ is continuous with respect to $u \vee v$, we have $w \subset u \vee v$. To prove the other inclusion $u \vee v \subset w$, we show that λ and ν are continuous at 0 with respect to w (see 3.1.3). For that, let W_0 be a 0–neighborhood in G and U, V, and W be, respectively, solid 0–neighborhoods with respect to u, v, and w with $\lambda(U) \cup \nu(V) \cup \mu(W) \subset W_0$. We show that $\lambda(W) \cup \nu(W) \subset W_0 + W_0 - W_0$. Let $x \in W$. Since $\lambda \perp \nu$, by 2.4 there are elements $y \in U$ and $z \in V$ with $x = y + z$. Since $x = y + (x \Delta y)$, $x \Delta y \leq z \in V$ and V is solid, we may assume that $z = x \Delta y$. Therefore $\lambda(x) = \lambda(y) + \lambda(z) = \lambda(y) + \mu(z) - \nu(z) \in W_0 + W_0 - W_0$ and analogously $\nu(x) \in W_0 + W_0 - W_0$. ■

The significance of 5.1 is that, under the assumption of 5.1, λ and ν are, respectively, exhaustive, locally exhaustive, σ–order continuous, order continuous iff so is μ.

5.2. Lemma ([23, 3.8 (b)]). *Let $\tilde{w}$ be a lattice uniformity on a lattice $\tilde{L}$, L a dense sublattice of $(\tilde{L}, \tilde{w})$, $\tilde{\mu} : \tilde{L} \to G$ a modular function, and $\mu = \tilde{\mu}|_L$ the restriction of $\tilde{\mu}$ to L. Then the $\tilde{\mu}$–uniformity (= weakest lattice uniformity which makes $\tilde{\mu}$ uniformly continuous) induces on L the μ–uniformity. Moreover, $\tilde{w}$ is the $\tilde{\mu}$–uniformity iff $\tilde{w}|_L$ is the μ–uniformity.*

5.3. Decomposition theorem. *Let $\mu : P \to G$ be an exhaustive measure and $u \in \mathcal{LUA}(P)$. Then there are unique measures $\lambda, \nu : P \to G$ such that $\mu = \lambda + \nu$, $\lambda \ll u$, and $\nu \perp u$, in particular, $\lambda \perp \nu$.*

Proof. Existence: Let w be the μ–uniformity and $u_1 := u \wedge w$. Since, by 4.5, $\mathcal{LUA}(P, w)$ is a Boolean algebra, u_1 has a complement v in $\mathcal{LUA}(P, w)$, i.e. $u_1 \perp v$ and $u_1 \vee v = w$. With the notation of 4.2 and 4.3, let a and b, respectively, be the images of u_1 and v under the dual order isomorphism $f := s \circ r^{-1} \circ q$. Let $\tilde{\mu} : (\tilde{P}, \tilde{w}) \to G$ be the continuous extension of the function $\hat{\mu} : \hat{P} \to G$ defined by $\hat{\mu}(\hat{x}) := \mu(x)$. Put $\tilde{\mu}_a(z) := \tilde{\mu}(a \wedge z)$ for

$z \in \tilde{P}$, $\hat{\mu}_a := \tilde{\mu}_a|_{\hat{P}}$, and $\mu_a(x) := \hat{\mu}_a(\hat{x})$ for $x \in P$. Since $I(\tilde{P}) = C(\tilde{P})$ (cf. the proof of 4.5), we have $a \in C(\tilde{P})$. It follows from 1.9 that $\tilde{\mu}_a$ is a measure; therefore $\hat{\mu}_a$ and μ_a are measures, too. With the aid of 5.2 one sees that $f(\mu_a$–uniformity) $= \sup N(\tilde{\mu}_a) \geq b = f(v)$, hence $\mu_a \ll v$. Analogously to $\tilde{\mu}_a$ and μ_a define $\tilde{\mu}_b$ and μ_b. Put $\lambda := \mu_b$ and $\nu := \mu_a$. Then $\lambda \ll u_1 \subset u$; $\nu \perp u$ since $\nu \ll v$ and $v \perp u$; in fact, $v \wedge u = (v \wedge w) \wedge u = v \wedge (w \wedge u) = v \wedge u_1$ and $v \perp u_1$. Finally, from $\tilde{\mu} = \tilde{\mu}_a + \tilde{\mu}_b$ follows $\mu = \lambda + \nu$.

Uniqueness: Let $\lambda_i, \nu_i : P \to G$ be measures with $\lambda_i \ll u$ and $\nu_i \perp u$ $(i = 1, 2)$. Then $\rho := \lambda_1 - \lambda_2 \ll u$ and $\rho = \nu_2 - \nu_1 \perp u$. But $\rho \ll u$ and $\rho \perp u$ imply that the ρ–uniformity is trivial, i.e. $\rho = 0$. ■

For measures on rings of sets, decomposition theorem 5.3 was first proved by Traynor [16]. Another proof of Traynor's theorem is given in [20]. Here we have used the method of [20]. As in the Boolean case [16, 20], different choices of u yield different decomposition theorems. If u is the ρ–uniformity for some measure ρ, then 5.3 becomes the Lebesgue decomposition theorem; $\lambda \ll u$ and $\nu \perp u$ can then be written as $\lambda \ll \rho$ and $\nu \perp \rho$.

Under additional assumptions, $\mu \ll \nu$ and $\mu \perp \nu$ can be characterized by means of $N(\mu)$ and $N(\nu)$ as in the Boolean case, as follows.

5.4. Remark. *Let $\mu, \nu : P \to G$ be measures.*

(a) *If w is an exhaustive complete uniformity of $\mathcal{LUA}(P)$, $\mu \ll w$, and $\nu \ll w$, then*

  (i) *$\mu \ll \nu$ iff $N(\mu) \supset N(\nu)$,*

  (ii) *$\mu \perp \nu$ iff $N(\mu) + N(\nu) = P$.*

(b) *If μ, ν are σ–order continuous, G is metrizable, and $(P, \leq)$ is σ–complete, then the supremum of the μ–uniformity and the ν–uniformity is complete, in particular,* (i) *and* (ii) *are true.*

Proof. (a) This follows immediately from 4.3 observing that, under the given assumptions, $\tilde{P} = \hat{P}$ (notation as in 4.3) and $\tilde{P}$ has a greatest element by 4.1.

(b) Let w be the supremum of the μ–uniformity and the ν–uniformity. Then w is σ–order continuous, it has a countable base and therefore is complete by [19, 2.2.3], since $(P, \leq)$ is σ–complete. Now apply (a). ■

Already in the Boolean case, a measure, which is not exhaustive, cannot be decomposed with respect to *any* $u \in \mathcal{LUA}(P)$ (in the sense of 5.3). We now show that locally exhaustive measures can be decomposed with respect to a uniformity u^*, where $u \in \mathcal{LUA}(P)$, and u^* is the finest uniformity of $\mathcal{LUA}(P)$ which induces for any $a \in P$ on $[0, a]$ the same uniformity as u.

5.5. Theorem. *Let $\mu : P \to G$ be a locally exhaustive measure and $u \in \mathcal{LUA}(P)$. Then there are unique measures $\lambda, \nu : P \to G$ such that $\mu = \lambda + \nu$, $\lambda \ll u^*$, and $\nu \perp u^*$, in particular, $\lambda \perp \nu$.*

Proof. The uniqueness of the decomposition can be proved as in 5.3. Let us prove the existence.

(i) For any $a \in P$, the interval $P_a := \{x \in P : x \leq a\}$ is a Δ-ℓ–semigroup with the induced partial order, the induced operation Δ, and the addition $\oplus$ defined by $x \oplus y := (x + y) \wedge a$. For any measure $\rho : P \to G$, the restriction $\rho_a := \rho|_{[0,a]}$ is a measure on P_a. We use in the following that the ρ–uniformity induces on P_a the ρ_a–uniformity. This fact immediately follows from the description of the ρ–uniformity by means of the submeasures defined in 3.1.2. For $v \in \mathcal{LUA}(P)$, we denote by v_a the uniformity induced by v on P_a.

(ii) For $a \in P$, μ_a has by 5.3 the unique decomposition $\mu_a = \lambda^a + \nu^a$, where $\lambda^a, \nu^a : P_a \to G$ are measures with $\lambda^a \ll u_a$ and $\nu^a \perp u_a$. If $b \in P$ and $b \geq a$, then $\mu_a = (\lambda^b|_{P_a}) + (\nu^b|_{P_a})$, $\lambda^b|_{P_a} \ll u_a$, and $\nu^b|_{P_a} \perp u_a$, hence $\lambda^a = \lambda^b|_{P_a}$ and $\nu^a = \nu^b|_{P_a}$ because of the uniqueness of the decomposition. It follows that there are measures λ and ν on P with $\lambda_a = \lambda^a$ and $\nu_a = \nu^a$ $(a \in P)$, i.e., the restrictions of λ and ν on P_a are λ^a and ν^a, respectively. Then $\mu = \lambda + \nu$ and $\lambda \ll u^*$, since $\lambda_a \ll u_a$ for any $a \in P$.

$\nu \perp u^*$: Let v be the ν–uniformity. Then $(v \wedge u^*)_a \subset v_a \wedge u^*_a = v_a \wedge u_a = 0$ (:= trivial uniformity), hence $(v \wedge u^*)_a = 0$ for any element $a \in P$ and therefore $v \wedge u^* = 0$, that is, $\nu \perp u^*$. ■

If P is a positive cone of a commutative ℓ–group, we get from 5.5 the following decomposition theorem of Basile and Traynor.

5.6. Corollary ([2, Theorem 3.3]). *Let E be a commutative ℓ–group and $T : E \to G$ a locally exhaustive homomorphism. Let τ be a locally solid group topology on E, and τ^* the finest locally solid group topology which agrees with τ on every interval $[a, b]$ of E. Then there are unique homomorphisms $T_1, T_2 : E \to G$ such that $T = T_1 + T_2$; T_1 is continuous with respect to τ^* and the infimum of τ^*, and the T_2–topology (that is the weakest locally solid group topology on E making T_2 continuous) is the trivial topology.*

To obtain 5.6 from 5.5, observe that there is a natural order isomorphism between the space of all locally solid group topologies on E and $\mathcal{LUA}(P)$, where P denotes the positive cone of E (cf. 2.1), and that any G–valued additive map on P has a unique extension to a G–valued homomorphism on E.

It follows from 5.1 that in 5.6 the T–topology is the supremum of the T_1–topology and the T_2–topology, in particular, T_1 and T_2 are locally exhaustive.

Basile and Traynor deduce from their result 5.6 a Hewitt–Yosida decomposition for homomorphisms on ℓ–groups. We now do the same for measures on Δ-ℓ–semigroups (using 5.5 instead of 5.6). We call a measure ν on P *purely non σ–order continuous* if the zero–measure is the only σ–order continuous measure ρ on P with $\rho \ll \nu$.

5.7. Hewitt–Yosida decomposition. *Let $\mu : P \to G$ be a locally exhaustive measure. Then there are unique measures $\lambda, \nu : P \to G$ such that $\mu = \lambda + \nu$, λ is σ–order continuous, and ν is purely non σ–order continuous. Moreover, $\lambda \perp \nu$.*

Proof. Apply 5.5 using for u the supremum of all σ–order continuous uniformities of $\mathcal{LUA}(P)$. For this uniformity we have $u = u^*$. ∎

Since $\lambda \perp \nu$ in 5.7, the measures λ and ν are locally exhaustive by 5.1.

Studying not only exhaustive but also locally exhaustive measures allows us to include the case of a homomorphism on ℓ–groups (or linear operators on Riesz spaces).

5.8. Remark. *Let E be a commutative ℓ–group.*

(a) *The trivial topology is the only exhaustive locally solid group topology on E.*

(b) *There is no nonzero exhaustive homomorphism on E.*

Proof. (a) If τ is an exhaustive locally solid group topology on E, then the uniform completion of $E/\overline{\{0\}}^{\tau}$ is a complete (not only Dedekind complete) ℓ–group (cf. 4.1 (b)) and therefore it has one element only, i.e. $\overline{\{0\}}^{\tau} = E$.

(b) If T is an exhaustive homomorphism on E, then the T–topology is trivial by (a), hence $T = 0$. ∎

6. Decomposition of MV–algebras and Hammer–Sobczyk's decomposition

6.1. Decomposition of complete MV–algebras

In this section, let L be an MV–algebra.

Here we present a decomposition theorem for complete MV–algebras (Theorem 6.1.6), which is an important tool for the characterization of connected and compact MV–algebras (see 6.1.9) and for the decomposition theorems for the measures of Section 6.2. Although decomposition theorem 6.1.6 is essentially a reformulation of the main result of [9], we will give the proof. We use in the proof of 6.1.6 partly the same tools as Jakubik (6.1.1 and Hölder's theorem), but our proof becomes simpler by using elements of the center of L to get the decomposition of L (cf. 6.1.5).

If $(H, \oplus, \leq)$ is a commutative ℓ–group and e a positive element of H, then the interval $\{x \in H : 0 \leq x \leq e\}$ becomes an MV–algebra with respect to the operations $x + y := (x \oplus y) \wedge e$ and $x' := e - x$; we denote this MV–algebra by $L_0(H, e)$. Chang [6] has shown that any MV–algebra is of this type; hereby observe that Chang's axioms of an MV–algebra and Mangani's axioms, which we use here, are equivalent by [15].

6.1.1. Theorem ([6], [9]). *There exists a commutative ℓ–group with the strong order unit e such that $L = L_0(H, e)$. Hereby, L is complete iff H is Dedekind complete, and L is totally ordered iff H is totally ordered.*

For the first statement, see [6] and [13], for the second see [9, Lemma 1.1 and Lemma 1.2].

We recall that an *atom* of a lattice L_0 with 0 is an element $a \in L_0 \setminus \{0\}$ such that whenever $x \in L_0$ and $0 \leq x \leq a$, we have either $x = 0$ or $x = a$. L_0 is *atomic* if for every $x \in L_0 \setminus \{0\}$ there is an atom a of L_0 with $a \leq x$; L_0 is *atomless* if it doesn't contain any atom.

6.1.2. Lemma. *Let L be complete. For any $a \in L$, put $\bar{a} := \sup\{na : n \in \mathbb{N}\}$. Then*

(a) *$\bar{a} \in C(L)$ for any $a \in L$,*

(b) *$\bar{a}$ is an atom of $C(L)$ iff $a \neq 0$ and $[0, a]$ is totally ordered.*

Proof. (a) was observed by Lacava [11, Proposition 15]. In fact, $\bar{a} \in I(L) = C(L)$.

(b)($\Rightarrow$) We represent L as an interval of an ℓ–group. Let $x, y \in [0, a]$. Put $b := x - x \wedge y$ and $c := y - x \wedge y$. Since $\bar{a}, \bar{b} \in C(L)$ by (a), $\bar{a} \geq \bar{b}$, and $\bar{a}$ is an atom of $C(L)$, we obtain either $\bar{b} = 0$ or $\bar{b} = \bar{a}$. If $\bar{b} = 0$, then $b = 0$ and therefore $x \leq y$. If $\bar{b} = \bar{a}$, then from $b \wedge c = (x - x \wedge y) \wedge (y - x \wedge y) = 0$ we deduce that $nb \wedge c = 0$ for $n \in \mathbb{N}$, therefore $0 = \bar{b} \wedge c = \bar{a} \wedge c = c$, thus $y \leq x$.

($\Leftarrow$) Let a_1, a_2 be disjoint elements of $C(L)$ with $\bar{a} = a_1 \vee a_2$. Then $a \wedge a_1$, $a \wedge a_2$ are disjoint elements of the totally ordered set $[0, a]$ and therefore one of these elements is 0. Assume that $a \wedge a_2 = 0$. Then $\bar{a} \wedge a_2 = 0$ and therefore $a_2 = 0$. ■

In the terminology of Jakubik [9], 6.1.2 (b) means precisely that $\bar{a}$ is an atom of $C(L)$ iff a is an α–atom of L.

We denote by $\mathcal{A}_\infty(L)$ and $\mathcal{A}_f(L)$, respectively, the set of all atoms a of $C(L)$ for which $[0,a]$ is infinite and finite. Let $Ł_\infty := L_0(\mathbb{R},1)$ be the real unit interval considered in the usual way as MV–algebra and $Ł_{n+1} := \{ i/n : i = 0,\dots,n\}$ the subspace of $Ł_\infty$ consisting of $n+1$ elements for $n \in \mathbb{N}$.

6.1.3. Proposition. *Let L be a complete MV–algebra and a an atom of $C(L)$. If $a \in \mathcal{A}_\infty(L)$, then $[0,a]$ is isomorphic to $Ł_\infty$. If $a \in \mathcal{A}_f(L)$, then $[0,a]$ is isomorphic to $Ł_{n+1}$ for some $n \in \mathbb{N}$.*

Proof. Since $a \in C(L)$, the interval $[0,a]$ is an MV–algebra with respect to the induced addition and the unary operation $x^* := x' \wedge a$ (see 1.6, 1.7 or [11]). Since $[0,a]$ is totally ordered by 6.1.2, $[0,a]$ is an interval of a Dedekind complete totally ordered ℓ–group H by 6.1.1. By Hölder's theorem [3, p. 300], H is isomorphic either to $\mathbb{R}$ or to $\mathbb{Z}$. In the first case, $[0,a]$ is isomorphic to $Ł_\infty$; in the second case, to $Ł_{n+1}$ for some $n \in \mathbb{N}$. ■

It is also possible to avoid the use of 6.1.1 by using the version of Hölder's theorem in [3, p. 322] to prove that an infinite totally ordered complete MV–algebra is isomorphic to $Ł_\infty$.

We now give a connection between atoms of L and atoms of $C(L)$.

6.1.4. Proposition. *Assume that L is complete.*

(a) *For any atom a in L there is an $\bar{a} \in \mathcal{A}_f(L)$ with $a \le \bar{a}$.*

(b) *For any $\bar{a} \in \mathcal{A}_f(L)$ there is an atom a of L with $a \le \bar{a}$.*

Proof. (a) Since $[0,a] = \{0,a\}$ is totally ordered, $\bar{a} := \sup\{na : n \in \mathbb{N}\}$ is an atom of $C(L)$ by 6.1.2. Since a is an atom of $[0,\bar{a}]$ but $Ł_\infty$ doesn't contain any atom, $[0,\bar{a}]$ and $Ł_\infty$ are not isomorphic. Hence $\bar{a} \in \mathcal{A}_f(L)$ by 6.1.3.

(b) is obvious. ■

6.1.5. Lemma. *Let L be complete and D be a set of pairwise disjoint elements of $C(L)$ with $\sup D = 1$. Then $f : x \mapsto (x \wedge d)_{d\in D}$ defines an isomorphism from L onto $\Pi_{d\in D}[0,d]$.*

Proof. As mentioned at the beginning of the proof of 6.1.3, the intervals $[0,d]$ are MV–algebras. Moreover, f is an MV–algebra homomorphism (see 1.6, 1.7 (c)). Since L is a distributive complete lattice, f is a lattice isomorphism. ■

6.1.6. Theorem. *Assume that L is a complete MV–algebra.*
Let $a_\infty := \sup \mathcal{A}_\infty(L)$, $a_f := \sup \mathcal{A}_f(L)$, and $c := (a_\infty \vee a_f)'$. Then

(a) *L is isomorphic to $[0,a_f] \times [0,a_\infty] \times [0,c]$.*

(b) $[0, a_f]$ *is isomorphic to a product* $\Pi_{\alpha \in A} \text{Ł}_{n_\alpha}$, *where* $A = \mathcal{A}_f(L)$.

(c) $[0, a_\infty]$ *is isomorphic to the power* Ł_∞^B, *where* $B = \mathcal{A}_\infty(L)$.

(d) *The center of* $[0, c]$ *is atomless.*

(e) L *is atomless iff* $a_f = 0$ *iff* $\mathcal{A}_f(L) = \emptyset$.

(f) L *is atomic iff* $a_\infty = c = 0$.

Proof. Since L is complete, $I(L)$ is complete, too. Using $C(P) = I(P)$, we get $a_f, a_\infty, c \in C(P)$. (a), (b), (c) now follow from 6.1.5 and 6.1.3.

(d) is obvious since $C([0, a]) = C(L) \cap [0, a]$ for any $a \in C(L)$.

(e) follows from 6.1.4.

(f) follows from (a), (b), and 6.1.4. ■

We now use 6.1.6 to study compactness and connectedness properties of MV–algebras.

6.1.7. Theorem. *Let u be an exhaustive Hausdorff uniformity of $\mathcal{LUA}(L)$ such that (L, u) is complete. Then*

(a) $(L, \leq)$ *is complete as a lattice. The isomorphisms of* 6.1.6 (a), (b), (c) *are also homeomorphisms. Hereby* $[0, a_f]$, $[0, a_\infty]$, $[0, c]$ *are endowed with the uniformity induced by u,* Ł_n *and* Ł_∞ *— with the uniformity induced by the usual uniformity of* $\mathbb{R}$, *and the products have the product uniformity.*

(b) *By the notation of* 6.1.6, $[0, a_f]$ *is totally disconnected and compact,* $[0, a_\infty]$ *is arcwise connected and compact,* $[0, c]$ *is arcwise connected and not compact if* $c \neq 0$.

Proof. (a) By 4.1, $(L, \leq)$ is complete and u order continuous. The second statement of (a) follows from the following consequence of [21, 5.10]: any order isomorphism between two complete lattices endowed with Hausdorff order continuous lattice uniformities is a homeomorphism.

(b) The product space $\Pi_{\alpha \in A} \text{Ł}_{n_\alpha}$ is totally disconnected and compact, moreover, homeomorphic to $[0, a_f]$ (cf. 6.1.6 (b)). Hence $[0, a_f]$ is totally disconnected and compact. Similarly we get with 6.1.6 (c) that $[0, a_\infty]$ is arcwise connected and compact. The center C of $[0, c]$ is by 6.1.6 (d) an atomless Boolean algebra, and closed in (L, u) since $C = I([0, c])$, hence complete (as uniform space). Therefore there is by [17, Satz 1.4] a continuous map $\gamma : I \to C$ defined on the real closed unit interval I with $\gamma(0) = 0$ and $\gamma(1) = c$. For any $a \in [0, c]$, $\alpha(t) := \gamma(t) \wedge a$ defines then a continuous map $\alpha : I \to [0, c]$ with $\alpha(0) = 0$ and $\alpha(1) = a$. Therefore $[0, c]$ is arcwise connected.

Assume that $[0, c]$ is compact. Then the closed subspace C is also compact. We have already seen that C is connected. It follows that $C = \{0\}$, i.e.

$c = 0$, since any compact Boolean algebra is totally disconnected by [18, Theorem 3]. ■

Under the assumption of 6.1.7, L is compact or totally disconnected or connected, respectively, iff $c = 0$ or $c = a_\infty = 0$ or $a_f = 0$. This leads to the following characterization of compact, totally disconnected and connected MV–algebras.

6.1.8. Corollary. *Let τ be a Hausdorff topology on L which makes the operations $+$ and $'$ of L continuous. Then (L, τ) is compact iff (L, τ) is isomorphic and homeomorphic to a product $Ł_\infty^B \times \Pi_{\alpha \in A} Ł_{n_\alpha}$, where $n_\alpha \in \mathbb{N} \setminus \{1\}$ for $\alpha \in A$.*

Proof of ($\Rightarrow$). Since τ is compact, τ is induced by a uniformity u, and $+$ and $'$ are uniformly continuous with respect to u. Therefore $u \in \mathcal{LUA}(L)$, see 2.8. Moreover, u is exhaustive by [21, 6.1] and complete. Therefore u satisfies the assumptions of 6.1.7, and so $L \simeq [0, c] \times Ł_\infty^B \times \Pi_{\alpha \in A} Ł_{n_\alpha}$, and $c = 0$ since u is compact. ■

6.1.9. Corollary. *Let u be an exhaustive Hausdorff uniformity of $\mathcal{LUA}(L)$ such that (L, u) is complete. Then*

(a) *L is atomic iff (L, u) is totally disconnected iff (L, u) is isomorphic and homeomorphic to a product $\Pi_{\alpha \in A} Ł_{n_\alpha}$, where $n_\alpha \in \mathbb{N} \setminus \{1\}$ for $\alpha \in A$. In particular, if (L, u) is totally disconnected, then (L, u) is compact.*

(b) *L is atomless iff (L, u) is arcwise connected iff (L, u) is connected.*

6.1.9 immediately follows from 6.1.7 and from the fact that L is atomless or atomic iff $a_f = 0$ or $a_\infty = c = 0$, respectively, see 6.1.6 (e), (f).

6.2. Hammer–Sobczyk's decomposition

In this section we prove Hammer–Sobczyk's decomposition for measures on Δ-ℓ–semigroups.

The original version of Hammer–Sobczyk's decomposition theorem says:

If A is a Boolean algebra with the greatest element e and $\mu : A \to [0, +\infty)$ is a finitely additive measure, then there are finitely additive two–valued measures $\mu_n : A \to [0, +\infty)$ such that

$$\mu = \lambda + \sum_{n \in I} \mu_n,$$

where $I \subset \mathbb{N}$, and measure λ is "continuous" in the following sense:

$$\forall \varepsilon > 0 \quad \exists n \in \mathbb{N} \quad \exists a_1, \ldots, a_n \in A \quad e = \vee_{i=1}^n a_i \quad \textit{and} \quad \lambda 11(a_i) < \varepsilon \quad (i = 1, \ldots, n).$$

In the case when μ is a σ–additive measure on a σ–algebra, this theorem produces a decomposition into the atomic part $\sum_{n\in I}\mu_n$ and the atomless part λ.

We will do the same in our setting. Hejcman's concept of chained uniform spaces plays an important role here. A uniform space (X, u) is called *chained* if for any $x, y \in X$ and $U \in u$ there is a finite number of elements $x_0, \dots, x_n \in X$ with $x_0 = x$, $x_n = y$, and $(x_{i-1}, x_i) \in U$ for $i = 1, \dots, n$.

6.2.1. Proposition. *For $u \in \mathcal{LUA}(P)$, the following conditions are equivalent:*

(1) *(P, u) is chained.*

(2) *For any $U \in u$ and $x, y \in P$ with $x \leq y$ there is a finite chain $x = x_0 \leq \cdots \leq x_n = y$ with $(x_{i-1}, x_i) \in U$ $(i = 1, \dots, n)$.*

(3) *For any 0–neighborhood U in (P, u) and $a \in P$ there is a finite number of elements $a_1, \dots, a_n \in U$ with $a = \sum_{i=1}^n a_i$.*

If P has a greatest element 1, *then a further equivalent condition is:*

(4) *For any 0–neighborhood U in (P, u) there is a finite number of elements $a_1, \dots, a_n \in U$ with $1 = \sum_{i=1}^n a_i$.*

Proof. (1)$\Leftrightarrow$(2) is easy to verify and holds even for lattice uniformities, see [23, 5.7].

(2)$\Rightarrow$(3): Let U be a 0–neighborhood in (P, u) and $a \in P$. Choose $x_i \in P$ with $0 = x_0 \leq \cdots \leq x_n = a$ and $(x_i, x_{i-1}) \in U^\Delta$ $(i = 1, \dots, n)$. Then $a_i := x_i \Delta x_{i-1} \in U$ and $a = \sum_{i=1}^n a_i$.

(3)$\Rightarrow$(2), (4)$\Rightarrow$(2) if P has a unit 1: Let x, y, $a \in P$ with $x \leq y$, $y \Delta x \leq a$, and $a = 1$ in case (4); moreover, let U, V be solid 0–neighborhoods in (P, u) such that $r \Delta s \in V$ implies $(r \wedge t)\Delta(s \wedge t) \in U$ for any r, s, $t \in P$. By (3) or (4), there are $a_1, \dots, a_n \in V$ with $\sum_{i=1}^n a_i = a$. Put $x_i := x + \sum_{j=1}^i a_j$ and $y_i := x_i \wedge y$ $(i = 0, \dots, n)$. Then $x_i \Delta x_{i-1} \leq a_i \in V$, hence $x_i \Delta x_{i-1} \in V$ and $y_i \Delta y_{i-1} \in U$ $(i = 1, \dots, n)$. Moreover, $x = y_0 \leq \cdots \leq y_n = y$. ∎

Let G be a Hausdorff topological Abelian group and $\mu : P \to G$ a measure. We say that P is *μ–chained* if (P, μ–uniformity) is chained. By 6.2.1, P is μ–chained iff for every 0–neighborhood W in G and for any $a \in P$ there are $a_1, \dots, a_n \in P$ such that $a = \sum_{i=1}^n a_i$ and $\mu([0, a_i]) \subset W$. We say that μ is *atomless* if $P/N(\mu)$ is atomless, and that μ is *atomic* if $P/N(\mu)$ is atomic.

6.2.2. Theorem. *Let G be a complete Hausdorff topological Abelian group and $\mu : P \to G$ an exhaustive measure. Let w be the μ–uniformity, $(\tilde{P}, \tilde{w})$ the* *uniform completion of* $(\hat{P}, \hat{w}) := (P, w)/N(w)$, *and* $\tilde{\mu} : \tilde{P} \to G$ *the continuous extension of the function* $\hat{\mu} : \hat{P} \to G$ *defined by* $\hat{\mu}(\hat{x}) := \mu(x)$, *where* $\hat{x} = x \Delta N(w)$. *Then the following conditions are equivalent:*

(1) P *is* μ*–chained.*

(2) $\tilde{P}$ *is* $\tilde{\mu}$*–chained.*

(3) $\tilde{\mu}$ *is atomless.*

(4) $(\tilde{P}, \tilde{w})$ *is arcwise connected.*

(5) $\tilde{\mu}([0, a])$ *is arcwise connected for any* $a \in \tilde{P}$.

(6) $\overline{\mu([0, a])}$ *is connected for any* $a \in P$.

Proof. First observe that w is the $\tilde{\mu}$–uniformity by 5.2 and is exhaustive by 4.1 (c). Moreover, $\tilde{P}$ is an MV–algebra by 1.8. Now (1) to (4) are equivalent by 6.1.9 (b) and [23, 5.8].

(4)⇒(5) follows from the continuity of $\tilde{\mu}$, and (5)⇒(6) from the fact that $\overline{\mu([0, a])} = \tilde{\mu}([0, \hat{a}])$ for $a \in P$.

(6)⇒(3): Since $\hat{\mu}([0, \hat{a}]) = \mu([0, a])$ for $a \in P$, we may assume that $\hat{P} = P$ and $\hat{\mu} = \mu$. Suppose that $\tilde{P}$ contains an atom a. Put $g := \tilde{\mu}(a)$. Let U be an open 0–neighborhood in G with $U \cap (g + U) = \emptyset$, and V be a closed 0–neighborhood in G contained in U. Choose a solid 0–neighborhood W in $(\tilde{P}, \tilde{w})$ such that $x, y \in \tilde{P}$ and $x \Delta y \in W$ imply $\tilde{\mu}(x) - \tilde{\mu}(y) \in V$. Let $b \in P$ with $a \Delta b \in W$. We show that $\mu([0, b] \cap P) \subset V \cup (g + V)$:

Let $x \in [0, b] \cap P$. If $x \geq a$, then $x \Delta a \leq b \Delta a \in W$ and therefore $\mu(x) - \tilde{\mu}(a) \in V$. Otherwise $x \wedge a = 0$, hence $x \Delta 0 = (x \wedge b) \Delta (x \wedge a) \leq b \Delta a \in W$ and so $\mu(x) \in V$.

It follows that $\overline{\mu([0, b] \cap P)} \subset V \cup (g + V) \subset U \cup (g + U)$, $\mu(0) \in U$, and $\mu(b) \in g + U$. Therefore $\overline{\mu([0, b] \cap P)}$ is not connected and so (6) doesn't hold. ■

6.2.3. Hammer–Sobczyk Decomposition. *Let* G *be a complete Hausdorff topological Abelian group and* $\mu : P \to G$ *an exhaustive measure. Then there exist measures* λ, μ_a $(a \in A)$ *such that*

(1) $\sum_{a \in A} \mu_a(x)$ *is summable uniformly in* $x \in P$.

(2) $\mu = \lambda + \sum_{a \in A} \mu_a$.

(3) *For any* $a \in A$, $P/N(\mu_a)$ *is isomorphic to* Ł_{n_a} *for some* $n_a \in \mathbb{N} \setminus \{1\}$.

(4) P *is* λ*–chained.*

Moreover, $\lambda(P)$ *is dense in an arcwise connected set, in particular,* $\overline{\lambda(P)}$ *is connected, and the range of* $\sum_{a \in A} \mu_a(P)$ *is relatively compact.*

Proof. We use the same notation as in 6.2.2. It is easy to see that Hammer–Sobczyk decomposition $\hat{\mu} = \hat{\lambda} + \sum \hat{\mu}_a$ yields the Hammer–Sobczyk decomposition for μ defining $\lambda(x) := \hat{\lambda}(\hat{x})$ and $\mu_a(x) := \hat{\mu}_a(\hat{x})$. Therefore we may assume that w is Hausdorff, i.e. $\mu = \hat{\mu}$. By 4.1 and 1.8, $\tilde{P}$ is a complete (as a lattice) MV–algebra, $\tilde{w}$ is order continuous and exhaustive. Let $A = \mathcal{A}_f(\tilde{P})$ (cf. 6.1), $a_f = \sup A$. For any $b \in C(\tilde{P})$, put $\tilde{\mu}_b(z) := \tilde{\mu}(b \wedge z)$, $\mu_b := \tilde{\mu}_b|_P$, $\tilde{\lambda} := \tilde{\mu}_{a'_f}$, and $\lambda := \mu_{a'_f}$. We now show that the measures λ and μ_a ($a \in A$) have the desired properties.

(2) Let $x \in \tilde{P}$. Since $\tilde{\mu}$ is order continuous, $\sum_{a\in A} \tilde{\mu}(a \wedge x)$ is summable and

$$\tilde{\mu}(\sup_{a\in A} x \wedge a) = \sum_{a\in A} \tilde{\mu}(a \wedge x) = \sum_{a\in A} \tilde{\mu}_a(x),$$

therefore $\tilde{\mu}(x) = \tilde{\mu}(x \wedge a'_f) + \tilde{\mu}(x \wedge a_f) = \tilde{\lambda}(x) + \tilde{\mu}(x \wedge \sup_{a\in A} a) = \tilde{\lambda}(x) + \tilde{\mu}(\sup_{a\in A} x \wedge a) = \tilde{\lambda}(x) + \sum_{a\in A} \tilde{\mu}_a(x)$.

(1) Let U be a 0–neighborhood in G. We will prove that there exists a finite set $A_0 \subset A$, such that $\sum_{a\in A} \tilde{\mu}_a(x) - \sum_{a\in A_1} \tilde{\mu}_a(x) \in U$ whenever A_1 is a finite set with $A_0 \subset A_1 \subset A$ and $x \in \tilde{P}$. Since $\tilde{\mu}$ is uniformly continuous, there is a solid 0–neighborhood W in $(\tilde{P}, \tilde{w})$ such that $\tilde{\mu}(x) - \tilde{\mu}(y) \in U$ whenever $x, y \in \tilde{P}$ and $x\Delta y \in W$. Because $\tilde{w}$ is order continuous, there is a finite subset A_0 of A such that $a_f \Delta \sup_{a\in A_0} a \in W$. Let $A_0 \subset A_1 \subset A$, A_1 be finite, $x \in \tilde{P}$, and $z = x \wedge a_f$. Then we have $z\Delta(z \wedge \sup_{a\in A_1} a) = (z \wedge a_f)\Delta(z \wedge \sup_{a\in A_1} a) \leq a_f \Delta \sup_{a\in A_0} a$, hence $z\Delta(z \wedge \sup_{a\in A_1} a) \in W$ and $\sum_{a\in A} \tilde{\mu}_a(x) - \sum_{a\in A_1} \tilde{\mu}_a(x) = \tilde{\mu}(z) - \tilde{\mu}(z \wedge \sup_{a\in A_1} a) \in U$.

(3) Let $a \in A$. $\tilde{P}/N(\tilde{\mu}_a)$ is isomorphic to $[0, a]$ and therefore, by 6.1.3, to Ł_{n_a} for some $n_a \in \mathbb{N} \setminus \{1\}$. Let $\tilde{w}_a$ be the $\tilde{\mu}_a$–uniformity and w_a be the restriction of $\tilde{w}_a$ to P. $P/N(w_a)$ can be identified with a subspace of $\tilde{P}/N(\tilde{w}_a)$. Since $\tilde{P}/N(\tilde{w}_a) = \tilde{P}/N(\tilde{\mu}_a)$ is finite and P is dense in $(\tilde{P}, \tilde{w}_a)$, we get $P/N(w_a) = \tilde{P}/N(\tilde{w}_a)$. Moreover, $N(w_a) = N(\mu_a)$ since w_a by 5.2 is a μ_a–uniformity. It follows that $P/N(\mu_a)$ is isomorphic to Ł_{n_a}.

(4) By 6.1.7, $[0, a'_f]$ is arcwise connected with respect to $\tilde{w}$ and therefore with respect to $\tilde{\lambda}$–uniformity, since $\tilde{\lambda} \ll \tilde{w}$. It follows that $\tilde{P}$ is $\tilde{\lambda}$–chained since $a_f \in N(\tilde{\lambda})$. By 5.2, the $\tilde{\lambda}$–uniformity induces on P the λ–uniformity. Moreover, P is dense in ($\tilde{P}$, $\tilde{\lambda}$–uniformity). Therefore P is λ–chained.

$\tilde{\lambda}(\tilde{P}) = \tilde{\lambda}([0, a'_f])$ is arcwise connected as a continuous image of the arcwise connected subset $[0, a'_f]$ of $(\tilde{P}, \tilde{w})$. Since $\tilde{\lambda}$ is the continuous extension of λ, the range $\lambda(P)$ of λ is dense in $\tilde{\lambda}(\tilde{P})$. For $\tilde{\nu} := \sum_{a\in A} \tilde{\mu}_a$, $\nu := \tilde{\nu}|_P$, one obtains similarly that $\tilde{\nu}(\tilde{P})$ is compact since $[0, a_f]$ is a compact subspace of $(\tilde{P}, \tilde{w})$ by 6.1.7 (b) and $\overline{\nu(P)} = \overline{\tilde{\nu}(\tilde{P})} = \tilde{\nu}(\tilde{P})$. ∎

We briefly compare the decomposition theorem 5.3 with the Hammer–Sobczyk decomposition 6.2.3. It is easy to see that there exists a finest uniformity $u \in \mathcal{LUA}(P)$ such that (P, u) is chained. Let $\mu = \lambda + \nu$ be the decomposition according to 5.3 with $\lambda \ll u$ and $\nu \perp u$. Then this measure λ coincides with the measure λ of the Hammer–Sobczyk decomposition $\mu = \lambda + \sum_{a \in A} \mu_a$. As an additional information, the Hammer–Sobczyk decomposition yields a precise description of the measure ν, namely $\nu = \sum_{a \in A} \mu_a$. If P is complete with respect to μ–uniformity, the measures λ and ν of the decomposition $\mu = \lambda + \nu$ are atomless and atomic, respectively.

For measures on Boolean rings, Theorem 6.2.2 is essentially contained in [17] and the Hammer–Sobczyk decomposition is contained in [20]. If $G = \mathbb{R}^n$ for some $n \in \mathbb{N}$, the statement about the range of measures in 6.2.2 ((1)$\Rightarrow$(6)) and in 6.2.3 can be sharpened by replacing the (arcwise) connectedness by the convexity. This follows from the Lyapunov's convexity theorem recently proved by G. Barbieri for measures on Δ-ℓ–semigroups (in particular, for measures on MV–algebras), generalizing the previous result of [1] for T_∞–valuations on clans of fuzzy sets.

References

1. A. Avallone and G. Barbieri, Range of finitely additive fuzzy measures, *Fuzzy Sets and Systems* 2102 (1996).
2. A. Basile and T. Traynor, Monotonely Cauchy locally solid topologies, *Order* 7 (1991), 407–416.
3. G. Birkhoff, *Lattice theory*, AMS Colloquium Publications, Vol. 25, Providence, Rhode Island, 1984.
4. D. Butnariu and E. P. Klement, *Triangular norm–based measures and games with fuzzy coalitions*, Kluwer Acad. Publ., Dordrecht, Holland, 1993.
5. C. C. Chang, Algebraic analysis of many valued logics, *Trans. Amer. Math. Soc.* 88 (1958), 467–490.
6. C. C. Chang, A new proof of the completeness of Łukasiewicz axioms, *Trans. Amer. Math. Soc.* 93 (1959), 74–80.
7. R. Cignoli, Complete and atomic algebras of the infinite valued Łukasiewicz logic, *Studia Logica* 50 (1991), 375–384.
8. J. Hejcman, Boundedness in uniform spaces and topological groups, *Czechoslovak J. Math.* 9 (1959), 544–562.
9. J. Jakubik, On complete MV–algebras, *Czechoslovak J. Math.* 45 (1995), 473–480.

10. I. Kluvanek, Integrale vectorielle de Daniell, *Mat. Fyz. Casopis Sloven. Akad. Vied* 15 (1965), 146–161.

11. F. Lacava, Sulla struttura delle Ł–algebre, *Accademia Nazionale dei Lincei*, *Estratto dai Rendiconti della Classe di Scienze Fisiche*, *Matematiche e Naturali*, serie VIII, vol. LXVII (1979), 275–281.

12. P. Mangani, Su certe algebre connesse con logiche a piú valori, *Boll. UMI* 8 (1973), 68–78.

13. D. Mundici, Interpretation of AF C^*–algebras in Łukasiewicz sentential calculus, *J. Funct. Anal.* 65 (1986), 15–63.

14. J. Pellaumail, Intégrale de Daniell à valeurs dans un groupe, *Rev. Roum. Math. Pures et Appl.*, Tome XVI (1971), 1227–1236.

15. D. Saeli, Problemi di decisione per algebre connesse a logiche a piú valori, *Accademia Nazionale dei Lincei*, *Rendiconti della Classe di Scienze Fisiche*, *Matematiche e Naturali* 59 (1975), 219–223.

16. T. Traynor, The Lebesgue decomposition for group–valued set functions, *Trans. Amer. Math. Soc.* 220 (1976), 307–319.

17. H. Volkmer and H. Weber, Der Wertebereich atomloser Inhalte, *Arch. Math.* 40 (1983), 464–474.

18. S. Warner, Compact rings and Stone–Čech compactifications, *Arch. Math.* 11 (1960), 327–332.

19. H. Weber, $\mathbb{R}$–freie Integrationstheorie I and II, *J. Reine Angew. Math.* 289 (1977), 30–54 and 290 (1977), 1–23.

20. H. Weber, Group– and vector–valued s–bounded contents, in: *Measure Theory* (*Oberwolfach, 1983*), Lecture Notes in Mathematics, Vol. 1089, Springer–Verlag, 1984, 181–198.

21. H. Weber, Uniform lattices I: A generalization of topological Riesz spaces and topological Boolean rings, *Annali di Matematica Pura e Applicata* 160 (1991), 347–370; and
Uniform lattices II: Order continuity and exhaustivity, *Annali di Matematica Pura e Applicata* 165 (1993), 133–158.

22. H. Weber, Metrization of uniform lattices, *Czechoslovak J. Math.* 43 (1993), 271–280.

23. H. Weber, On modular functions, *Funct. et Approx.* 24 (1996), 35–52.

24. H. Weber, An abstraction of clans of fuzzy sets, *Ricerche di Matematica*, to appear.

25. H. Weber, manuscript.

Giuseppina Barbieri
Dipartimento di Matematica
e Applicazioni
Università Federico II
Via Cintia, Complesso Monte S. Angelo
Napoli 80126
Italy
email: barbieri@matna2.dma.unina.it

Hans Weber
Dipartimento di Matematica
e Informatica
Universitá degli studi di Udine
Via delle Scienze 206
Udine 33100
Italy
email: weber@dimi.uniud.it

ON THE RANGES OF ADDITIVE CORRESPONDENCES*

ACHILLE BASILE

Abstract. In this work we consider a variety of finitely additive set–valued set–functions with infinite–dimensional range–spaces. The results are *à la Lyapunov*, namely they furnish some conditions for the range to have compact/convex closure. Some material concerning set–valued set–functions with values in semigroups is also given both because of its own interest and because it is preparatory for the topological vector space situation.

1. Introduction

The core of this paper is mainly concerned with variations on the celebrated theorem of Lyapunov, which states that the range of an arbitrary countably additive ($\mathbb{R}^\ell$–valued) vector measure is compact, and if in addition the measure is atomless, the range is also convex. The large variety of applications of this theorem makes it one of the best loved from the Measure Theory. Applications run from Statistical Decision Theory and Game Theory (papers in the early fifties by people like D. Blackwell and A. Dvoretzky, A. Wald, J. Wolfowitz — see, for example, [11] and [17]) to the Bang–Bang Principle ([29, p. 444]), passing through Economics, where numerous results are based, or can be based, on the Lyapunov theorem (a source for several references is the book [1] by C. Aliprantis and K. Border).

Among all possible "variations" (that is, directions of generalization, in more standard terminology), we shall be especially dealing with the possibility of a multiplicity of values, finite additivity, and the possibility that the range–space is infinite–dimensional. In other words, we shall investigate the range, denoted by $R\Phi$, of a correspondence

$$\Phi : \mathcal{F} \to 2^X \setminus \{\emptyset\},$$

where

$\mathcal{F}$ is a field of subsets of a given set Ω,

X is a metrizable topological vector space,

$\Phi(\emptyset) = \{0\}$,

$\Phi(E \cup F) = \Phi(E) + \Phi(F) = \{x + y : \ x \in \Phi(E),\ y \in \Phi(F)\}$ for disjoint $E, F \in \mathcal{F}$.

*Thanks to my friend Prof. H. Weber for the fruitful discussions on this paper.

Each of the cited variations has its own motivation in the Economic Theory.

• It was K. Vind [26] who first based on convexity properties of special atomless *set–measures* the fundamental Core–Equivalence Theorem. Even before, with reference to statistical and game–theoretical applications, H. Richter [22] generalized Blackwell's result by proving the convexity of the range of integrals of certain multifunctions.

It is because of Richter's paper that we will refer to multivalued variations of the Lyapunov theorem as the Lyapunov–Richter theorem.

• In the measure–theoretical modeling of perfect competition, T. Armstrong and M. Richter [5, 6] have pointed out the role finite additivity plays in avoiding some paradoxical situations.

• It is a relatively recent phenomenon[1] that we have the increasing usage of infinite–dimensional spaces for modeling of economic systems with infinitely many commodities. The list of contributors is rather long. Excellent sources for references are: the book [3] by C. Aliprantis, D. Brown and O. Burkinshaw; the forthcoming paper [2] by Aliprantis, Border and Burkinshaw; Chapter 34, written by A. Mas–Colell and W. Zame, of the fourth volume of the Handbook of Mathematical Economics [21]; and the book [20] whose editors are M. Khan and N. Yannelis.

Of course, as a consequence of our idea of dealing with finitely additive correspondences having infinite–dimensional range–space, we cannot expect to prove a general theorem that looks exactly like the classical Lyapunov theorem. Indeed, even if for countably additive $\mathbb{R}^\ell$–valued correspondences, D. Schmeidler proved a theorem[2] which is quite Lyapunov–like, in the the single–valued case (i.e., the case of a vector measure) we have weaker conclusions:

• the range of a nonatomic, finitely additive, $\mathbb{R}^\ell$–measure defined on a σ–field is convex but not necessarily closed, moreover, the range is relatively convex if the domain is only assumed to be a field (Armstrong and Prikry [4]);[3]

• in the classical book [15] by J. Diestel and J. J. Uhl, one can find an ℓ_2–valued countably additive atomless measure of bounded variation whose range is neither compact nor convex, though a result by Uhl [15, Theorem 10, p. 266] would ensure that the closure of the range is both compact and convex.

On the other hand, it is the convexity part of the Lyapunov theorem to

[1] Basic papers, however, already appeared in the early seventies.

[2] Schmeidler proved, in [23], that the values and the range of Φ are convex and compact under the following assumptions: $X = \mathbb{R}^\ell$, the domain $\mathcal{F}$ is a σ–field, the set $\Phi(\Omega)$ is compact, Φ is null on λ–null members of $\mathcal{F}$ for a given atomless probability λ on $\mathcal{F}$, and $\Phi(\cup_{n=1}^\infty F_n) = \{x : \ x = \sum_n x_n,$ where $x_n \in \Phi(F_n)$ and $\sum x_n$ converges absolutely$\}$.

[3] A finitely additive extension of the convexity part of the cited theorem by Schmeidler is in the paper [7] by Avallone and Basile.

which most of the applications are due. Moreover, what one often is really needed is the convexity of the closure of the range.

Let us refer to the convexity of the closure of a set as its *relative convexity*. Evidently, the above considerations make us feel undiminished in proving theorems mainly ensuring, under suitable hypotheses, the relative convexity of the range of nonatomic finitely additive correspondences.

The rest of the paper is structured in the following way. The main results have been stated in Sections 4 and 5; preparatory stuff is contained in Section 3, where, except for Theorem 3.2, the correspondences with values in semigroups are considered. In its turn, part of the material of Section 3 rests on the interpretation of additive correspondences as the contents (that is, finitely additive measures) with values in suitable semigroups. Because of this, Section 2 contains the machinery of semigroup–valued contents, which we shall use in Sections 3 and 4.

2. Semigroup–valued contents

A few theorems concerning semigroup–valued set functions will be used in this paper as a tool for deriving our results of Sections 3–5. For the sake of completeness, we shall briefly recall what we need.

Suppose S is an abelian semigroup (additive notation) with neutral element 0. A finitely additive mapping $\mu : \mathcal{F} \to S$ such that $\mu(\emptyset) = 0$ is called an S–valued *content*. An atom for the content μ is defined in the usual way: it is a set F of the domain not belonging to $\mathcal{N}(\mu) := \{G \in \mathcal{F} : \mu(H) = 0$ for $H \subseteq G,\ H \in \mathcal{F}\}$ and such that either E or $F \setminus E$ belongs to $\mathcal{N}(\mu)$ whenever $E \in \mathcal{F}$ and $E \subseteq F$. Several other basic notions from ordinary Measure Theory may be transferred to the abstract case by assuming that S is a *metric semigroup*. The latter means that we also have a metric d on S, which is *semi-invariant*, i.e., $d(a + c, b + c) \leq d(a, b)$ for all $a, b, c \in S$. Naturally, the semi-invariance guarantees the uniform continuity of the sum in S. If S is a metric semigroup and μ is a content with values in S, then μ is said to be

a *measure* if it is countably additive and defined on a σ–field;

exhaustive if $\mu(F_n) \to 0$ whenever the sets $F_n \in \mathcal{F}$ form a disjoint sequence;

σ*–order continuous* if $\mu(F_n) \to 0$ whenever $F_n \in \mathcal{F}$ and $F_n \searrow \emptyset$;

nonatomic if for any positive ε there is a finite partition $\{F_i\} \subset \mathcal{F}$ of Ω such that $E \in \mathcal{F}$ and $E \subseteq F_i \Rightarrow d(\mu(E), 0) < \varepsilon$.

Naturally, a nonatomic content is atomless. Combining the following theorem with the results that can be obtained in the finitely additive framework by assuming the nonatomicity, we understand the reason why the latter concept is considered as a finitely additive analogue of the lack of atoms for measures.

2.1. Theorem. *If μ is σ–order continuous and its domain is a σ–field, then μ is nonatomic if and only if it is atomless.*

The next decomposition theorem is stated exactly in the form which we need to apply. The only reference to its real nature is given here: it is the semigroup generalization of the atomic–nonatomic decomposition theorem.

2.2. Theorem. *Assume that $\mu : \mathcal{F} \to S$ is exhaustive and the metric semigroup S is complete. Then there are two exhaustive contents $\mu_1, \mu_2 : \mathcal{F} \to S$ such that $\mu = \mu_1 + \mu_2$; the content μ_1 is nonatomic, and the range of μ_2 is a relatively compact subset of S.*

Let μ be an S–valued content. The variation $v(\mu, F)$ is defined as the supremum of all sums $\sum_{E\in\pi} d(0, \mu(E))$, where $\pi \subseteq \mathcal{F}$ runs over all finite partitions of F. There are no surprises: $v(\mu)$ is the smallest among all contents $\lambda : \mathcal{F} \to [0, \infty]$ such that $d(0, \mu(\cdot)) \leq \lambda(\cdot)$. We may also introduce the classical pseudometric $d_\mu(F, E) := \sup_{G\subseteq E\triangle F} d(0, \mu(G))$ on $\mathcal{F}$ and say that whenever d_μ is equivalent to d_λ for some nonnegative real–valued content λ, λ is a *control* for μ. Another useful tool, the classical Stone space argument, can be registered in a way described in the next theorem. Therein, the absolute continuity we refer to has to be meant in $\varepsilon - \delta$ terms.

2.3. Theorem. *Assume that Ω is the Stone space of a given Boolean algebra and $\mathcal{F}$ is the field of clopen subsets of Ω. Let Σ be the σ–field generated by $\mathcal{F}$. Given an arbitrary complete metric semigroup S, the set of all exhaustive S–valued contents μ on $\mathcal{F}$ may be mapped onto the set of all σ–order continuous S–valued contents on Σ in a one-to-one way. If the bijective mapping above is denoted by $\mu \mapsto \tilde{\mu}$, its inverse mapping is the restriction $\tilde{\mu} \mapsto \mu = \tilde{\mu}|_{\mathcal{F}}$ and the range of μ is a dense subset of the range of $\tilde{\mu}$.*

Moreover, if $\nu : \mathcal{F} \to T$ is an exhaustive content (where T is a complete metric semigroup), then

$$\mu \ll \nu \iff \tilde{\mu} \ll \tilde{\nu} \iff \mathcal{N}(\tilde{\mu}) \supseteq \mathcal{N}(\tilde{\nu}).$$

Here is a first straightforward application of the Stone space argument.

2.4. Proposition. *Any S–valued content μ of bounded variation admits the variation $v(\mu)$ as a control.*

Proof. Obviously, $d_\mu(E,F) \le v(\mu, E\Delta F)$. So μ and its variation are exhaustive, and so to prove the remaining part $v(\mu) \ll \mu$ we may use Theorem 2.3. Therefore our task is to show $\mathcal{N}(\tilde{v}) \supseteq \mathcal{N}(\tilde{\mu})$, letting v be an abbreviation for $v(\mu)$. In actuality we will prove more; namely, that $\tilde{v}$ is the variation of $\tilde{\mu}$, so that equality holds between the $\tilde{v}$–null sets and the $\tilde{\mu}$–null sets. Let us take a set $E \in \Sigma$ and a sequence $F_n \in \mathcal{F}$ such that $\mu(F_n) \to \tilde{\mu}(E)$ and $v(F_n) \to \tilde{v}(E)$.[4] From $d(0, \mu(F_n)) \le v(F_n)$, we recognize that $d(0, \tilde{\mu}(E)) \le \tilde{v}(E)$. This is the same as $\tilde{v} \ge v(\tilde{\mu})$. A further use of Theorem 2.3 tells us that we get equality provided $\tilde{v}|_{\mathcal{F}} = v(\tilde{\mu})|_{\mathcal{F}}$. The latter can be easily proved. ∎

In the case when μ is exhaustive without being of finite variation, the fact that the variation of $\tilde{\mu}$ is the extension of $v(\mu)$ is replaced by the fact that $(\mathcal{F}, d_\mu)$ is a subspace of the pseudometric space $(\Sigma, d_{\tilde{\mu}})$. Moreover, μ is nonatomic if and only if $\tilde{\mu}$ is nonatomic.

By means of the Stone space argument and Theorem 2.1, the Lyapunov–type results due to J. Uhl, V. Kadets [18] and V. Kadets and G. Shekhtman [19], admit the formulation indicated in Theorem 2.5.

In accordance with [19] we say that a topological vector space X is a *Lyapunov space* if each X–valued atomless measure has a relatively convex range. RNP is an abbreviation for Radon–Nikodym property [15]; for B–convex spaces see [14].

2.5. Theorem. *Let X be a Banach space and $\mu : \mathcal{F} \to X$ a content.*

(Uhl) *If X has the RNP and $v(\mu) < \infty$, then $R\mu$ is relatively compact.*

(Uhl, Kadets) *If $v(\mu) < \infty$ and X is either B–convex or has the RNP, then the nonatomicity of μ entails the relative convexity of $R\mu$.*

(Kadets–Shekhtman) *If μ is exhaustive and nonatomic, then its range is relatively convex provided X is c_0, or ℓ_p (for $p \in [1,\infty) \setminus \{2\}$), or a direct sum of any Lyapunov space with one of the previous spaces.*

We have in mind concrete semigroups to deal with. Let us assume that the range space X of our additive correspondences is a complete metrizable topological vector space. We denote by U_r the closed ball in X with radius r and center at the origin, and let

$$\delta(X_1, X_2) := \inf\{r > 0 : \quad X_i \subseteq X_j + U_r,\ i,j = 1,2,\ i \ne j\}$$

[4]Of course, this is a bit more than the density indicated in Theorem 2.3. Nevertheless, the sequence exists as the proof of Theorem 2.3 would show.

be the Hausdorff distance between two subsets X_1 and X_2 of X. By $\mathcal{C}(X)$ we denote the complete metric semigroup consisting of the closed non-empty subsets of X when they are summed according to the operation

$$\stackrel{*}{+}: \ (X_1, X_2) \mapsto \mathrm{cl}(X_1 + X_2),$$

and the distance is given by δ. The collection $\mathcal{Q}(X)$ of all compact members of $\mathcal{C}(X)$ is a closed subsemigroup of $\mathcal{C}(X)$. In $\mathcal{Q}(X)$, we do not distinguish between $+$ and $\stackrel{*}{+}$. Moreover, it is a routine matter to check the following.

2.6. Proposition. *If a subset of $\mathcal{Q}(X)$ is relatively compact, then the union of its elements is relatively compact in X.*

Remark. What we have just said does not really depend on the assumption that X is a linear space. If we simply assume that X is a complete metric semigroup, then everything goes almost in the same way. Indeed, we have only to modify the definition of δ as follows:

$$\delta(X_1, X_2) := \inf \left\{ r > 0 : \quad X_i \subseteq \bigcup_{a \in X_j} \{b : \ d(a, b) \leq r\}, \ i, j = 1, 2, \ i \neq j \right\}.$$

Clearly, the two formulas coincide for groups.

3. Set correspondences

Let Φ be a correspondence defined on $\mathcal{F}$, with values in X and such that $\Phi(\emptyset) = \{0\}$. The space X is assumed to be a metric semigroup $(X, +, 0, d)$. Let us write $\|x\|$ instead of $d(0, x)$. According to Schmeidler [24], the variation of a correspondence is a straightforward extension of that for measures. For $F \in \mathcal{F}$, we set

$$v(\Phi, F) := \sup \left\{ \sum_{E \in \pi} \|x_E\| : \quad \begin{array}{l} \pi \text{ is a finite partition of } F, \ \pi \subseteq \mathcal{F} \\ x_E \in \Phi(E) \end{array} \right\}.$$

Obviously, the set function defined above vanishes on the empty set, is increasing, superadditive, and coincides with $v(\bar{\Phi})$ for $\bar{\Phi}$ being the correspondence $F \mapsto \mathrm{cl}(\Phi(F))$. The finite additivity of the variation is ensured if the inclusion $\Phi(E \cup F) \subseteq \mathrm{cl}(\Phi(E) + \Phi(F))$ holds whenever E and F are disjoint. So $v(\Phi)$ is a content if Φ is finitely additive, or in a little bit more general case if Φ is $\stackrel{*}{+}$–finitely additive, which means that $\mathrm{cl}(\Phi(E \cup F)) = \mathrm{cl}(\Phi(E) + \Phi(F))$ for disjoint E and F.

The terminology is transparent: "Φ *is* $\overset{*}{+}$*–finitely additive*" means that "$\bar{\Phi} : \mathcal{F} \to \mathcal{C}(X)$ *is a* $\mathcal{C}(X)$*–valued content.*" Consequently, for $\overset{*}{+}$–finitely additive correspondences we have at our disposal the machinery for semigroup–valued contents. For example, the variation of Φ, as defined here, coincides with that of the $\mathcal{C}(X)$–valued content $\bar{\Phi}$; we may use the pseudometric $\delta_{\bar{\Phi}}$, that we prefer to denote by d_Φ, and

$$d_\Phi(E,F) = \sup_{G \subseteq E \triangle F} \inf\{r > 0 : \ \Phi(G) \subseteq U_r\} = \sup\left\{ \|x\| : x \in \bigcup_{G \subseteq E \triangle F} \Phi(G) \right\}.$$

Proposition 2.4 says the following.

3.1. Corollary. *Let* Φ *be* $\overset{*}{+}$*–finitely additive. If* Φ *is of finite variation, then* $v(\Phi)$ *is a control for* Φ*.*

In the special case when X is a linear space we may try to weaken the finite variation assumption by assuming that Φ is exhaustive. This is done in the next theorem à la Bartle–Dunford–Schwartz (the working argument is that of [27, (4.1), (4.2), (4.3)]; compare also with [16, (8.3)]). Notice that removing in the statement of Theorem 3.2 the hypothesis that there are enough functionals on X, would lead us to face the famous Maharam problem.

3.2. Theorem. *Let* X *be a metrizable topological vector space with a separating dual* X'*. If* Φ *is* $\overset{*}{+}$*–finitely additive and exhaustive, then it admits a control.*

Let us go back to semigroups in general. For the correspondence Φ, we define $\mathcal{S}(\Phi)$ as the set of its selections which also are X–valued contents.

3.3. Lemma. *Assume that* Φ *is closed–valued and* $\overset{*}{+}$*–finitely additive. Given two finite subfields* $\mathcal{G} \subseteq \mathcal{H}$ *of* $\mathcal{F}$*, we have that for any* $\mu \in \mathcal{S}(\Phi|_{\mathcal{G}})$ *and any* $\varepsilon > 0$ *there is* $\sigma \in \mathcal{S}(\Phi|_{\mathcal{H}})$ *such that* $\sup_{\mathcal{G}} d(\mu(\cdot), \sigma(\cdot)) \leq \varepsilon$*.*

The next result was proved by Costé [12]. We think it is worthwhile to give here a plain proof of this result which, contrary to [12], avoids the usage of a relatively "sophisticated" tool such as the Mittag–Leffler Theorem on projective systems of metric spaces.

3.4. Theorem. *Let* Φ *be closed–valued,* $\overset{*}{+}$*–finitely additive and defined on a countable field* $\mathcal{F}$*. Then the equality* $\Phi(F) = \mathrm{cl}\,\{\mu(F) : \mu \in \mathcal{S}(\Phi)\}$ *holds for all* $F \in \mathcal{F}$*.*

Proof. We must prove that $\Phi(F) \subseteq \mathrm{cl}\,\{\mu(F) : \mu \in \mathcal{S}(\Phi)\}$. Let us adopt the following notation: $\mathcal{F}_0$ is the field generated by F, the remaining sets in

$\mathcal{F}$ are $F_1, F_2, \ldots$, and the field generated by $\mathcal{F}_n \cup \{F_{n+1}\}$ is $\mathcal{F}_{n+1}$. Naturally, $\mathcal{F}_n \uparrow \mathcal{F}$. Assume $x \in \Phi(F)$ and $\varepsilon > 0$. If we define $\mu_0(F^c)$ to be any of the points in $\Phi(F^c)$, $\mu_0(F) = x$, and $\mu_0(\emptyset) = 0$, we will have $\mu_0 \in \mathcal{S}(\Phi|_{\mathcal{F}_0})$. By means of Lemma 3.3, we can find for any natural number n a set function μ_n such that $\mu_n \in \mathcal{S}(\Phi|_{\mathcal{F}_n})$ and $\sup_{\mathcal{F}_{n-1}} d(\mu_n(\cdot), \mu_{n-1}(\cdot)) \leq \varepsilon/2^n$. The sequence of set functions $\mu_n \chi_{\mathcal{F}_n} \in X^{\mathcal{F}}$ converges pointwise to a set function μ that one promptly checks to be in $\mathcal{S}(\Phi)$. Finally, $d(x, \mu(F)) \leq \varepsilon$. ■

With an argument quite similar to that above, it is possible to prove the following theorem.

3.5. Theorem. *If Φ is defined on a countable $\mathcal{F}$ and is finitely additive, then for any F we have $\Phi(F) = \{\mu(F) : \mu \in \mathcal{S}(\Phi)\}$.*

As we shall show later, the assumption on X in the next theorem is satisfied if $X = \ell_p$ and $p < 1$.

3.6. Theorem. *Assume that each $\overset{*}{+}$–finitely additive, exhaustive, nonatomic, X–valued correspondence has a relatively compact range. Then the range of any $\overset{*}{+}$–finitely additive correspondence Φ which is exhaustive and relatively compact–valued is relatively compact.*

Proof. Let us work with the content $\bar{\Phi} : \mathcal{F} \to \mathcal{Q}(X)$. By Theorem 2.2, we have the decomposition $\bar{\Phi} = \Phi_1 + \Phi_2$, where $\Phi_i : \mathcal{F} \twoheadrightarrow \mathcal{Q}(X)$ is a finitely additive, exhaustive correspondence; Φ_1 is nonatomic and $\{\Phi_2(F) : F \in \mathcal{F}\}$ is a relatively compact subset of $\mathcal{Q}(X)$. Evidently, $R\Phi \subseteq R\Phi_1 + R\Phi_2$ and so, by our assumption on X and by Proposition 2.6, we can conclude that the range of Φ is relatively compact. ■

3.7. Lemma (9, Lemma 3.1). *Let ϕ be a nonnegative real–valued function defined on $\mathcal{F}$. Suppose the inequality $|\phi(F) - \phi(E)| \leq \phi(F \setminus E)$ holds whenever $E, F \in \mathcal{F}$ and $E \subseteq F$. Then for ϕ to be bounded it is sufficient to be bounded over disjoint sequences from $\mathcal{F}$.*

Applying the above lemma to $\phi(F) := \sup_{x \in \Phi(F)} \|x\|$, we obtain the following useful theorem.

3.8. Theorem. *The range of Φ is bounded if Φ is $\overset{*}{+}$–finitely additive, bounded–valued and exhaustive.*

4. Banach space–valued correspondences

Throughout this section X will be a Banach space. There is a classical result in measure theory which states that the exhaustivity of a vector–valued content may be equivalently described in terms of relative weak compactness of the range of the content. A generalization of this result to the multivalued case follows (see [10, 8]).

4.1. Theorem. *Let Φ be $\overset{*}{+}$-finitely additive. Then the following are equivalent:*

(1) *Φ is exhaustive and its values are relatively weakly compact.*

(2) *The range of Φ is relatively weakly compact.*

It is worth observing that the proof of Theorem 4.1 is based on two well known results of functional analysis. One (used for proving (1)⇒(2)) is due to James, and characterizes the weakly compact subsets of a Banach space, as those on which each continuous linear form attains its supremum. The other implication can be proved by means of the Orlicz–Pettis theorem on the norm convergence of a series which is weakly subseries convergent.

Now we present an infinite–dimensional version of the Lyapunov–Richter theorem.

4.2. Theorem. *For a nonatomic $\overset{*}{+}$-finitely additive Φ the values and the range are relatively convex provided one of the following hypotheses is satisfied:*

(a) *The variation of Φ is finite and X is either B–convex or has the RNP.*

(b) *Φ is exhaustive and X is one of the sequence spaces c_0 or ℓ_p with $2 \neq p \geq 1$.*

Proof. Without loss of generality we may assume $\mathcal{F}$ is as in Theorem 2.3. Let $\tilde{\bar{\Phi}}$ be the extension of $\bar{\Phi}$, according to Theorem 2.3. Proving that the values and the range of $\tilde{\bar{\Phi}}$ are convex, we will be done. Given a control λ for $\tilde{\bar{\Phi}}$ (see Corollary 3.1 and Theorem 3.2), we can find a tree[5] $\{\Omega_{i_1,\dots,i_n}\colon i_j = 0,1;\ n \in \mathbb{N}\}$ in Σ such that $\lambda(\Omega_{i_1,\dots,i_n}) = \frac{1}{2^n}\lambda(\Omega)$. Assume $\mathcal{A}$ is the field generated by the tree. Because of Theorem 3.4, applied to $\tilde{\bar{\Phi}}\big|_{\mathcal{A}}$, we have

$$\tilde{\bar{\Phi}}(A) = \mathrm{cl}\left\{\mu(A) : \quad \mu \in \mathcal{S}(\tilde{\bar{\Phi}}\big|_{\mathcal{A}})\right\} \quad \text{for all } A \in \mathcal{A}. \tag{4.1}$$

We show that $\tilde{\bar{\Phi}}(\Omega)$ is convex, then the rest shall be straightforward. Assume $x, y \in \tilde{\bar{\Phi}}(\Omega)$ and $\varepsilon, \theta \in (0,1)$; set $z = x\theta + y(1-\theta)$. Using (4.1), we can find two contents μ and ν from $\mathcal{A}$ to X such that $\|x - \mu(\Omega)\| < \varepsilon$, $\|y - \nu(\Omega)\| < \varepsilon$,

[5]The structure of a tree is described by the sequence of disjoint unions: $\Omega = \Omega_0 \cup \Omega_1$, $\Omega_0 = \Omega_{00} \cup \Omega_{01}$, $\Omega_1 = \Omega_{10} \cup \Omega_{11}$, $\Omega_{00} = \Omega_{000} \cup \Omega_{001}$, ...

and $\mu(A), \nu(A) \in \tilde{\Phi}(A)$ for all $A \in \mathcal{A}$. Evidently, μ and ν are λ–absolutely continuous and therefore nonatomic. Applying Theorem 2.5 to the content $\varrho : \mathcal{A} \to \mathbb{R} \times X \times X$ given by $\varrho(\cdot) = \big(\lambda(\cdot), \mu(\cdot), \nu(\cdot)\big)$, we obtain the relative convexity of its range. Let us choose now a set $A \in \mathcal{A}$ such that both $\big\|\mu(A) - \theta\mu(\Omega)\big\|$ and $\big\|\nu(A) - \theta\mu(\Omega)\big\|$ are smaller than ε. Then

$$\mu(A) + \nu(\Omega \setminus A) \in \tilde{\Phi}(A) + \tilde{\Phi}(\Omega \setminus A) \subseteq \tilde{\Phi}(\Omega)$$

and $\big\|z - [\mu(A) + \nu(\Omega \setminus A)]\big\| < 3\varepsilon$, whence $z \in \mathrm{cl}\,\tilde{\Phi}(\Omega) = \tilde{\Phi}(\Omega)$. ∎

Due to the fact that in normed spaces, weak compactness is of a sequential nature, the result which says that in ℓ_1 we cannot distinguish between weak and strong convergence of sequences tells us that in ℓ_1 we cannot distinguish between (relative) weak compactness and (relative) compactness. Consequently, an alternative statement of the Lyapunov–Richter theorem for ℓ_1–valued correspondences could be registered as follows.

4.3. Corollary. *Let Φ be a $\overset{*}{+}$–finitely additive ℓ_1–valued correspondence. If Φ is exhaustive and relatively compact–valued, then $R\Phi$ is relatively compact. Moreover, if Φ is also nonatomic, then $R\Phi$ (and all values) are relatively convex.*

Let us consider the formula

$$\Phi(F) = \mathrm{cl}\,\{\mu(F) : \quad \mu \in \mathcal{S}(\Phi)\} \qquad \text{for all } F \in \mathcal{F}. \tag{4.2}$$

In [13], the formula has been proved under the assumptions that the values of Φ are closed, bounded and convex subsets of X, the correspondence Φ is $\overset{*}{+}$–countably additive on a σ–field, and X is separable or has the RNP. As an immediate consequence of Theorem 4.2, we have the next result.

4.4. Corollary. *Let Φ be $\overset{*}{+}$–finitely additive, closed–valued, exhaustive, and nonatomic. Then formula (4.2) holds in each of the following cases:*

X has the RNP and $v(\Phi)$ is finite;

X is B–convex and separable, and $v(\Phi)$ is finite;

$X = c_0$ or ℓ_p $(p \geq 1,\ p \neq 2)$.

5. ℓ_p–valued correspondences, $p \in (0,1)$

The aim of this section is to carry over some of the results we have seen before to the non-locally convex case of spaces ℓ_p with $p \in (0,1)$. So throughout this section let $X = \ell_p$ with its F–norm $\|x\|^p = \sum_{i\in\mathbb{N}} |x_i|^p$. Obviously, ℓ_p can be continuously embedded in ℓ_1. Since for all our spaces ℓ_p the set of continuous linear forms is ℓ_∞, we may consider the weak topology of ℓ_p that coincides with the restriction to ℓ_p of the weak topology of ℓ_1, that is, $\sigma(\ell_p, \ell_\infty) = \sigma(\ell_1, \ell_\infty)|_{\ell_p}$. Moreover, we remember that boundedness in ℓ_p can be described in terms of boundedness in F–norm. The next theorem is from [28, Theorem 3.7].

5.1. Theorem. *Each convex closed and bounded subset of ℓ_p, $p \in (0,1)$, is necessarily compact.*

5.2. Lemma. *For a bounded subset B of ℓ_p the closure $\mathrm{cl}_{\sigma(\ell_1,\ell_\infty)}(B)$ is still a subset of ℓ_p and is bounded.*

We have all the elements for giving the Lyapunov–Richter theorem for correspondences with values in ℓ_p $(p < 1)$.

5.3. Theorem. *Let Φ be $\overset{*}{+}$–finitely additive, exhaustive, and nonatomic. Then the values and range of Φ are both relatively convex and relatively compact.*

Proof. Because of the continuous embedding of ℓ_p into ℓ_1, we may consider Φ as ℓ_1–valued. It follows then from Theorem 4.2 that $\mathrm{cl}_{\ell_1}(R\Phi)$ and any $\mathrm{cl}_{\ell_1}(\Phi(F))$ are convex. Let the set B be either $R\Phi$ or $\Phi(F)$. Since B is a bounded subset of ℓ_p, Lemma 5.2 tells us that $\mathrm{cl}_{\ell_1}(B)$ is still a bounded subset of ℓ_p. Certainly, it is also closed in ℓ_p, and so Theorem 5.1 gives us the compactness of $\mathrm{cl}_{\ell_1}(B)$. To conclude, we need only observe that $\mathrm{cl}_{\ell_1}(B) = \mathrm{cl}_{\ell_p}(B)$. ■

The implication (2)⇒(1) of Theorem 4.1 can be preserved in the present context. This is because the argument in [8] still applies if we replace the Orlicz–Pettis theorem with the following result due to C. Swartz [25, Theorem 13]: in the sequence spaces ℓ_p, a series is convergent if it is subseries convergent with respect to the topology of pointwise convergence.

5.4. Theorem. *Let Φ be $\overset{*}{+}$–finitely additive. If $R\Phi$ is a relatively weakly compact subset of ℓ_p, then Φ is exhaustive.*

Our final result presents a statement that is similar to the implication (2)⇒(1) in Theorem 4.1.

5.5. Theorem. *Let Φ be $\overset{*}{+}$–finitely additive, exhaustive, and relatively compact–valued. Then the range of Φ is relatively compact.*

Proof. Use Theorems 3.6 and 5.3. ■

Remark. A different argument for proving a variation of Theorem 5.5 is possible. In a sense, we can say that it is more strictly related to the special nature of the range space we are dealing with. Here is the argument. Assume that Φ is $\overset{*}{+}$–finitely additive, exhaustive, and with values that are relatively weakly compact bounded subsets of ℓ_p. As in the proof of Theorem 5.3, Φ can be considered ℓ_1–valued. By Theorem 4.1, we know that $R\Phi$ is a subset of ℓ_p which is ℓ_1–relatively weakly compact. Applying Theorem 3.8 we see that $R\Phi$ is also a bounded subset of ℓ_p and therefore (by Lemma 5.2) it is ℓ_p–relatively weakly compact. We have obtained a result that is similar to (1)$\Rightarrow$(2) of Theorem 4.1.

References

1. C. D. Aliprantis and K. C. Border, *Infinite Dimensional Analysis*, Springer, Berlin, 1994.
2. C. D. Aliprantis, K. C. Border, and O. Burkinshaw, Economies with many commodities, *Journal of Economic Theory* 74 (1997), 62–105.
3. C. D. Aliprantis, D. J. Brown, and O. Burkinshaw, *Existence and optimality of competitive equilibria*, Springer, New York, 1990.
4. T. E. Armstrong and K. Prikry, Lyapunov's theorem for nonatomic, finitely additive, bounded, finite–dimensional, vector–valued measures, *Trans. Amer. Math. Soc.* 266 (1981), 499–514.
5. T. E. Armstrong and M. K. Richter, The core–Walras equivalence, *Journal of Economic Theory* 33 (1984), 116–151.
6. T. E. Armstrong and M. K. Richter, Existence of nonatomic core–Walras allocations, *Journal of Economic Theory* 38 (1986), 137–159.
7. A. Avallone and A. Basile, Lyapunov–Richter theorem in the finitely additive setting, *Journal of Mathematical Economics* 22 (1993), 557–561.
8. A. Avallone and A. Basile, Lyapunov–Richter theorem in B–convex spaces, *Journal of Mathematical Economics* 29 (1998).
9. W. G. Bade and P. C. Curtis, The Wedderburn decomposition of commutative Banach algebras, *American Journal of Mathematics* 82 (1960), 851–866.
10. A. Basile, Finitely additive correspondences, *Proc. Amer. Math. Soc.* 121 (1994), 883–891.
11. D. Blackwell, The range of certain vector integrals, *Proc. Amer. Math. Soc.* 2 (1951), 390–395.

12. M. A. Costé, Sur les multimesures a valeurs fermees bornees d'un espace de Banach, *Compte Rendue de l'Academie des Sciences de Paris* 280 (1975), 567–570.

13. M. A. Costé, Densite des selecteurs d'une multimesures a valeurs convexes fermees bornees d'un espace de Banach separable, *Compte Rendue de l'Academie des Sciences de Paris* 282 (1976), 967–969.

14. J. Diestel, H. Jarchow, and A. Tonge, *Absolutely Summing Operators*, Cambridge University Press, 1995.

15. J. Diestel and J. J. Uhl, *Vector Measures*, American Mathematical Society, Providence–Rhode Island, 1977.

16. L. Drewnowski, Additive and countably additive correspondences, *Annales Societatis Mathematicae Polonae, Series* I: *Commentationes Mathematicae* XIX (1976), 25–54.

17. A. Dvoretzky, A. Wald and J. Wolfowitz, Relations among certain ranges of vector measures, *Pacific Journal of Mathematics* 1 (1951), 59–74.

18. V. M. Kadets, A remark on Lyapunov's theorem on a vector measure, *Functional Analysis and Applications* 25 (1991), 295–297.

19. V. M. Kadets and G. Shekhtman, The Lyapunov theorem for ℓ_p–valued measures, *Saint Petersburg Mathematical Journal* 4 (1993), 961-966.

20. M. A. Khan and N. C. Yannelis (eds.), *Equilibrium Theory in Infinite Dimensional Spaces*, Springer–Verlag, Berlin, 1991.

21. A. Mas–Colell and W. R. Zame, Equilibrium theory in infinite dimensional spaces, in: *Handbook of Mathematical Economics*, Vol. IV, North–Holland, 1991, 1835–1898.

22. H. Richter, Verallgemeinerung eines in der Statistik benötigten Satzes der Masstheorie, *Mathematische Annalen* 150 (1963), 85–90.

23. D. Schmeidler, Convexity and compactness in countably additive correspondences, in: *Differential games and related topics* (eds. H. W. Kuhn and G. P. Szego), North–Holland, 1971.

24. D. Schmeidler, On set correspondences into uniformly convex Banach spaces, *Proc. Amer. Math. Soc.* 34 (1972), 97–101.

25. C. Swartz, A generalized Orlicz–Pettis theorem and applications, *Mathematische Zeitschrift* 163 (1978), 283–290.

26. K. Vind, Edgeworth allocations in exchange economy with many traders, *International Economic Review* 5 (1964), 165–177.

27. H. Weber, Group and vector valued s–bounded contents, in: *Measure Theory Oberwolfach 1983*, *LNM 1089*, Springer–Verlag, 1984.

28. H. Weber, Compact convex sets in non-locally convex spaces, *Note di Matematica* 12 (1992), 271–289.

29. E. Zeidler, *Nonlinear Functional Analysis and its Applications*, Vol. III, Springer–Verlag, New York, 1985.

Achille Basile
Facoltà di Economia
Dipartimento di Matematica e Statistica
Università Federico II
via Cintia, Complesso Monte S. Angelo
Napoli 80126
Italy
email: basile@matna2.dma.unina.it

EXTREMAL VECTORS FOR A CLASS OF LINEAR OPERATORS

PER ENFLO

Abstract. We prove a qualitative result characterizing the behavior of backward minimal vectors introduced in [1].

We start by recalling the definition of backward minimal vectors from Ansari–Enflo [1]. H will denote a separable Hilbert space over the real or complex numbers. $R(T)$ will denote the range of T.

1. Definition. *Let $T : H \to H$ be a bounded operator with dense range. Let $x_0 \in H$ and $\epsilon > 0$ with $\epsilon < \|x_0\|$. There is a unique vector y^{ϵ}_{n,x_0} such that $\left\|T^n y^{\epsilon}_{n,x_0} - x_0\right\| \le \epsilon$ and $\left\|y^{\epsilon}_{n,x_0}\right\| = \inf\left\{\|y\| : \ \|T^n y - x_0\| \le \epsilon\right\}$. y^{ϵ}_{n,x_0} are called backward minimal vectors.*

When there is no ambiguity, we will drop x_0 and ϵ in y^{ϵ}_{n,x_0}. The minimality of $y^{\epsilon}_{n,x_0} = y_n$ immediately gives the following.

2. Orthogonality Relation. *If $r_n \perp y_n$, then $T^n r_n \perp T^n y_n - x_0$.*

In [1] backward minimal vectors and other types of extremal vectors were studied and their connection to invariant subspaces was established. In this paper we will give results on backward minimal vectors — or rather $T^n y_n$ — for a class of linear operators. We conjecture that this class includes multiplication by $1 - z$ on H^2.

We say that an operator T on H has P1, if there is a sequence of polynomials $\ell_m(T)$ such that $\left\|T^m \ell_m(T)\right\|_{op} = 1$ and $T^m \ell_m(T) \to I$ in strong operator topology. We have the following result.

3. Proposition. *Multiplication by $1 - z$ on H^2 has* P1.

Proof. Let T_0 denote multiplication by $1 - z$ on H^2. We first observe that

$$\text{For every polynomial } P(z) \text{ there is a polynomial } P_0 \text{ such that } P_0(T_0) \text{ is a multiplication by } P(z) \text{ on } H^2. \tag{1}$$

To verify (1), we see that it is true for deg $P(z) = 0$ and that it follows easily by induction on deg $P(z)$.

Now let V_m denote multiplication by $\frac{(1-z)^m}{(b_m - z)^m}$ on H^2 where $b_m = 1 + \frac{1}{2^m}$. It is easy to see that $\|V_m\| < 1$ and $V_m \to I$ in strong operator topology. Obviously V_m can be approximated in norm topology by operators of the form "multiplication by $(1-z)^m P_m(z)$," with a polynomial P_m. Since they, by (1), have the form $T_0^m \ell_m(T_0)$, the proposition is proved. ■

From now on we will assume that T has dense range, $R(T) \neq H$, $\|x_0\| = 1$, $x_0 \notin R(T)$, and $0 < \epsilon < \frac{1}{2}$. We say that T has P2 if for all x_0 and ϵ as above, with $y_n = y_{n,x_0}^{\epsilon}$, we have $\frac{\|y_n\|}{\|y_{n-1}\|} \to \infty$ as $n \to \infty$.

In [2] it is proved that for T_0 being multiplication by $1-z$ on H^2 and x_0 being the constant function $\equiv 1$, $y_{n,x_0}^{\epsilon} = y_n$ has the form $\frac{A_n}{(a_1-z)(a_2-z)\ldots(a_n-z)}$ where $A_n, a_1, a_2, \ldots, a_n$ are complex numbers, $|a_i| > 1$ for $1 \leq i \leq n$. It is conjectured in [2] that

$$T_0^n y_n \textit{ converges in } H^2 \textit{ to a multiple of a singular inner function.} \tag{2}$$

To prove (2), one step would be to show that T_0 has P2 and that $T_0^n y_n$ converges in H^2. Then it follows from the Proposition 4 below that $T_0(\lim T_0^n y_n)$ is non-cyclic. Obviously this function has no zero inside the unit disk.

4. Proposition. *Assume that T has* P2. *Assume that $T^n y_n \to z$ for some x_0 and ϵ. Then Tz is non-cyclic for all operators commuting with T.*

Proof. Let V commute with T. Put $Vy_{n-1} = a_n y_n + r_n$, $r_n \perp y_n$. Then $VT^n y_{n-1} = a_n T^n y_n + T^n r_n$. And so

$$\left\langle VTT^{n-1} y_{n-1},\ x_0 - T^n y_n \right\rangle = a_n \langle T^n y_n,\ x_0 - T^n y_n\rangle + \langle T^n r_n,\ x_0 - T^n y_n\rangle.$$

When $n \to \infty$, the scalars $a_n \to 0$ and so the expression on the right–hand side converges to 0, while the expression on the left–hand side converges to $\langle VTz,\ x_0 - z\rangle$. This proves the proposition. ■

We do not know if $T^n y_n$ converges for operators satisfying P1 and P2. The main result of this paper will show, however, that the sequence $T^n y_n$ varies slowly.

5. Theorem. *Assume that T satisfies* P1 *and* P2. *Assume T has dense range, $x_0 \notin R(T)$, $\|x_0\| = 1$, and $0 < \epsilon < \frac{1}{2}$. Then $\left\|T^n y_n - T^{n+1} y_{n+1}\right\| \to 0$ as $n \to \infty$.*

The proof of this theorem will follow from the lemmas below.

6. Lemma. *Assume that T satisfies the assumptions of Theorem 5. Then for every $\delta > 0$ and every integer m, there is $K = K(m,\delta)$ such that for all n*

$$\left\|y_{n+m,x_0}^{\epsilon+\delta}\right\| \leq K \left\|y_{n,x_0}^{\epsilon}\right\|.$$

Proof. Let $T^j\ell_j(T) \to I$ strongly, $\left\|T^j\ell_j(T)\right\|_{op} \leq 1$. Choose m such that $\left\|T^m\ell_m(T)x_0 - x_0\right\| \leq \delta$. Put $y^{\epsilon}_{n,x_0} = y_n$. Then

$$\begin{aligned}
&\left\|T^{m+n}\ell_m(T)y_n - x_0\right\| = \left\|T^m\ell_m(T)(T^n y_n) - x_0\right\| \\
&\leq \left\|T^m\ell_m(T)(T^n y_n - x_0)\right\| + \left\|(T^m\ell_m(T) - I)x_0\right\| \\
&\leq \epsilon + \delta \quad \text{for all } n.
\end{aligned}$$

Moreover,

$$\left\|y^{\epsilon+\delta}_{n+m,x_0}\right\| \leq \left\|\ell_m(T)y_n\right\| \leq \left\|\ell_m(T)\right\|_{op}\left\|y_n\right\|.$$

With $K = \left\|\ell_m(T)\right\|_{op}$ the lemma is proved. ∎

7. Lemma. *Assume that T satisfies the assumptions of Theorem 5. Then, if $\epsilon < \epsilon'$,*

$$\frac{\left\|y^{\epsilon}_{n,x_0}\right\|}{\left\|y^{\epsilon'}_{n,x_0}\right\|} \to \infty \quad \text{as} \quad n \to \infty.$$

Proof. Put $\epsilon' = \epsilon + \delta$. Lemma 6 gives

$$\frac{\left\|y^{\epsilon'}_{n+1,x_0}\right\|}{\left\|y^{\epsilon}_{n+1,x_0}\right\|} < \frac{K\left\|y^{\epsilon}_{n,x_0}\right\|}{\left\|y^{\epsilon}_{n+1,x_0}\right\|} \quad \text{for all } n.$$

Since T has P2, the right–hand side converges to 0 as $n \to \infty$ and so the lemma is proved. ∎

Let θ_n denote the angle between $T^n y_n$ and $x_0 - T^n y_n$. It is also easy to see that if $\theta_n \geq \frac{\pi}{2} + \gamma$, $0 < \gamma < \frac{1}{10}$, then $\left\|T^n(1 + \epsilon \sin\gamma)y_n - x_0\right\| \leq \frac{\epsilon(\cos\gamma + 1)}{2}$ and so, with $\epsilon_1 = \frac{\epsilon(\cos\gamma + 1)}{2}$, we get

$$\left\|y^{\epsilon_1}_{n,x_0}\right\| \leq (1 + \epsilon)\left\|y^{\epsilon}_{n,x_0}\right\|. \tag{3}$$

Equation (3) and Lemma 7 give the next result.

8. Lemma. *If T satisfies the assumptions of Theorem 5, then*

$$\theta_n \to \frac{\pi}{2} \quad \text{as} \quad n \to \infty.$$

With Lemma 8 we can now complete the proof of Theorem 5. Put $T^n y_n = \alpha_n x_0 + \beta_n s_n$ where $s_n \perp x_0$, $\left\|s_n\right\| = 1$. Since, as $\theta_n \to \frac{\pi}{2}$, $T^n y_n$ approaches to be tangential to the ϵ–sphere around x_0, Lemma 8 gives

$$\alpha_n \to 1 - \epsilon^2 \quad \text{and} \quad \beta_n \to \epsilon\sqrt{1 - \epsilon^2} \quad \text{as} \quad n \to \infty. \tag{4}$$

Now, Lemma 6 gives, with $m = 1$, $\left\|y^{\epsilon+\delta}_{n+1,x_0}\right\| < K\left\|y^{\epsilon}_{n,x_0}\right\|$. Thus, we have $Ty^{\epsilon+\delta}_{n+1,x_0} = Cy^{\epsilon}_{n,x_0} + r_n$ where $r_n \perp y^{\epsilon}_{n,x_0}$ and $C \leq K\|T\|$. And so

$$\begin{aligned} T^{n+1}y^{\epsilon+\delta}_{n+1,x_0} &= CT^n y^{\epsilon}_{n,x_0} + T^n r_n \\ \text{where} \quad T^n r_n &\perp x_0 - T^n y^{\epsilon}_{n,x_0}. \end{aligned} \tag{5}$$

Now (4) and Lemma 8 give with $T^{n+1}y^{\epsilon+\delta}_{n+1,x_0} = \alpha^1_{n+1}x_0 + \beta'_{n+1}s'_{n+1}$, $\left\|s'_{n+1}\right\| = 1$, and $s'_{n+1} \perp x_0$, that

$$\alpha'_{n+1} \to 1 - (\epsilon + \delta)^2 \quad \text{as} \quad n \to \infty. \tag{6}$$

Consider the orthogonal projection $z^{(n)}$ of the element $T^{n+1}y^{\epsilon+\delta}_{n+1,x_0}$ onto Span $\left\{x_0, T^n y^{\epsilon}_{n,x_0}\right\}$. Lemma 8 and (5) give that $\left\langle z^{(n)},\ x_0 - T^n y^{\epsilon}_{n,x_0}\right\rangle \to 0$ as $n \to \infty$. With (6) this gives that there is a positive function $g(\delta)$, $g(\delta) \to 0$ as $\delta \to 0$, such that

$$\left\|T^{n+1}y^{\epsilon}_{n+1,x_0} - T^n y_{n,x_0}\right\| < g(\delta) \quad \text{if } n \text{ is sufficiently large.} \tag{7}$$

We now put $y^{\epsilon+\delta}_{n+1}, x_0 = C'y^{\epsilon}_{n+1,x_0} + r_{n+1}$ where $\left|C'\right| < 1$ and $r_{n+1} \perp y^{\epsilon}_{n+1,x_0}$. In the same way as we prove (7) we get

$$\left\|T^{n+1}y^{\epsilon+\delta}_{n+1,x_0} - T^{n+1}y^{\epsilon}_{n+1,x_0}\right\| < g(\delta) \quad \text{if } n \text{ is sufficiently large.} \tag{8}$$

Since δ is an arbitrary positive number, (7) and (8) complete the proof of Theorem 5.

References

1. S. Ansari and P. Enflo, Extremal vectors and invariant subspaces, *Trans. Amer. Math. Soc.* 350 (1998), 539–558.
2. A. Spalsbury, Ph.D. Thesis, Kent State University, Kent, OH, 1996.

Per Enflo
Department of Mathematics and Computer Science
Kent State University
Kent, OH 44242
USA
email: enflo@mcs.kent.edu

TOWARDS A THEORY OF NONLINEAR ORTHOMORPHISMS

ZAFER ERCAN* and ANTONY W. WICKSTEAD

Abstract. The concepts of orthomorphisms and central operators are well known and established in the theory of linear operators on vector lattices. In this work we introduce versions of the definitions of orthomorphisms and of central operators for nonlinear operators. We investigate the order structure of the corresponding spaces of these nonlinear operators. If the vector lattice is Dedekind complete, then both spaces are also Dedekind complete vector lattices, but in general they are not even vector lattices.

1. Introduction

Among the classes of linear operators on Archimedean vector lattices which have been studied in great detail are the classes of *band preserving* operators and of *orthomorphisms*, which are simply those band preserving operators which are also order bounded. The space $\mathrm{Orth}(E)$, of all orthomorphisms on E, is known to be an Archimedean vector lattice under the usual operator order, and moreover, for $S, T \in \mathrm{Orth}(E)$ and $x \in E_+$ the formulae

$$(S \vee T)x = Sx \vee Tx \quad \text{and} \quad (S \wedge T)x = Sx \wedge Tx$$

hold. If E is Dedekind complete then so is $\mathrm{Orth}(E)$. An important subspace of $\mathrm{Orth}(E)$ is $\mathcal{Z}(E)$, the (order) ideal generated by the identity operator on E. $\mathcal{Z}(E)$ is called the *center* of E, it is order dense in $\mathrm{Orth}(E)$ and its elements are known as *central* operators. This space is order and algebra isomorphic to a sublattice and subalgebra of $C(K)$ for some compact Hausdorff space K. Many results for $\mathcal{Z}(E)$ were first proved using a representation of this form although representation–free proofs are now available as well. A convenient reference for these facts about $\mathrm{Orth}(E)$ and $\mathcal{Z}(E)$ is [4].

In this work we introduce versions of the definitions of orthomorphisms and of central operators for nonlinear operators. We investigate the order structure of these spaces. If E is Dedekind complete then both spaces are Dedekind complete vector lattices, but in general they are not even vector lattices. The smaller space that we define is an ideal in the larger, but not order dense. Finally we initiate a study of representations for the smaller space analogous to that stated for $\mathcal{Z}(E)$ above.

*This author gratefully acknowledges the support of TUBITAK (the Scientific and Technical Research Council of Turkey) during the preparation of this work.

2. Nonlinear orthomorphisms

If E is an Archimedean vector lattice and $\phi : E \to E$ is a mapping with $\phi(0) = 0$ then we say that ϕ is *order bounded* if whenever $a, b \in E$ with $a \leq b$ the set

$$\left\{ \sum_{k=1}^{n} \left|\phi(x_k) - \phi(x_{k-1})\right| : \quad a = x_0 \leq x_1 \leq \cdots \leq x_n = b \right\}$$

is order bounded. We call $\phi : E \to E$ *increasing* if $x \geq y \Rightarrow \phi(x) \geq \phi(y)$. If ϕ is increasing and $a = x_0 \leq x_1 \leq \cdots \leq x_n = b$ then

$$\begin{aligned}\sum_{k=1}^{n} \left|\phi(x_k) - \phi(x_{k-1})\right| &= \sum_{k=1}^{n} (\phi(x_k) - \phi(x_{k-1})) \\ &= \phi(b) - \phi(a),\end{aligned}$$

so that increasing maps are automatically order bounded.

We call the mapping $\phi : E \to E$ *locally band preserving* if $x - y \perp z \Rightarrow \phi(x) - \phi(y) \perp z$. We denote by $\mathcal{ORTH}(E)$ the vector space of all order bounded locally band preserving mappings on E, with the pointwise vector operations. We call the elements of $\mathcal{ORTH}(E)$ (*nonlinear*) *orthomorphisms*. We order $\mathcal{ORTH}(E)$ by defining $\phi \geq \psi$ to mean that $\phi - \psi$ is increasing. Below we show that if E is Dedekind complete then the increasing mappings span $\mathcal{ORTH}(E)$. Note that if ϕ is any locally band preserving mapping on E and $x, y \in E$ then the fact that $x - x \wedge y \perp y - x \wedge y$ implies that $\phi(x) - \phi(x \wedge y) \perp \phi(y) - \phi(x \wedge y)$. Hence

$$\begin{aligned}(\phi(x) \wedge \phi(y) - \phi(x \wedge y))^+ &= \Big((\phi(x) - \phi(x \wedge y)) \wedge (\phi(y) - \phi(x \wedge y))\Big)^+ \\ &= (\phi(x) - \phi(x \wedge y))^+ \wedge (\phi(y) - \phi(x \wedge y))^+ = 0,\end{aligned}$$

so that $\phi(x) \wedge \phi(y) \leq \phi(x \wedge y)$. Similarly it can be shown that $\phi(x) \vee \phi(y) \geq \phi(x \vee y)$ for all $x, y \in E$. Thus the a locally band preserving mapping is increasing if and only if it is a lattice homomorphism.

For a linear operator $T : E \to E$ the condition of being locally band preserving is equivalent to the usual definition of being band preserving. Also, for linear operators, our definition of order boundedness reduces to the standard one. Furthermore a linear operator is increasing if and only if it is positive, so that the linear mappings in $\mathcal{ORTH}(E)$ form precisely $\mathrm{Orth}(E)$ and the orders on the two are the same.

Before proving the main result of this section, we need some technical properties of nonlinear locally band preserving mappings. The first of these asserts that locally band preserving mappings have the *Hammerstein property* introduced by Koldunov in [3].

2.1. Proposition. *Let E be an Archimedean vector lattice, $\phi : E \to E$ a mapping with $\phi(0) = 0$ which is locally band preserving, $u, v, z \in E$ and $u \perp v$, then we have*

$$\phi(u+v+z) - \phi(u+z) = \phi(v+z) - \phi(z).$$

Proof. As we have

$$(u+v+z) - (u+z) = (v+z) - z = v \perp u,$$

we have

$$\phi(u+v+z) - \phi(u+z), \quad \phi(v+z) - \phi(z) \perp u,$$

and hence

$$[\phi(u+v+z) - \phi(u+z)] - [\phi(v+z) - \phi(z)] \perp u.$$

Similarly

$$[\phi(u+v+z) - \phi(u+z)] - [\phi(v+z) - \phi(z)] \perp v,$$

so the disjointness of u and v forces

$$[\phi(u+v+z) - \phi(u+z)] - [\phi(v+z) - \phi(z)] = 0,$$

which establishes the claim. ∎

2.2. Corollary. *If $\phi : E \to E$ is locally band preserving, $u, v \in E$ with $u \perp v$, then $\phi(u+v) = \phi(u) + \phi(v)$.*

Proof. Use the preceding proposition with $z = 0$ and use the fact that $\phi(0) = 0$. ∎

2.3. Corollary. *If $\phi : E \to E$ is locally band preserving and $x, y \in E$, then $\phi(x \vee y) + \phi(x \wedge y) = \phi(x) + \phi(y)$.*

Proof. Put $z = x \wedge y$, $u = (x-y)^+$, and $v = (x-y)^-$ in Proposition 2.1. ∎

In fact it is not difficult to show that, for an arbitrary mapping ϕ between vector lattices, the property that $\phi(x \vee y) + \phi(x \wedge y) = \phi(x) + \phi(y)$ for all $x, y \in E$ characterizes the Hammerstein property.

2.4. Theorem. *If E is a Dedekind complete vector lattice then the space $\mathcal{ORTH}(E)$ is a Dedekind complete vector lattice when ordered by the cone of increasing mappings.*

Proof. Given $\phi \in \mathcal{ORTH}(E)$ and $a, b \in E$, let

$$m(a,b) = \sup\left\{\sum_{k=1}^{n} \left|\phi(x_k) - \phi(x_{k-1})\right| : \quad a = x_0 \le x_1 \le \cdots \le x_n = b\right\}$$

if $0 \le a \le b$,

and

$$m(a,b) = -\sup\left\{\sum_{k=1}^{n} \left|\phi(x_k) - \phi(x_{k-1})\right| : \quad a = x_0 \le x_1 \le \cdots \le x_n = b\right\}$$

if $a \le b \le 0$.

Next define, for each $x \in E$, $m(x) = m(0, x^+) + m(-x^-, 0)$. The proof that $\mathcal{ORTH}(E)$ is a lattice proceeds in three steps, first showing that $m \in \mathcal{ORTH}(E)$, then that both $m - \phi$ and $m + \phi$ are increasing and then that m is the least upper bound of ϕ and $-\phi$ in $\mathcal{ORTH}(E)$. As a final step we prove the Dedekind completeness.

Step 1. If $0 \le a \le b$ then $m(0,b) = m(0,a) + m(a,b)$. To see this, first note that

$$\begin{aligned}
m(0,b) &= \sup\left\{\sum_{k=1}^{n} \left|\phi(x_k) - \phi(x_{k-1})\right| : \quad 0 = x_0 \le x_1 \le \cdots \le x_n = b\right\} \\
&\ge \sup\left\{\sum_{k=1}^{n} \left|\phi(x_k) - \phi(x_{k-1})\right| : \quad 0 = x_0 \le x_1 \le \cdots \le x_n = a\right\} \\
&+ \sup\left\{\sum_{k=1}^{n} \left|\phi(x_k) - \phi(x_{k-1})\right| : \quad a = x_0 \le x_1 \le \cdots \le x_n = b\right\} \\
&= m(0,a) + m(a,b).
\end{aligned}$$

If $0 = x_0 \le x_1 \le \cdots \le x_n = b$ then $0 = x_0 \wedge a \le x_1 \wedge a \le \cdots \le x_n \wedge a = a$ and $a = x_0 \vee a \le x_1 \vee a \le \cdots \le x_n \vee a = b$, so we have

$$\begin{aligned}
&\sum_{k=1}^{n} \left|\phi(x_k) - \phi(x_{k-1})\right| = \\
&\quad = \sum_{k=1}^{n} \left|\phi(x_k \vee a) - \phi(x_{k-1} \vee a) + \phi(x_k \wedge a) - \phi(x_{k-1} \wedge a)\right| \\
&\quad \le \sum_{k=1}^{n} \left|\phi(x_k \vee a) - \phi(x_{k-1} \wedge a)\right| + \sum_{k=1}^{n} \left|\phi(x_k \wedge a) - \phi(x_{k-1} \wedge a)\right| \\
&\quad \le m(a,b) + m(0,a),
\end{aligned}$$

where the first equality follows from Corollary 2.3. Taking the supremum now shows that $m(0,b) \le m(0,a) + m(a,b)$ completing proof of the equality. Similarly if $a \le b \le 0$ then $m(a,0) = m(a,b) + m(b,0)$.

If $0 \le x, y$ and $x - y \perp z$ then

$$\begin{aligned}\left|m(x) - m(y)\right| &= \left|m(0,x) - m(0,y)\right| \\ &= \left|[m(0, x\wedge y) + m(x\wedge y, x)] - [m(0, x\wedge y) + m(x\wedge y, y)]\right| \\ &= \left|m(x\wedge y, x) - m(x\wedge y, y)\right| \\ &\le m(x\wedge y, x) + m(x\wedge y, y).\end{aligned}$$

If $x \wedge y = x_0 \le x_1 \le \cdots \le x_n = x$ then for each k we have

$$0 \le x_k - x_{k-1} \le x - x\wedge y \le \left|x - y\right| \perp z,$$

so that

$$\sum_{k=1}^{n} \left|\phi(x_k) - \phi(x_{k-1})\right| \perp z,$$

and taking the supremum shows that $m(x\wedge y, x) \perp z$. Similarly $m(x\wedge y, y) \perp z$ and hence we have $m(x) - m(y) \perp z$. A similar proof establishes this if $x, y \le 0$ and $x - y \perp z$. If we merely have $x, y \in E$ with $x - y \perp z$ then certainly $x^+ - y^+ \perp z$ and $x^- - y^- \perp z$ so that $m(x^+) - m(y^+)$, $m(-x^-) - m(-y^-) \perp z$, and hence $m(x) - m(y) \perp z$.

If $x \ge y$ then $x^+ \ge y^+$ and $-x^- \ge -y^-$ so that we certainly have $m(0,x^+) \ge m(0,y^+)$ and $m(-x^-,0) \ge m(-y^-,0)$ and hence we have $m(x) \ge m(y)$. I.e. m is increasing and therefore order bounded so certainly lies in $\mathcal{ORTH}(E)$.

Step 2. Note that if $0 \le a \le b$ then taking $n = 1$ in the definition of $m(a,b)$ we see that $m(a,b) \ge \left|\phi(a) - \phi(b)\right|$. Hence we have

$$m(b) = m(0,b) = m(0,a) + m(a,b) \ge m(0,a) + \left|\phi(a) - \phi(b)\right|,$$

so we have $(m \pm \phi)(b) \ge (m \pm \phi)(a)$. The same inequalities hold, with similar proof, if $a \le b \le 0$. In general, if $x \ge y$, we may apply the facts that $x^+ \ge y^+$ and $-x^- \ge -y^-$, these special cases and the general definition of m to conclude that $(m \pm \phi)(x) \ge (m \pm \phi)(y)$. Thus $m \ge \pm\phi$ in the order of $\mathcal{ORTH}(E)$.

Step 3. We show that if $\psi \ge \pm\phi$ then $\psi \ge m$. This will show that m is the modulus of ϕ, which will certainly guarantee that $\mathcal{ORTH}(E)$ is a vector lattice. If $0 \le a = x_0 \le x_1 \le \cdots \le x_n = b$ then the fact that $\psi \pm \phi$ are both increasing shows that $(\psi \pm \phi)(x_{k-1}) \le (\psi \pm \phi)(x_k)$ for $1 \le k \le n$. Hence

$$\sum_{k=1}^{n} \left|\phi(x_k) - \phi(x_{k-1})\right| \le \sum_{k=1}^{n} \left(\psi(x_k) - \psi(x_{k-1})\right) = \psi(b) - \psi(a)$$

and taking a supremum we see that $m(a,b) = m(b) - m(a) \le \psi(b) - \psi(a)$, so that $(\psi - m)(a) \le (\psi - m)(b)$. The same conclusion is similarly seen to hold if $a \le b \le 0$. The general case follows from the definition of m and Corollary 2.2 for ψ. Thus $\psi - m$ is increasing and hence m is least amongst the upper bounds of ϕ as required.

Step 4. It suffices to consider an increasing nonnegative family in $\mathcal{ORTH}(E)$ which is bounded above and then show that this family has a supremum. Suppose that $0 \le \phi_\alpha \uparrow \le \phi$ then define $\psi : E \to E$ by

$$\psi(x) = \sup_\alpha \phi_\alpha(x^+) + \inf_\alpha \phi_\alpha(-x^-),$$

which certainly exists as each $\phi_\alpha \le \phi$ and E is Dedekind complete. It is clear that $\psi(0) = 0$ and that ψ is increasing. Suppose that $x, y, z \in E$ and that $x - y \perp z$. For each β we have

$$\begin{aligned}\phi_\beta(x^+) &= \phi_\beta(x^+) - \phi_\beta(y^+) + \phi_\beta(y^+)\\ &\le \bigvee_\alpha (\phi_\alpha(x^+) - \phi_\alpha(y^+)) + \bigvee_\alpha \phi_\alpha(y^+)\\ &\le \bigvee_\alpha \left|\phi_\alpha(x^+) - \phi_\alpha(y^+)\right| + \bigvee_\alpha \phi_\alpha(y^+)\end{aligned}$$

and taking the supremum over all β we see that

$$\psi(x^+) - \psi(y^+) \le \bigvee_\alpha \left|\phi_\alpha(x^+) - \phi_\alpha(y^+)\right|.$$

From this, and the similar inequality with the roles of x and y reversed, we see that

$$\left|\psi(x^+) - \psi(y^+)\right| \le \bigvee_\alpha \left|\phi_\alpha(x^+) - \phi_\alpha(y^+)\right|.$$

As $x - y \perp z$, we certainly have $x^+ - y^+ \perp z$, so for each α we have $\left|\phi_\alpha(x^+) - \phi_\alpha(y^+)\right| \perp z$. It follows that the supremum of all such terms is also orthogonal to z and hence so is $\psi(x^+) - \psi(y^+)$. Similarly we see that $\psi(-x^-) - \psi(-y^-) \perp z$ and in a manner that is by now familiar we have $\psi(x) - \psi(y) \perp z$. We have now established that $\psi \in \mathcal{ORTH}(E)$.

We show next that ψ is an upper bound for the collection of all ϕ_α. Fix β and suppose that $x \le y$, so that $x^+ \le y^+$ and $-x^- \le -y^-$. For each α we may pick $\gamma \ge \alpha, \beta$ and note that $(\phi_\gamma - \phi_\beta)(y^+) \ge (\phi_\gamma - \phi_\beta)(x^+)$ so that

$$\phi_\alpha(x^+) + \phi_\beta(y^+) \le \phi_\gamma(x^+) + \phi_\beta(y^+) \le \phi_\gamma(y^+) + \phi_\beta(x^+) \le \bigvee_\gamma \phi_\gamma(y^+) + \phi_\beta(x^+).$$

Taking a supremum over α we see that

$$\psi(x^+) + \phi_\beta(y^+) \le \psi(y^+) + \phi_\beta(x^+).$$

Similarly we see that

$$\psi(-x^-) + \phi_\beta(-y^-) \leq \psi(-y^-) + \phi_\beta(-x^-),$$

which shows that

$$(\psi - \phi_\beta)(x) \leq (\psi - \phi_\beta)(y),$$

so that ψ is indeed an upper bound for the family of all ϕ_α.

It remains only to show that if τ is another upper bound for the collection of all ϕ_α then $\tau - \psi$ is increasing, which will show that ψ is indeed the least upper bound for the family, completing the proof that $\mathcal{ORTH}(E)$ is Dedekind complete. If $x \leq y$ then for each α we have $(\tau - \phi_\alpha)(x^+) \leq (\tau - \phi_\alpha)(y^+)$ and $(\tau - \phi_\alpha)(-x^-) \leq (\tau - \phi_\alpha)(-y^-)$. It follows that

$$\bigvee_\alpha \phi_\alpha(y^+) + \tau(x^+) \leq \bigvee_\alpha \phi_\alpha(x^+) + \tau(y^+)$$

and

$$\bigvee_\alpha \phi_\alpha(-y^-) + \tau(-x^-) \leq \bigvee_\alpha \phi_\alpha(-x^-) + \tau(-y^-),$$

so that $\psi(y) + \tau(x) \leq \psi(x) + \tau(y)$. I.e. $(\tau - \psi)(x) \leq (\tau - \psi)(y)$, completing the proof. ∎

Since the order bounded mappings on E clearly form a vector space under the pointwise operations, it follows that a locally band preserving mapping is order bounded if and only if it is the difference of two increasing locally band preserving maps. More significantly, in order to prove that a locally band preserving operator on a Dedekind complete vector lattice is order bounded, it suffices to consider in the definition of order boundedness only order intervals with one end being 0 as this suffices for the above construction of the modulus to work.

Clearly the Dedekind completeness of E is needed for the last part of the proof to work, but it is not clear that it is needed for $\mathcal{ORTH}(E)$ to be a vector lattice. However the following example shows that the condition of Dedekind completeness cannot just be omitted from the hypotheses of Theorem 2.4.

2.5. Example. Let $\sigma_n : \mathbb{R} \to \mathbb{R}$ be the continuous function which has a graph consisting of straight line segments of slope alternatively $+1$ and -1, each of length $2/n$, with $\sigma_n(0) = 0$ and with $\sigma_n(t) = t$ on $[-1/n, 1/n]$. It is not difficult to see that $\sigma_n \in \mathcal{ORTH}(\mathbb{R})$ and that σ_n^+ has graph consisting of straight line segments of slope alternately 1 and 0. As $n \to \infty$ we have σ_n^+ converging uniformly to half the identity function on $\mathbb{R}$.

Define $\phi : c \to c$ by

$$\phi(x)_n = \begin{cases} \sigma_n(x_n) & \text{if } n \text{ is even,} \\ 0 & \text{if } n \text{ is odd} \end{cases}$$

and observe that $\left|\sigma_n(x_n)\right| \leq 1/n$, so that $\phi(x) \in c_0$. It is not difficult to show that $\phi \in \mathcal{ORTH}(E)$. If ϕ^+ did exist in $\mathcal{ORTH}(E)$ then we would certainly have $\phi^+(x)_{2n-1} = 0$ for all $n \in \mathbb{N}$. On the other hand we certainly have $\phi^+(x)_{2n} \geq \sigma_{2n}^+(x_{2n})$ for all $n \in \mathbb{N}$. Considering the action of ϕ^+ on the constantly one sequence $\mathbf{1}$, we see that $\phi(\mathbf{1})_{2n-1} = 0$ while $\phi(\mathbf{1})_{2n} \geq \sigma_{2n}^+(\mathbf{1}) \to {}^1\!/_2$, which is impossible if $\phi^+(\mathbf{1}) \in c$. ∎

We have already observed that $\mathrm{Orth}(E)$ is a subspace of $\mathcal{ORTH}(E)$ and we know that it is a lattice. It is indeed a sublattice of $\mathcal{ORTH}(E)$.

2.6. Proposition. *If E is a Dedekind complete vector lattice then* $\mathrm{Orth}(E)$ *is a sublattice of $\mathcal{ORTH}(E)$.*

Proof. If $\phi \in \mathrm{Orth}(E)$ we let ψ be the modulus of ϕ calculated in $\mathrm{Orth}(E)$ and τ be the modulus of ϕ calculated in $\mathcal{ORTH}(E)$. We certainly have $\psi \geq \tau$. If $0 \leq y \leq x$ then putting $x_0 = 0$, $x_1 = y$, and $x_2 = x$, we have

$$\begin{aligned} \phi(y) &= \phi(y) - \phi(0) \\ &\leq \left|\phi(y) - \phi(0)\right| + \left|\phi(x) - \phi(y)\right| \\ &= \left|\phi(x_1) - \phi(x_0)\right| + \left|\phi(x_2) - \phi(x_1)\right| \\ &\leq \tau(x). \end{aligned}$$

It is well known that for $x \geq 0$ we have

$$\psi(x) = \sup\{\phi(y) : 0 \leq y \leq x\} \leq \tau(x)$$

so that $\psi(x) = \tau(x)$ for all $x \geq 0$. A similar proof establishes equality when $x \leq 0$ and the general case follows as usual. ∎

The use of the cone of increasing functions to order $\mathcal{ORTH}(E)$ seems to be the most sensible choice, but it does take some effort to get used to it. Here are a few examples to help the reader get a feel for this order.

2.7. Proposition. *Suppose that E is a Dedekind complete vector lattice and that $T : E \to E$ is a positive linear orthomorphism. If we define $\pi(x) = \left|T(x)\right|$ then $\pi \in \mathcal{ORTH}(E)$ and $|\pi| = T$.*

Proof. Clearly $\pi(0) = 0$. If $x, y, z \in E$ and $x - y \perp z$ then $Tx - Ty = T(x - y) \perp z$. As $\big||Tx| - |Ty|\big| \leq |Tx - Ty|$ we must have $\pi(x) - \pi(y) =$

$|Tx| - |Ty| \perp z$. Thus π is locally band preserving. The argument that we use below to show that $|\pi|$ exists will also show that it is order bounded, so lies in $\mathcal{ORTH}(E)$.

If $0 \le x$ then

$$\begin{aligned}
|\pi|(x) &= \sup\left\{\sum_{k=1}^{n} \big|\pi(x_k) - \pi(x_{k-1})\big| : \quad 0 = x_0 \le x_1 \le \cdots \le x_n = x\right\} \\
&= \sup\left\{\sum_{k=1}^{n} \Big||T(x_k)| - |T(x_{k-1})|\Big| : \quad 0 = x_0 \le x_1 \le \cdots \le x_n = x\right\} \\
&= \sup\left\{\sum_{k=1}^{n} \big|T(x_k) - T(x_{k-1})\big| : \quad 0 = x_0 \le x_1 \le \cdots \le x_n = x\right\} \\
&= \sup\left\{\sum_{k=1}^{n} T(x_k) - T(x_{k-1}) : \quad 0 = x_0 \le x_1 \le \cdots \le x_n = x\right\} \\
&= Tx.
\end{aligned}$$

Similarly, if $x \le 0$ we can show that

$$\begin{aligned}
|\pi|(x) &= -\sup\left\{\sum_{k=1}^{n} \big|\pi(x_k) - \pi(x_{k-1})\big| : \quad x = x_0 \le x_1 \le \cdots \le x_n = 0\right\} \\
&= T(x),
\end{aligned}$$

so that in general

$$|\pi|(x) = |\pi|(x^+) + |\pi|(-x^-) = T(x^+) + T(-x^-) = T(x^+ - x^-) = T(x),$$

as claimed. ■

Taking T to be the identity, we see that the mapping $x \mapsto |x|$ is in $\mathcal{ORTH}(E)$ and has the identity mapping as its modulus. On the other hand the mapping $x \mapsto x^+$ (or, more generally, $x \mapsto (a \vee x) \wedge b$, for fixed $a, b \in E$) is already increasing so is its own modulus.

3. Nonlinear central mappings

We call $\phi : E \to E$ a *(nonlinear) central mapping* if $\phi(0) = 0$ and there exists a real number α such that $\big|\phi(x) - \phi(y)\big| \le \alpha|x - y|$ for all $x, y \in E$. We denote the space of all such maps on E by $\mathcal{Z}(E)$ and give it the pointwise linear operations and order it using the cone of increasing functions. This space may also be normed by defining $\|\phi\| = \inf\left\{\alpha : \ \big|\phi(x) - \phi(y)\big| \le \alpha|x - y| \ \ \forall x, y \in E\right\}$. It is again easy to verify that the linear mappings in $\mathcal{Z}(E)$ are precisely $\mathrm{Z}(E)$

and that the two orders and norms coincide on $\mathcal{Z}(E)$. The following result is exactly what one would expect from the linear theory.

3.1. Theorem. *If E is a Dedekind complete vector lattice then $\mathcal{Z}(E)$ is the lattice ideal generated in $\mathcal{ORTH}(E)$ by the identity mapping on E.*

Proof. If $\phi \in \mathcal{Z}(E)$ and $x, y, z \in E$ with $x - y \perp z$ then the fact that $|\phi(x) - \phi(y)| \leq \alpha|x - y|$ shows that $\phi(x) - \phi(y) \perp z$ so that ϕ is locally band preserving. To see that ϕ is order bounded, simply observe that if $a = x_0 \leq x_1 \leq \cdots \leq x_n = b$ then

$$\sum_{k=1}^{n} |\phi(x_k) - \phi(x_{k-1})| \leq \sum_{k=1}^{n} \alpha|x_k - x_{k-1}| = \sum_{k=1}^{n} \alpha(x_k - x_{k-1}) = \alpha(b - a).$$

Thus we certainly have $\mathcal{Z}(E) \subseteq \mathcal{ORTH}(E)$ and we see that if either $0 \leq a \leq b$ or $a \leq b \leq 0$ then

$$\begin{aligned} |\phi|(b) - |\phi|(a) &= \sup\left\{\sum_{k=1}^{n} |\phi(x_k) - \phi(x_{k-1})| : \quad a = x_0 \leq x_1 \leq \cdots \leq x_n = b\right\} \\ &\leq \alpha(b - a). \end{aligned}$$

Suppose that $x \geq y$, then we have $x^+ \geq y^+ \geq 0$ and $0 \geq -x^- \geq -y^-$, so that

$$\begin{aligned} |\phi|(y^+) - |\phi|(x^+) &\leq \alpha(y^+ - x^+) \\ |\phi|(-y^-) - |\phi|(-x^-) &\leq \alpha(-y^- + x^-), \end{aligned}$$

so that $|\phi|(y) - |\phi|(x) \leq \alpha(y - x)$. I.e. $(\alpha i - |\phi|)(x) \leq (\alpha i - |\phi|)(y)$, where i denotes the identity on E, so that $|\phi| \leq \alpha i$ in the ordering of $\mathcal{ORTH}(E)$.

Now suppose that $\phi \in \mathcal{ORTH}(E)$ and that $0 \leq \phi \leq \alpha i$ for some $\alpha \in \mathbb{R}$. If $x, y \in E$ then

$$\begin{aligned} \phi(y) - \phi(x) &\leq \phi(x \vee y) - \phi(x) \\ &= (\alpha i - \phi)(x) - (\alpha i - \phi)(x \vee y) + (\alpha i)(x \vee y) - (\alpha i)(x) \\ &\leq (\alpha i)(x \vee y) - (\alpha i)(x) \\ &= \alpha(x \vee y - x) \\ &= \alpha(y - x)^+ \\ &\leq \alpha|y - x| \end{aligned}$$

and similarly $\phi(x) - \phi(y) \leq \alpha|y - x|$, so that $|\phi(x) - \phi(y)| \leq \alpha|x - y|$ and ϕ certainly lies in $\mathcal{Z}(E)$. In general, if $|\phi| \leq \alpha i$ then $0 \leq \phi^+, \phi^- \leq \alpha i$ so that $\phi^+, \phi^- \in \mathcal{Z}(E)$ and hence $\phi \in \mathcal{Z}(E)$. This completes the proof that $\mathcal{Z}(E)$ is precisely the ideal generated in $\mathcal{ORTH}(E)$ by i. ∎

The norm on $\mathcal{Z}(E)$ is precisely the order unit norm induced by the identity mapping. It follows, of course, that if E is Dedekind complete then $\mathcal{Z}(E)$ will also be a Dedekind complete vector lattice. In the linear case, $Z(E)$ is order dense in $\mathrm{Orth}(E)$. This is no longer so in the nonlinear setting — this is another major difference between the linear and the nonlinear theory.

3.2. Example. Take $\mathbb{R}$ as our vector lattice and define $\phi : \mathbb{R} \to \mathbb{R}$ by $\phi(x) = 1$ if $x > 0$ and $\phi(x) = 0$ if $x \le 0$. Any map of $\mathbb{R}$ into itself which leaves 0 fixed is locally band preserving and this map is order bounded as if $a = x_0 \le x_1 \le \cdots \le x_n = b$ then either all terms $\phi(x_k)$ are zero or all are 1 or all up to some point are zero and then all are 1. It follows that the terms $\left|\phi(x_k) - \phi(x_{k-1})\right|$ are all zero, except possibly for one taking the value 1. Thus certainly the set

$$\left\{\sum_{k=1}^{n} \left|\phi(x_k) - \phi(x_{k-1})\right| : \quad a = x_0 \le x_1 \le \cdots \le x_n = b\right\}$$

is order bounded in $\mathbb{R}$. Thus $\phi \in \mathcal{ORTH}(E)$ and clearly ϕ is increasing.

If we let i denote the identity on $\mathbb{R}$, then if $m, n \in \mathbb{N}$ with $m \le k$, we have

$$\begin{aligned}
&(\phi \wedge mi)(1) = \\
&\quad = \sum_{k=1}^{n} (\phi \wedge mi)\left(\frac{k}{n}\right) - (\phi \wedge mi)\left(\frac{k-1}{n}\right) \\
&\quad \le \sum_{k=1}^{n} \left(\phi\left(\frac{k}{n}\right) - \phi\left(\frac{k-1}{n}\right)\right) \wedge \left(\frac{mk}{n} - \frac{m(k-1)}{n}\right) \\
&\quad = \left(\phi\left(\frac{1}{n}\right) - \phi(0)\right) \wedge \left(\frac{m}{n}\right) + \sum_{k=2}^{n} \left(\phi\left(\frac{k}{n}\right) - \phi\left(\frac{k-1}{n}\right)\right) \wedge \left(\frac{m}{n}\right) \\
&\quad = \left(\phi\left(\frac{1}{n}\right) - \phi(0)\right) \wedge \left(\frac{m}{n}\right) \le \frac{m}{n}.
\end{aligned}$$

Thus we have $(\phi \wedge mi)(1) = 0$ for all $m \in \mathbb{N}$. From step 4 in the proof of Theorem 2.4 we saw that $\bigvee_{m\in\mathbb{N}}(\phi \wedge mi)$ exists in $\mathcal{ORTH}(\mathbb{R})$ and that its value at 1 is $\bigvee_{m\in\mathbb{N}}(\phi \wedge mi)(1) = 0$, so this supremum is certainly not ϕ and $\mathcal{Z}(E)$ is not order dense in $\mathcal{ORTH}(E)$. ∎

A possible alternative candidate for a nonlinear analogue of the space of linear orthomorphisms would be the linear span of the order continuous increasing members of $\mathcal{ORTH}(E)$. It is unknown to the authors whether or not $\mathcal{Z}(E)$ is always order dense in this space. It is clear that $\mathcal{Z}(E)$ will certainly be an ideal in it.

4. Some concrete examples

We commence by looking at the very simplest case, namely when $E = \mathbb{R}$. Unlike the linear case, the spaces that arise are far from trivial. For example the space $\mathcal{ORTH}(\mathbb{R})$ is easily seen to be precisely the space of functions from $\mathbb{R}$ into itself which leave 0 fixed and such that the restriction to every bounded interval is of bounded variation. The order continuous maps in $\mathcal{ORTH}(\mathbb{R})$ may similarly be identified with the space of continuous functions from $\mathbb{R}$ into $\mathbb{R}$, leaving 0 fixed, which are of bounded variation on each bounded interval. Similarly the space $\mathcal{Z}(\mathbb{R})$ may be identified with the space of functions from $\mathbb{R}$ into $\mathbb{R}$ which leave 0 fixed and which satisfy a Lipschitz condition on the whole of $\mathbb{R}$. An alternative description of this latter space in terms which may be rather more familiar to modern analysts is easily given.

4.1. Theorem. *The space $\mathcal{Z}(\mathbb{R})$ is isometrically order isomorphic to $L_\infty(\mathbb{R})$.*

Proof. In Chapter 19, Section 5, of [2] it is established that for absolutely continuous functions the identity $f(x) = \int_0^x f'(t)\,dt$ holds for all $x \in \mathbb{R}$. If f satisfies a lipschitz condition on $\mathbb{R}$, then f' lies in $L_\infty(\mathbb{R})$. The order isomorphism merely states that f is increasing if and only if f' is nonnegative and the isometry states that the best lipschitz constant is precisely that essential supremum norm of f'. ■

We conclude by describing $\mathcal{Z}(C_0(\Sigma, E))$ where Σ is a locally compact Hausdorff space and E a Dedekind complete Banach lattice in a manner analogous to the description of $\mathrm{Z}(C_0(\Sigma, E))$ given in [1]. We let $\mathcal{Z}(E)_s$ denote $\mathcal{Z}(E)$ equipped with the strong operator topology in which $f_\alpha \to f$ if and only if $f_\alpha(e) \to f(e)$ for all $e \in E$. By $C^b(\Sigma, \mathcal{Z}(E)_s)$ we mean the space of all functions from Σ into $\mathcal{Z}(E)$ which are continuous for the strong operator topology on $\mathcal{Z}(E)$ and which are bounded in the norm of $\mathcal{Z}(E)$. We order $C^b(\Sigma, \mathcal{Z}(E)_s)$ by defining $\phi_1 \le \phi_2$ to mean that, for all $\sigma \in \Sigma$, $\phi_1(\sigma) \le \phi_2(\sigma)$ in $\mathcal{Z}(E)$.

4.2. Theorem. *If Σ is a locally compact Hausdorff space and E is a Banach lattice, then $\mathcal{Z}(C_0(\Sigma, E))$ is isometrically order isomorphic to $C^b(\Sigma, \mathcal{Z}(E)_s)$.*

Proof. Define a map $\phi \mapsto \hat{\phi}$ of $C^b(\Sigma, \mathcal{Z}(E)_s)$ into $\mathcal{Z}(C_0(\Sigma, E))$ by $\hat{\phi} f(\sigma) = \phi(\sigma) f(\sigma)$, for $f \in C_0(\Sigma, E)$ and $\sigma \in \Sigma$. As $\phi(\Sigma)$ is bounded in $\mathcal{Z}(E)$ there is a constant M such that $\left|\phi(\sigma)(e_1) - \phi(\sigma(e_2)\right| \le M|e_1 - e_2|$ for all $\sigma \in \Sigma$, $e_1, e_2 \in E$. Clearly $\hat{\phi} f : \Sigma \to E$ and it is equally clear that it vanishes at infinity. The continuity is also immediate from the strong operator continuity of ϕ, so that $\hat{\phi} f \in C_0(\Sigma, E)$.

To see that $\hat{\phi} \in \mathcal{Z}(C_0(\Sigma, E))$ note first that we clearly have $\hat{\phi}(0) = 0$. Also for arbitrary $f, g \in C_0(\Sigma, E)$ and $\sigma \in \Sigma$ we have

$$\begin{aligned}\left|\hat{\phi}f - \hat{\phi}g\right|(\sigma) = \left|\hat{\phi}f(\sigma) - \hat{\phi}g(\sigma)\right| = \left|\phi(\sigma)f(\sigma) - \phi(\sigma)g(\sigma)\right| \leq \\ \leq M\left|f(\sigma) - g(\sigma)\right| = M\left|f - g\right|(\sigma),\end{aligned}$$

so that $\left|\hat{\phi}f - \hat{\phi}g\right| \leq M\left|f - g\right|$.

The map $\phi \mapsto \hat{\phi}$ is one-to-one as if $\hat{\phi} = \hat{\psi}$, $\sigma \in \Sigma$, and $e \in E$, choose $f \in C_0(\Sigma)$ with $f(\sigma) = 1$. Then

$$\begin{aligned}\phi(\sigma)(e) = \phi(\sigma)\big((f \otimes e)(\sigma)\big) = \hat{\phi}(f \otimes e)(\sigma) = \\ = \hat{\psi}(f \otimes e)(\sigma) = \psi(\sigma)\big((f \otimes e)(\sigma)\big) = \psi(\sigma)(e),\end{aligned}$$

where $f \otimes e(\tau) = f(\tau)e$. Thus $\phi(\sigma) = \psi(\sigma)$ for all $\sigma \in \Sigma$ and hence $\phi = \psi$.

Now suppose that $\Phi \in \mathcal{Z}(C_0(\Sigma, E))$. We show that the map $\phi \mapsto \hat{\phi}$ is onto by constructing $\phi \in C^b(\Sigma, \mathcal{Z}(E))$ with $\hat{\phi} = \Phi$. For each $\sigma \in \Sigma$, let f_σ denote a function in $C_0(\Sigma)$ which is identically one on some compact neighborhood U_σ of σ, which vanishes off some other relatively compact open neighborhood of σ and with $0 \leq f_\sigma \leq 1$. Define $\phi : \Sigma \to \mathcal{Z}(E)$ by $\phi(\sigma)(e) = \phi(f_\sigma \otimes e)(\sigma)$. We see that $\phi(\sigma) \in \mathcal{Z}(E)$ as for $e_1, e_2 \in E$ we have

$$\begin{aligned}\left|\phi(\sigma)(e_1) - \phi(\sigma)(e_2)\right| &= \left|\Phi(f_\sigma \otimes e_1)(\sigma) - \Phi(f_\sigma \otimes e_2)(\sigma)\right| \\ &\leq \|\Phi\|\left|(f_\sigma \otimes e_1)(\sigma) - (f_\sigma \otimes e_2)(\sigma)\right| \\ &= \|\Phi\|\left|e_1 - e_2\right|.\end{aligned}$$

This also shows that ϕ is a bounded function into $\mathcal{Z}(E)$.

To see that ϕ is continuous for the strong operator topology on $\mathcal{Z}(E)$, suppose that $\sigma_\gamma \to \sigma$. Choose γ_0 so that $\gamma \geq \gamma_0$ implies that $\sigma_\gamma \in U_\sigma$. Then for each $\gamma \geq \gamma_0$ we have

$$\begin{aligned}\left|\Phi(f_{\sigma_\gamma}) \otimes e)(\sigma_\gamma) - \Phi(f_\sigma \otimes e)(\sigma_\gamma)\right| &= \|\Phi\|\left|(f_{\sigma_\gamma} \otimes e)(\sigma_\gamma) - (f_\sigma \otimes e)(\sigma_\gamma)\right| \\ &= \|\Phi\|\left|e - e\right| = 0,\end{aligned}$$

so that $\Phi(f_{\sigma_\gamma} \otimes e)(\sigma_\gamma) = \Phi(f_\sigma \otimes e)(\sigma_\gamma)$ for all $\gamma \geq \gamma_0$. If we now fix $e \in E$ and suppose that $\gamma \geq \gamma_0$ then

$$\begin{aligned}\left\|\phi(\sigma_\gamma)(e) - \phi(\sigma)(e)\right\| &= \left\|\Phi(f_{\sigma_\gamma} \otimes e)(\sigma_\gamma) - \Phi(f_\sigma \otimes e)(\sigma)\right\| \\ &= \left\|\Phi(f_\sigma \otimes e)(\sigma_\gamma) - \Phi(f_\sigma \otimes e)(\sigma)\right\| \to 0.\end{aligned}$$

Hence $\phi(\sigma_\gamma)(e) \to \phi(\sigma)(e)$ as $\sigma_\gamma \to \sigma$, so that ϕ does indeed have the required continuity.

We now show that $\hat{\phi} = \Phi$. Fixing $f \in C_0(\Sigma, E)$ and $\sigma \in \Sigma$ we have

$$\begin{aligned}\left|\hat{\phi}f(\sigma) - \Phi(f)(\sigma)\right| &= \left|\phi(\sigma)(f(\sigma)) - \Phi(f)(\sigma)\right| \\ &\le \left|\Phi(f_\sigma \otimes f(\sigma))(\sigma) - \Phi(f)(\sigma)\right| \\ &\le \|\Phi\|\left|(f_\sigma \otimes f(\sigma))(\sigma) - f(\sigma)\right| \\ &= \|\Phi\|\left|f_\sigma(\sigma)f(\sigma) - f(\sigma)\right| = 0.\end{aligned}$$

Thus the mapping $\phi \mapsto \hat{\phi}$ is also onto.

It is routine to show that the map $\phi \mapsto \hat{\phi}$ is an order isomorphism and that the function that is constantly the identity operator on E correspond to the identity operator on $C_0(\Sigma, E)$ which suffices, given that the norms are the order unit norms induced by these respective order units, to establish that the map is an isometry. ■

If $C_0(\Sigma, E)$ is Dedekind complete (see [1] for conditions under which this occurs) then $\mathcal{Z}(C_0(\Sigma, E))$ and hence $C^b(\Sigma, \mathcal{Z}(E)_s)$ will also be Dedekind complete vector lattices. In particular if $E = \mathbb{R}$ and Ω is a compact Hausdorff extremally disconnected space then $C^b(\Omega, \mathcal{Z}(\mathbb{R})_s)$ will be a Dedekind complete vector lattice. This should be contrasted with the situation for the norm continuous functions from Ω into $\mathcal{Z}(\mathbb{R}) = L_\infty(\mathbb{R})$ which is only Dedekind complete if Ω is a finite set.

References

1. Z. Ercan and A. W. Wickstead, Banach lattices of continuous Banach lattice–valued functions, *J. Math. Anal. Appl.* 198 (1996), 121–136.
2. C. Goffman, *Real Functions*, Rinehart, New York, 1953.
3. A. V. Koldunov, Hammerstein operators preserving disjointness, *Proc. Amer. Math. Soc.* 123 (1995), 1083–1095.
4. A. C. Zaanen, *Riesz Spaces* II, North–Holland, Amsterdam, New York, Oxford, 1983.

Zafer Ercan
Department of Mathematics
Middle East Technical University
06531 Ankara
Turkey
email: zercan@rorqual.cc.metu.edu.tr

Antony W. Wickstead
Department of Pure Mathematics
The Queen's University of Belfast
Belfast, BT7 1NN
Northern Ireland
email: a.wickstead@qub.ac.uk

FINITELY GENERATED VECTOR SUBLATTICES

CHARLES B. HUIJSMANS

Abstract. Let E^+ be the positive cone of an Archimedean vector lattice E. It is shown in [4, Theorem 2.1] that for arbitrary $u, v \in E^+$ the vector sublattice $R(u,v)$ of E generated by u and v can be described as follows:

$$R(u,v) = \text{Span}\,\{(\alpha u + \beta v)^+ : \ \alpha, \beta \in \mathbb{R}\}.$$

We will demonstrate in this paper that this result no longer holds for three or more positive elements. It also ceases to be true for the vector sublattice generated by two (or more) arbitrary elements. Moreover, we will show that every finitely generated vector sublattice of E is finite-dimensional if and only if E is hyper-Archimedean.

1. Introduction

Throughout, E denotes an Archimedean vector lattice (= Riesz space). For unexplained terminology and the elementary theory of vector lattices, normed vector lattices and f-algebras we refer to [1, 2, 13, 15, 17].

If D is a non-empty subset of E, then the linear span of D, i.e., the linear subspace of E generated by D, consists of all finite linear combinations of elements of D and will be denoted by Span $\{D\}$. By $R(D)$ we denote the vector sublattice (= Riesz subspace) generated by D; so $R(D)$ is the intersection of all vector sublattices of E containing D. Notice that Span $\{D\} \subset R(D)$. This inclusion may be proper, even in the case when D is a linear subspace of E (in which case, of course, $D = \text{Span}\,\{D\}$). For later purposes observe that for each non-empty subset D of E, we have $R(D) = R(\text{Span}\,\{D\})$. Furthermore, if D itself is a vector sublattice of E, then $D = \text{Span}\,\{D\} = R(D)$.

The order ideal generated by an arbitrary non-empty subset D of E is denoted by $A(D)$. Hence, $A(D)$ is the smallest ideal of E containing D, i.e., the intersection of all ideals in E containing D. Notice that in general

$$D \subset \text{Span}\,\{D\} \subset R(D) \subset A(D)$$

with equality everywhere when D itself is an ideal. Contrary to $R(D)$, the ideal $A(D)$ is very easy to describe (see [1, Section 1] or [13, Section 17]). Indeed, $A(D)$ consists of all $f \in E$ such that

$$|f| \leq |a_1 g_1| + \cdots + |a_n g_n|$$

for appropriate $a_1, \ldots, a_n \in \mathbb{R}$, $g_1, \ldots, g_n \in D$, and $n = n(f)$ depending on f. If D consists of a single element, say g, then $A(\{g\})$ is in short denoted by A_g.

This principal order ideal consists of all $f \in E$ satisfying $|f| \leq |ag|$ for some real number a.

It is much more difficult to characterize the elements of $R(D)$. To the best of my knowledge, the only general result in this direction is due to G. Jameson in [10, 2.2.11], and it reads as follows. If L is a linear subspace of E and the set of all finite infima (finite suprema) of elements of L is denoted by $\wedge L$ ($\vee L$, respectively), then

$$R(L) = \wedge(\vee L) = \vee(\wedge L).$$

The latter equality is evidently a consequence of the distributivity of E considered as a lattice (see e.g. [1, Section 1] or [13, Section 12]). A nice application of the G. Jameson result (which is not used very often in the literature) can be found in a paper by D. R. Hart [8, proof of Lemma 1.5]. It is proved there that if L is a linear subspace of E and $T : L \to L$ is a positive linear mapping with the property that $x \perp y$ ($x \in L$, $y \in E$) entails $Tx \perp y$, then T has a (unique) extension to a positive operator $\tilde{T} : R(L) \to R(L)$ such that $x \perp y$ ($x \in R(L)$, $y \in E$) still implies $\tilde{T}x \perp y$. The extended operator $\tilde{T}$ is therefore a positive orthomorphism on $R(L)$ (for the concept of orthomorphism, consult [2, Section 8] or [17, Section 139]).

Back to the general situation: if D is an arbitrary non-empty subset of E, then $R(D) = R(\text{Span}\,\{D\})$ consists, by the above result of G. Jameson, of all finite infima of all finite suprema of finite linear combinations of elements of D. In particular, if D is a finite subset of E, say $D = \{f_1, \dots, f_n\}$, then the vector sublattice $R(f_1, \dots, f_n)$ generated by $f_1, \dots, f_n$ consists of all finite infima of all finite suprema of linear combinations of $f_1, \dots, f_n$. The order of finite infima and finite suprema may be interchanged herein. This is not a simple description and, in practice, it is difficult to handle. It came therefore as a surprise in 1991 that in the case of *two positive* elements $u, v \in E^+$, the elements of the vector sublattice $R(u, v)$ generated by u and v are rather easy to characterize. Indeed, the following result was proved then by D. J. Brown, B. de Pagter and the author [4, Theorem 2.1].

1.1. Theorem. *Let E be an Archimedean vector lattice and $u, v \in E^+$. Then the vector sublattice $R(u, v)$ generated by u and v satisfies the equality*

$$R(u, v) = \text{Span}\,\{(\alpha u + \beta v)^+ : \ \alpha, \beta \in \mathbb{R}\}.$$

Closely related are the following equalities:

$$\begin{aligned} R(u, v) &= \text{Span}\,\{u, \ (v - \alpha u)^+ : \ \alpha \geq 0\} \\ &= \text{Span}\,\{u, \ (v - \alpha u)^+ : \ \alpha \in \mathbb{R}\}, \end{aligned}$$

and dually (by interchanging the roles of u and v):

$$R(u,v) = \operatorname{Span}\{v,\ (u-\alpha v)^+ :\ \alpha \geq 0\}$$
$$= \operatorname{Span}\{v,\ (u-\alpha v)^+ :\ \alpha \in \mathbb{R}\}.$$

This result has an interesting application in mathematical economics. Indeed, let E be an Archimedean f–algebra with the unit element e (for this concept, see [2, Definition 8.19] or [17, Definition 140.8]). If $\lambda \geq 0$, then a *call option* on $u \in E^+$ with *striking price* λ, will be denoted by $c_\lambda(u)$ and is, by definition,

$$c_\lambda(u) = (u - \lambda e)^+.$$

By Theorem 1.1 (cf. [4, Theorem 4.1]), the vector sublattice $R(u,e)$ generated by u and e satisfies

$$R(u,e) = \operatorname{Span}\{e,\ c_\lambda(u) :\ \lambda \geq 0\}.$$

This characterization of $R(u,e)$ as *portfolios of calls* on u generalizes an earlier result due to D. J. Brown and S. A. Ross [5, Theorem 1] who showed it for the $C(X)$–case. It is worthwhile to note in this respect that the class of Archimedean unital f–algebras is much wider than the class of f–algebras of type $C(X)$. Take for instance f–algebras of measurable functions.

In order to make this paper more or less self–contained, we present for the reader's convenience a sketch of the proof of Theorem 1.1 using representations. It is still an open problem whether this result can be accomplished without the axiom of choice (or some weaker version of it). The proof is divided into three steps.

In the first step it is shown that the theorem is true for the vector lattice $E = C([0,1])$ of all real continuous functions on $[0,1]$ and the functions $u = \mathbf{1}$ and $v = \mathbf{i}$, where $\mathbf{1}$ is the constant one function (that is, $\mathbf{1}(x) = 1,\ 0 \leq x \leq 1$) and $\mathbf{i}$ is the identity function (that is, $\mathbf{i}(x) = x,\ 0 \leq x \leq 1$). It turns out that $R(\mathbf{1},\mathbf{i})$ consists of all continuous piecewise linear functions on $[0,1]$ with finitely many pieces.

Secondly, the $C(X)$–case, with X a compact Hausdorff space, $u = \mathbf{1}_X$, the constant one function on X, and $0 \leq v \leq \mathbf{1}_X$, is reduced to the first case by means of an argument due to A. C. M. van Rooij (private communication, cf. [7, Section 4] and [10, Theorem 6.3]).

Finally, the general case is proven then by means of the Yosida–Kakutani representation theorem for principal order ideals [13, Section 45]. For more details of the proof of Theorem 1.1, see [4, Section 2].

Incidentally, instead of considering $C([0,1])$, $u = \mathbf{1}$, and $v = \mathbf{i}$ in the first step above, one can just as well take $E = C(\mathbb{R}^2_+)$, the vector lattice of all real continuous functions on the first quadrant and the coordinate functions u and v given by

$$u(x,y) = x, \qquad v(x,y) = y \qquad (x, y \geq 0).$$

In this case, $R(u,v)$ consists of all continuous functions defined on the first quadrant that are piecewise homogeneous linear in (finitely many) conical polyhedral sectors of the first quadrant with common vertex 0 (cf. the section on free vector lattices in the book of G. Birkhoff [3, Chapter XV, Section 5, Theorem 7]). A similar argument as in the second and third step above leads then to the desired result.

It is one of the purposes of the present paper to show that the situation in Theorem 1.1 is very special. To be more precise, we will demonstrate in Section 2 by means of counterexamples that this result is no longer valid for two (or more) arbitrary elements of E, nor does it hold in general for three or more positive elements! Apart from this, we will exhibit some more examples closely related to the topic under consideration.

In the final Section 3, we will prove that every finitely generated vector sublattice of E is finite–dimensional if and only if E is hyper-Archimedean, and also we will present there a survey of characterizations of (uniformly complete) hyper-Archimedean vector lattices.

2. Examples

We start this section with an example of a finitely generated vector sublattice that properly contains the span of its generators.

2.1. Example. Consider $E = \mathbb{R}^3$ with the coordinatewise vector space operations and partial ordering. For $u = (1,1,0)$ and $v = (0,1,1)$ we have $(u-v)^+ = (1,0,0)$. Since

$$(\alpha, \beta, \gamma) = (\beta - \gamma)u + \gamma v + (\alpha + \gamma - \beta)(u-v)^+,$$

the vector sublattice $R(u,v)$ generated by u and v satisfies $R(u,v) = \mathbb{R}^3$, whereas Span $\{u,v\}$ is given by the plane with equation $x - y + z = 0$. ■

In general, there is a substantial difference between the vector sublattice generated by finitely many positive elements and finitely many arbitrary elements, even in the case of a single element. Indeed, if $0 < u \in E$, then the vector sublattice $R(u)$ generated by u satisfies

$$R(u) = \text{Span}\,\{u\} = \{\alpha u : \ \alpha \in \mathbb{R}\},$$

so $\dim R(u) = 1$ (the same being true, of course, for $-u$). This situation, however, changes drastically for an arbitrary element $f \in E$.

2.2. Proposition. *Let E be an Archimedean vector lattice. Then*

(i) *If $f \in E$, then*

$$R(f) = \operatorname{Span}\{f^+, f^-\} = \{\alpha f^+ + \beta f^- : \alpha, \beta \in \mathbb{R}\}.$$

(ii) *If $f \notin E^+$ and $-f \notin E^+$, then $\dim R(f) = 2$ (whereas $\dim \operatorname{Span}\{f\} = 1$).*

Proof. (i) Obviously, $\operatorname{Span}\{f^+, f^-\} \subset R(f)$. In addition, $\operatorname{Span}\{f^+, f^-\}$ is a linear subspace of E containing f (take $\alpha = 1$, $\beta = -1$). Therefore, once we can show that $\operatorname{Span}\{f^+, f^-\}$ is a vector sublattice of E, the smallest vector sublattice of E containing f, viz. $R(f)$, is then included in $\operatorname{Span}\{f^+, f^-\}$ and we have equality. To this end, take $g \in \operatorname{Span}\{f^+, f^-\}$, say $g = \alpha f^+ + \beta f^-$. Since $\alpha f^+ \perp \beta f^-$, we get $|g| = |\alpha| f^+ + |\beta| f^-$ (see e.g. [13, Theorem 14.4]), whence $|g| \in \operatorname{Span}\{f^+, f^-\}$. This shows that $\operatorname{Span}\{f^+, f^-\}$ is a vector sublattice of E. We will consider an extension of this situation in Lemma 3.3.

(ii) Assume now that both $f \notin E^+$ and $-f \notin E^+$. If $\gamma f^+ + \delta f^- = 0$ and $\gamma \neq 0$, then $f^+ = -\frac{\delta}{\gamma} f^-$, so

$$f = f^+ - f^- = (-\frac{\delta}{\gamma} - 1) f^-.$$

This yields that $f \in E^+$ or $-f \in E^+$, contradicting the hypothesis. It follows that $\gamma = 0$. Similarly, it is proven that $\delta = 0$, so $\{f^+, f^-\}$ is linearly independent. Consequently, $\dim R(f) = 2$ and so $\operatorname{Span}\{f\}$ is a proper subset of $R(f)$. ∎

Based on the considerations concerning the difference between $R(u)$ with $u \in E^+$ and $R(f)$ with $f \in E$ it is to be expected that the equality

$$R(f, g) = \operatorname{Span}\{(\alpha f + \beta g)^+ : \alpha, \beta \in \mathbb{R}\}$$

will *not* hold for arbitrary $f, g \in E$. Below we will present a counterexample showing that in the equality

$$R(u, v) = \operatorname{Span}\{(\alpha u + \beta v)^+ : \alpha, \beta \in \mathbb{R}\}$$

positivity of u and v is essential.

2.3. Example. Consider the function $f \in C(\mathbb{R}^2)$ defined by

$$f(x, y) = x \vee |y| \qquad (x, y \in \mathbb{R}).$$

It is an easy exercise to verify that f is not differentiable on the negative x–axis

$$\{(x,y) \in \mathbb{R}^2 : \ x \le 0, \ y = 0\}$$

and the half lines

$$\{(x,y) \in \mathbb{R}^2 : \ y = x, \ x \ge 0\}, \qquad \{(x,y) \in \mathbb{R}^2 : \ y = -x, \ x \ge 0\}.$$

However, any function of type $(\alpha x + \beta y)^+$ is not differentiable on the whole line

$$\{(x,y) \in \mathbb{R}^2 : \ \alpha x + \beta y = 0\}.$$

Thus the function f cannot be a finite linear combination of functions of type $(\alpha x + \beta y)^+$. Therefore,

$$x \vee |y| \notin \operatorname{Span}\{(\alpha x + \beta y)^+ : \ \alpha, \beta \in \mathbb{R}\},$$

although $x \vee |y| \in R(x,y)$, the vector sublattice generated by the coordinate functions x and y. Hence, $\operatorname{Span}\{(\alpha x + \beta y)^+ : \ \alpha, \beta \in \mathbb{R}\}$ is a proper subset of $R(x,y)$. ■

We will show next that, rather surprisingly, the characterization in Theorem 1.1 of the vector sublattice generated by two positive elements ceases to hold for three (and more) positive elements, i.e., if u, v, w are positive elements of an Archimedean vector lattice E and $R(u,v,w)$ is the vector sublattice generated by u, v, and w, then

$$\operatorname{Span}\{(\alpha u + \beta v + \gamma w)^+ : \ \alpha, \beta, \gamma \in \mathbb{R}\}$$

is, in general, a proper subset of $R(u,v,w)$. An analogous observation holds then, of course, for four and more positive elements.

2.4. Example. Choose $E = C(\mathbb{R}^3_+)$, the Archimedean vector lattice of all real continuous functions on the first octant of $\mathbb{R}^3$ and consider the function f defined by

$$f(x,y,z) = (x \wedge y) - z \qquad (x,y,z \ge 0),$$

so

$$f(x,y,z) = \begin{cases} x - z & (y \ge x \ge 0, \ z \ge 0) \\ y - z & (x \ge y \ge 0, \ z \ge 0). \end{cases}$$

The part of the intersection of the planes with equations $y = z$ and $x+y+z = 1$ in the first octant has parametric form

$$\begin{pmatrix} 1 \\ 0 \\ 0 \end{pmatrix} + \lambda \begin{pmatrix} -2 \\ 1 \\ 1 \end{pmatrix} \qquad (0 \le \lambda \le \tfrac{1}{2}).$$

Hence, if $0 \leq y \leq x$, then $0 \leq \lambda \leq 1 - 2\lambda$ shows that $0 \leq \lambda \leq 1/3$. An analogous observation holds in the case $x = z$ and $0 \leq x \leq y$. Consequently, the projections of the null set of $f(x, y, z)$ onto the part of the plane with equation $x + y + z = 1$ $(0 \leq x, y, z \leq 1)$ within the first octant consist of the two line segments

$$\begin{cases} y = z \\ x + y + z = 1 \end{cases} \qquad (0 \leq y,\ z \leq 1/3)$$

and

$$\begin{cases} x = z \\ x + y + z = 1 \end{cases} \qquad (0 \leq x,\ z \leq 1/3).$$

The parametric forms of these two line segments are

$$\begin{pmatrix} 1 \\ 0 \\ 0 \end{pmatrix} + \lambda \begin{pmatrix} -2 \\ 1 \\ 1 \end{pmatrix} \qquad (0 \leq \lambda \leq 1/3)$$

and

$$\begin{pmatrix} 0 \\ 1 \\ 0 \end{pmatrix} + \lambda \begin{pmatrix} 1 \\ -2 \\ 1 \end{pmatrix} \qquad (0 \leq \lambda \leq 1/3).$$

Therefore, the function $f^+(x, y, z) = (x \wedge y - z)^+$ is not differentiable on these two line segments. It follows immediately that f^+ cannot be a finite linear combination of functions of type $(\alpha x + \beta y + \gamma z)^+$, as the intersection in the first octant of the zero set of a function of the form $\alpha x + \beta y + \gamma z$ with the plane with equation $x + y + z = 1$ is either empty, a point, or a full line segment (i.e., from boundary to boundary in the first octant) inside the planar part $x + y + z = 1$; $0 \leq x, y, z \leq 1$. Hence, although the function $(x \wedge y - z)^+$ is a member of the vector sublattice $R(x, y, z)$ of E generated by the coordinate functions x, y, and z, this function does not belong to

$$\operatorname{Span}\left\{(\alpha x + \beta y + \gamma z)^+ : \ \alpha, \beta, \gamma \in \mathbb{R}\right\}.$$

■

3. Hyper-Archimedean vector lattices

In this section we will tackle the problem of determining the class of vector lattices in which all finitely generated vector sublattices are finite-dimensional. We begin with the following simple characterization of such vector lattices.

3.1. Lemma. *Let E be an Archimedean vector lattice. Then the following are equivalent:*

(i) *Every finitely generated vector sublattice of E is finite dimensional.*

(ii) *Every finite-dimensional linear subspace of E is contained in a finite-dimensional vector sublattice of E.*

Proof. (i)⇒(ii): Let L be a finite–dimensional linear subspace of E with the basis $f_1, \ldots, f_n$. Hence, $L = \operatorname{Span}\{f_1, \ldots, f_n\}$. Since $L \subset R(f_1, \ldots, f_n)$ and, by hypothesis, $\dim R(f_1, \ldots, f_n) < \infty$, statement (ii) is immediate.

(ii)⇒(i): Consider a finitely generated vector sublattice $R(f_1, \ldots, f_n)$ of E. By (ii), the finite–dimensional linear subspace $M = \operatorname{Span}\{f_1, \ldots, f_n\}$ of E is contained in a finite dimensional vector sublattice S of E. Now, $R(f_1, \ldots, f_n) \subset S$, since $R(f_1, \ldots, f_n)$ is the smallest vector sublattice of E containing $f_1, \ldots, f_n$ and S is some vector sublattice of E containing $f_1, \ldots, f_n$. It follows that

$$\dim R(f_1, \ldots, f_n) \le \dim S < \infty,$$

which is the desired result. ■

Condition (ii) of Lemma 3.1 was introduced by L. C. Moore Jr. in an unpublished note [14].

Recall that a lattice (or a Riesz) norm on a vector lattice E is any norm that is absolute $\left(\|f\| = \big\||f|\big\| \text{ for all } f \in E\right)$ and monotone ($0 \le u \le v$ in E^+ implies $\|u\| \le \|v\|$). Equivalently, $|f| \le |g|$ implies $\|f\| \le \|g\|$. A vector lattice E equipped with a lattice norm is simply called a normed vector lattice (or a normed Riesz space). If E is, in addition, complete with respect to this norm, then E is termed a Banach lattice. For more on these concepts, see e.g. [2, Section 12] and [13, Section 62].

In [6], D. Cozart and L. C. Moore Jr. consider normed vector lattices E in which every finite–dimensional linear subspace of E is so to say "*close to a finite dimensional vector sublattice of E*." More precisely, for every finite subset $\{f_1, \ldots, f_n\}$ of E and every $\epsilon > 0$ there exists a finite dimensional vector sublattice S of E and $g_1, \ldots, g_n \in S$ such that $\|f_i - g_i\| < \epsilon$ ($i = 1, \ldots, n$). Such normed vector lattices are said to have the *finite–dimensional vector sublattice property.* A sufficient condition for this property is that E have the principal projection property (for the definition of this notion, see [1, Definition 2.8] or [13, Definition 24.8]). Normed vector lattices with the finite–dimensional vector sublattice property are important in the study of the nonstandard hulls of normed vector lattices (for details we refer to [6]).

In this connection it was therefore a natural question to ask which Archimedean vector lattices have property (ii) (or, equivalently, (i)) of Lemma 3.1.

If such a vector lattice E carries a lattice norm, then E obviously has the finite–dimensional vector sublattice property.

In [14], L. C. Moore Jr. shows, by means of a very difficult proof, that (ii) (and hence (i)) of Lemma 3.1 is equivalent to the property that E is hyper-Archimedean. The main reason that his proof is so involved is that he tries to avoid representations (and thus the axiom of choice). Apart from the length of his proof, it is, in my opinion, also very untransparent. I believe that with the representation theory the proof becomes much simpler and more clear. I wish to express my gratitude to S. J. Bernau (private communication) for pointing out this idea to me.

Let us recall some of the relevant notions. A vector lattice E is termed *hyper-Archimedean* [17, p. 140] if the quotient vector lattice E/A is Archimedean for each order ideal A in E. By [13, Theorem 60.2] this is equivalent to saying that all ideals in E are uniformly closed. We list some other characterizations (see [13, Theorem 37.6, 61.1, and 61.2]). In these references it is shown that the following statements are equivalent:

(iii) *E is hyper-Archimedean.*

(iv) *For each pair $u, v \in E^+$ there exists $m \in \mathbf{N}$ (depending on u and v) such that*

$$v \wedge nu = v \wedge mu$$

for all $n \geq m$ (*or* $v \wedge (m+1)u = v \wedge mu$, *or* $v \wedge (m+1)u \leq mu$).

(v) *Every principal ideal of E is a projection band.*

By (v), every weak order unit in a hyper-Archimedean vector lattice is a strong order unit. In [9] many more equivalences are presented. We mention one striking characterization. Let $\mathcal{A}_p(E)$ be the distributive lattice of all principal ideals in E. Then *E is hyper-Archimedean if and only if $\mathcal{A}_p(E)$ is a Boolean ring (and hence a Boolean algebra as soon as E has a strong order unit)*. The equivalence of (iii) and (iv) above easily shows that statement (i) of Lemma 3.1 implies the hyper-Archimedean property. We formulate this in the next lemma.

3.2. Lemma. *Let E be an Archimedean vector lattice in which every finitely generated vector sublattice is finite–dimensional. Then E is hyper-Archimedean.*

Proof. It suffices to show (iv) above. To this end, choose $u, v \in E^+$. The vector sublattice $R(u, v)$ generated by u and v is Archimedean and finite–dimensional. Let $\dim R(u, v) = k < \infty$. By a well known result of A. I. Yudin [13, Theorem 26.11], $R(u, v)$ is linearly and lattice isomorphic to $\mathbb{R}^k$ with the usual coordinatewise ordering. Since $\mathbb{R}^k$ is hyper-Archimedean there exists $m \in \mathbf{N}$ such that

$$v \wedge nu = v \wedge mu \qquad (n = m+1,\ m+2, \dots).$$

By the equivalence of (iii) and (iv) above, E is hyper-Archimedean. ■

Next, we aim at the converse of Lemma 3.2. In the case that E is, in addition, uniformly complete, this is very easy. Indeed, by a result due to W. A. J. Luxemburg and L. C. Moore Jr. [12, Theorem 7.5] or [13, Theorem 61.4], the following are equivalent:

(a) *E is hyper-Archimedean and uniformly complete.*

(b) *Every principal ideal in E is finite–dimensional.*

Furthermore, and this yields much more insight in this class of vector lattices, each of the conditions (a) and (b) is equivalent to the following nice representation result:

(c) *E is linearly and lattice isomorphic to the vector lattice of all finitely supported functions on some non-empty set X (that is, all functions on X that vanish outside a finite subset of X, the subset depending on the function).*

Although, strictly speaking, the next three equivalences are beyond the scope of this paper, they are interesting in their own right and yield some extra information on the class of vector lattices under discussion.
Each of the conditions (a), (b), and (c) is equivalent to

(d) *Every ideal of E is a (projection) band.*

(e) *The distributive lattice $\mathcal{A}(E)$ of all ideals in E is a Boolean algebra.*

(f) *Every linear functional on E is order bounded.*

The equivalences (a)⇔(d)⇔(e) are due to the author [9]. The proof of (a)⇔(f) is presented in [17, Theorem 85.7] and is also due to the author. It can also be found as Exercise IV.1 in [15] (with a different proof). For many more equivalences of (a) through (f) consult [9] again.

Now, suppose that E is hyper-Archimedean and uniformly complete. If $f_1, \dots, f_n \in E$, then

$$R(f_1, \dots, f_n) \subset A(f_1, \dots, f_n) = A_{|f_1| \vee \dots \vee |f_n|} (= A_{|f_1| + \dots + |f_n|})$$

shows that the finitely generated vector sublattice $R(f_1, \dots, f_n)$ is contained in the finite–dimensional principal ideal $A_{|f_1| \vee \dots \vee |f_n|}$ (use (b) above), whence $\dim R(f_1, \dots, f_n) < \infty$.

As we will demonstrate next, the latter is true even without the extra assumption that E is uniformly complete, albeit at the cost of a more difficult proof. Before stating and proving the main result of this section, we will make some preparations. The first step is formulated in a lemma and generalizes Proposition 2.2 (i) of the present paper.

3.3. Lemma. *Let E be an Archimedean vector lattice and $\{u_1, \ldots, u_n\}$ a disjoint subset of E^+ (that is, $u_i \wedge u_j = 0$; $i, j = 1, \ldots, n$, $i \neq j$). Then*

$$\operatorname{Span}\{u_1, \ldots, u_n\} = R(u_1, \ldots, u_n).$$

Proof. It is enough to verify that the linear span of $u_1, \ldots, u_n$ is a vector sublattice of E. To this end, take any

$$f = \alpha_1 u_1 + \cdots + \alpha_n u_n \in \operatorname{Span}\{u_1, \ldots, u_n\}.$$

By [13, Theorem 14.4], $|f| = |\alpha_1| u_1 + \cdots + |\alpha_n| u_n$. That is, we see that $|f| \in \operatorname{Span}\{u_1, \ldots, u_n\}$ as well and the proof is complete. ∎

Secondly, it follows from the combination of Theorems 37.1 (i), 37.6, and 37.7 of [13] that an Archimedean vector lattice E is hyper-Archimedean and has a strong order unit (= weak order unit in this case) if and only if E is linearly and lattice isomorphic to a vector lattice of step functions with respect to an algebra (= field) Λ of subsets of some non-empty point set X. Compare this with the situation when E is hyper-Archimedean and uniformly complete. In that case (see (c) above), Λ is the *ring* of all finite subsets of X.

Remark. As a sidestep we notice that in the aforementioned representation theorems of hyper-Archimedean vector lattices which are uniformly complete or have a strong order unit, respectively, the following results play a key role. If E is an Archimedean vector lattice and $\mathcal{P}$ denotes the set of all proper prime (order) ideals of E equipped with the so-called hull–kernel (Zariski) topology ($\mathcal{P}$ is sometimes called the *structure space* of E; see [13, Section 36]), then

(α) *$\mathcal{P}$ is compact if and only if E has a strong order unit* [13, Theorem 37.1 (i)],

(β) *$\mathcal{P}$ is Hausdorff if and only if E is hyper-Archimedean* [13, Theorem 37.6],

(γ) *$\mathcal{P}$ is discrete if and only if E is hyper-Archimedean and uniformly complete.*

Equivalence (γ) is due to the author [9]. It also occurs in [17, the end of Section 85].

We are now in a position to state and prove the main result of this section.

3.4. Theorem. *Let E be an Archimedean vector lattice. Then the following are equivalent:*

(i) *Every finitely generated vector sublattice of E is finite dimensional.*

(iii) *E is hyper-Archimedean.*

Proof. (i)$\Rightarrow$(iii) by Lemma 3.2.

(iii)$\Rightarrow$(i): Suppose that E is hyper-Archimedean and take $f_1, \ldots, f_n \in E$. Let

$$F = A(f_1, \ldots, f_n) = A_{|f_1| \vee \cdots \vee |f_n|}$$

and note that $R(f_1, \ldots, f_n) \subset F$. The Archimedean vector lattice F is hyper-Archimedean in its own right and has, in addition, a strong order unit, for example, $|f_1| \vee \cdots \vee |f_n|$ or $|f_1| + \cdots + |f_n|$. Consider the aforementioned representation of F as a vector lattice of step functions with respect to an algebra Λ of subsets of a non-empty point set X. Every element $f \in F$ can therefore be written in the form

$$f = \sum_{k=1}^{n} a_k \chi_{A_k} \qquad (a_k \in \mathbb{R};\ A_k \in \Lambda,\ k = 1, \ldots, n),$$

where $n = n(f)$ depends on f and χ denotes the characteristic (indicator) function. Without loss of generality we can assume that the real numbers a_k are mutually distinct and the sets A_k are non-empty and pairwise disjoint. Each of the elements $f_1, \ldots, f_n \in F$ takes therefore only finitely many values on sets of the algebra Λ (let us call these the "*level sets*" of $f_1, \ldots, f_n$). Consider all possible non-empty intersections of the level sets of $f_1, \ldots, f_n$ and (to simplify notations) denote them by $C_1, \ldots, C_m$. Since the sets $C_k \in \Lambda$ ($k = 1, \ldots, m$) are pairwise disjoint, the characteristic functions $\chi_{C_1}, \ldots, \chi_{C_m}$ are pairwise disjoint in $F \subset E$, so by Lemma 3.3,

$$\operatorname{Span}\left\{\chi_{C_1}, \ldots, \chi_{C_m}\right\} = R(\chi_{C_1}, \ldots, \chi_{C_m}).$$

This shows that $\operatorname{Span}\left\{\chi_{C_1}, \ldots, \chi_{C_m}\right\}$ is a vector sublattice. Moreover, the above construction shows that

$$f_k \in \operatorname{Span}\left\{\chi_{C_1}, \ldots, \chi_{C_m}\right\} \qquad (k = 1, \ldots, n)$$

and hence

$$R(f_1, \ldots, f_n) \subset \operatorname{Span}\left\{\chi_{C_1}, \ldots, \chi_{C_m}\right\}.$$

Since $\dim \operatorname{Span}\left\{\chi_{C_1}, \ldots, \chi_{C_m}\right\} < \infty$, the vector sublattice $R(f_1, \ldots, f_n)$ is finite–dimensional as well, and we are done.

The essential ingredient in the proof of (iii)⇒(i) above is the fact that we may assume without loss of generality that E is hyper-Archimedean and possesses a strong order unit (otherwise, pass to F). ∎

By using the above representation theorem of a hyper-Archimedean vector lattice with a strong order unit, it is not difficult to show either that a vector lattice E is hyper-Archimedean and uniformly complete if and only if E is hyper-Archimedean and disjoint–σ–complete. Recall that a subset G of E^+ is said to be disjoint whenever the elements of G are pairwise disjoint; the vector lattice E is termed *disjoint–σ–complete* (or, in another terminology, conditionally laterally σ–complete) whenever every countable disjoint set in E^+ that is bounded from above has a supremum.

3.5. Theorem. *Let E be an Archimedean vector lattice. Then the following are equivalent:*

(a) *E is hyper-Archimedean and uniformly complete.*

(g) *E is hyper-Archimedean and disjoint–σ–complete.*

Proof. (a)⇒(g): By (a)⇔(b), every principal ideal in E is finite–dimensional. Take a countable disjoint set $\{u_n\}_{n=1}^{\infty}$ in E^+ (so $u_n \wedge u_m = 0,\ n \neq m$), bounded from above by some element w. Since A_w is finite–dimensional, A_w is Dedekind σ–complete, so $0 \leq u_n \leq w\ (n = 1, 2, \ldots)$ implies that $v = \sup u_n$ exists in A_w. It is not difficult to show that $v = \sup u_n$ in E as well. Hence, E is disjoint–σ–complete.

(g)⇒(a): In order to show that every principal ideal in E is of finite–dimension, we argue by means of contradiction. Suppose on the contrary that A_u is *not* finite–dimensional for some $u \in E^+$. The principal ideal A_u is disjoint–σ–complete and hyper-Archimedean in its own right and has, in addition, a strong order unit u. Hence, A_u is linearly and lattice isomorphic to a vector lattice H of all step functions with respect to an algebra of subsets of a non-empty set X, the element u corresponding to the step function identically 1 on X. Clearly, H is also disjoint–σ–complete. Since A_u is Archimedean and of infinite–dimension, there exists, by Theorem 26.10 of [13], a countable disjoint set $\{w_n\}_{n=1}^{\infty}$ in A_u^+, corresponding to a disjoint sequence $\{w_n(x)\}_{n=1}^{\infty}$ of step functions in H^+. Let $\{r_n\}_{n=1}^{\infty}$ be the sequence of rational numbers in $(0, 1)$. Multiplying by a suitable factor, we may assume that

$\max_{x \in X} w_n(x) = r_n$ $(n = 1, 2, \ldots)$. Write every $w_n(x)$ in its standard representation

$$w_n(x) = \sum_{i=1}^{k_n} b_i^{(n)} \chi_{B_i^{(n)}}(x) \qquad (n = 1, 2, \ldots)$$

(for n fixed, $b_i^{(n)} > 0$ mutually distinct; all $B_i^{(n)} \in \Lambda$ non-empty, mutually disjoint; one of the $b_i^{(n)}$, say $b_1^{(n)}$, is equal to r_n). Since H is disjoint–σ–complete and $0 \leq w_n(x) \leq \chi_X(x) = 1$ for all $x \in X$, the supremum $w(x) = \sup w_n(x)$ exists in H and $w(x) \leq 1$ for all $x \in X$. Let us represent w in its standard form:

$$w(x) = \sum_{i=1}^{k} a_i \chi_{A_i}(x).$$

For every n there exists $x_n \in B_1^{(n)}$ such that $w_n(x_n) = b_1^{(n)} = r_n$ and it follows from $w_{n_1} \wedge w_{n_2} = 0$ for $n_1 \neq n_2$ that $B_1^{(n_1)} \cap B_1^{(n_2)} = \phi$, so for $n_1 \neq n_2$ we have $x_{n_1} \neq x_{n_2}$. Moreover, $r_n = w_n(x_n) \leq w(x_n)$ implies $w(x_n) > 0$ $(n = 1, 2, \ldots)$. Hence, at least one of the sets A_i from the standard representation of $w(x)$, say A_1, must contain an infinite number of the points x_n. For each x_n we have $w(x_n) = a_1 \geq w_n(x_n) = r_n$, so $a_1 \geq r_n$ for infinitely many n. There exists therefore an index N such that

$$x_N \in A_1 \cap B_1^{(N)}, \qquad w(x_N) = a_1 > r_N = b_1^{(N)} = w_N(x_N).$$

Note that for all $x \in A_1 \cap B_1^{(N)}$

$$w(x) = a_1, \qquad w_N(x) = r_N.$$

Consider now the element $z \in H$ defined by

$$z(x) = w(x) - (a_1 - r_N)\chi_{A_1 \cap B_1^{(N)}}(x)$$

for all $x \in X$. It is now an easy matter to check that $z \geq w_N$. If $n \neq N$, then $w_n \wedge w_N = 0$, so for $x \in A_1 \cap B_1^{(N)}$ we have

$$w_n(x) = 0 < r_N = z(x),$$

and for $x \notin A_1 \cap B_1^{(N)}$ we get

$$w_n(x) \leq w(x) = z(x).$$

Consequently, $z \geq w_n$ for all $n \neq N$ holds as well. Therefore we have on the one hand that $z(x) \geq w(x)$ for all $x \in X$. On the other hand, however, for $x \in A_1 \cap B_1^{(N)}$ we have

$$z(x) = r_N < a_1 = w(x),$$

a contradiction. Hence, every principal ideal in E is finite–dimensional and we are done. ■

As a corollary of Theorem 3.5 we obtain that a disjoint–σ–complete hyper-Archimedean vector lattice is Dedekind σ–complete (and even super Dedekind complete; cf. [12, Theorem 7.5] or [13, Theorem 61.4]). Statement (g) has an analogue in the theory of Boolean rings: every disjoint–σ–complete Boolean ring is Dedekind σ–complete (cf. [16, Theorem 3.5]). In connection with this, we mention one final equivalence to (a)–(g) which follows easily from a combination of (b) and Theorem 26.10 of [13]:

(h) *Every disjoint set in E^+ that is bounded from above is finite.*

Incidentally, it is well known [13, Theorem 29.3] that for a vector lattice E the following two properties are equivalent: (i) each set having a supremum contains an at most countable subset having the same supremum, and (ii) every disjoint set in E^+ that is bounded from above is (at most) countable.

The equivalence of (a) and (g) above shows that a hyper-Archimedean vector lattice E is uniformly complete if and only if E is disjoint–σ–complete. It is worthwhile to observe that although in hyper-Archimedean vector lattices uniform completeness and disjoint–σ–completeness are equivalent, these two properties are independent in arbitrary vector lattices. Indeed, the uniformly complete vector lattice $C([0,1])$ is not disjoint–σ–complete. Conversely, for an example of a disjoint–σ–complete vector lattice that is not uniformly complete, consider the vector lattice E of all countable step functions on $[0,1]$ (with Lebesgue measure); a member of E is therefore of the form

$$\sum_{n=1}^{\infty} a_n \chi_{A_n}(x),$$

where $a_n \in \mathbb{R}$ and A_n are non-empty mutually disjoint Lebesgue measurable subsets of $[0,1]$.

Notice in conclusion that the paper shows that finitely generated vector sublattices play a certain role in several branches of mathematics, such as mathematical economics [4, 5], operator theory on vector lattices [8] and to some extent even in nonstandard analysis [6, 14].

References

1. C. D. Aliprantis and O. Burkinshaw, *Locally Solid Riesz Spaces*, Pure and Applied Mathematics Series, Vol. 76, Academic Press, New York London, 1978.

2. C. D. Aliprantis and O. Burkinshaw, *Positive Operators*, Pure and Applied Mathematics Series, Vol. 119, Academic Press, New York London, 1985.

3. G. Birkhoff, *Lattice Theory*, 3rd edition, Amer. Math. Soc. Coll. Publ. Providence, 1967.

4. D. J. Brown, C. B. Huijsmans, and B. de Pagter, Approximating Derivative Securities in f–Algebras, in: *Positive Operators, Riesz Spaces and Economics* (eds. C. D. Aliprantis, K. C. Border, and W. A. J. Luxemburg), Studies in Economic Theory, Vol. 2, Springer, Berlin Heidelberg New York, 1991, 171–177.

5. D. J. Brown and S. A. Ross, Spanning, Valuation and Options, *Economic Theory* 1 (1991), 3–12.

6. D. Cozart and L. C. Moore Jr., The nonstandard hull of a normed Riesz space, *Duke Math. J.* 41 (1974), 263–275.

7. P. G. Dodds, B. de Pagter, and W. Ricker, Reflexivity and order properties of scalar–type spectral operators in locally convex spaces, *Trans. Amer. Math. Soc.* 293 (1986), 355–380.

8. D. R. Hart, Some properties of disjointness preserving operators, *Indag. Math.* 47 (1985), 183–197.

9. C. B. Huijsmans, Riesz spaces for which every ideal is a projection band, *Indag. Math.* 38 (1976), 30–34.

10. C. B. Huijsmans and B. de Pagter, Subalgebras and Riesz subspaces of an f–algebra, *Proc. London Math. Soc.* 48 (1984), 161–174.

11. G. Jameson, *Ordered Linear Spaces*, Lecture Notes in Mathematics, Vol. 141, Springer, Berlin Heidelberg New York, 1970.

12. W. A. J. Luxemburg and L. C. Moore Jr., Archimedean quotient Riesz spaces, *Duke Math. J.* 34 (1967), 725–739.

13. W. A. J. Luxemburg and A. C. Zaanen, *Riesz Spaces* I, North–Holland, Amsterdam, 1971.

14. L. C. Moore Jr., A note on hyper-Archimedean Riesz spaces, 1973 (unpublished).

15. H. H. Schaefer, *Banach Lattices and Positive Operators*, Die Grundlehren der Mathematischen Wissenschaften in Einzeldarstellungen, Vol. 215, Springer, Berlin Heidelberg New York, 1974.

16. E. C. Smith Jr. and A. Tarski, Higher degrees of distributivity and completeness in Boolean algebras, *Trans. Amer. Math. Soc.* 84 (1957), 230–257.

17. A. C. Zaanen, *Riesz Spaces* II, North–Holland, Amsterdam, 1983.

Charles B. Huijsmans
Department of Mathematics
Leiden University
The Netherlands

DUALITY IN OPERATOR SPACES*

VICTOR LOMONOSOV

Abstract. It is well known that for any Hilbert space H the second dual of the space of all compact operators $K(H)$ on H coincides with the space of all bounded operators $L(H)$. In this note we generalize this statement to Banach spaces.

We begin by fixing some notation and terminology regarding the Banach spaces and operators on them. If B is a Banach space, then B^* denotes its dual space, and $L(B)$ the space of all bounded operators on B. The subspace of $L(B)$ consisting of all finite rank operators will be denoted by $F(B)$.

Let $T(B) = \{t = x \otimes y : \ x \in B, \ y \in B^*\}$ be the collection of all elementary tensors in $F(B)$, where as usual a tensor $t = x \otimes y \in T(B)$ acts on $z \in B$ as $t(z) = (z, y)x$, and (z, y) denotes the standard duality between B and its dual B^*.

We denote by $\operatorname{tr}(A)$ the trace functional on $F(B)$; in particular, for each $t = x \otimes y \in T(B)$ we have $\operatorname{tr}(t) = (x, y)$. Below, we consider dualities between the spaces of operators with respect to the duality $(R, S) = \operatorname{tr}(R, S)$, where either the operator R or the operator S is finite–dimensional. On the space $L(B)$ we consider two norms; namely, the usual operator norm

$$\|A\|_O = \sup \{\|Ax\| : \ x \in B, \ \|x\| \leq 1\}$$

and the integral norm

$$\|A\|_I = \sup \left\{ \left|\operatorname{tr}(AS)\right| : \quad S \in F(B), \ \|S\|_O \leq 1 \right\}, \tag{1}$$

which can be infinite. On the space $F(B)$ we consider as well the nuclear norm

$$\|S\|_N = \inf \left\{ \sum_i \|x_i\| \|y_i\| : \quad S = \sum_i x_i \otimes y_i \right\}. \tag{2}$$

Following A. Pietsch [4] we say that an operator $A \in L(B)$ is integral if $\|A\|_I < \infty$. The collection of all integral operators on B is denoted by $I(B)$. From (1) and (2) it follows that for each $A \in L(B)$ and $S \in F(B)$ we have

$$\begin{aligned} \left|\operatorname{tr}(AS)\right| &\leq \|A\|_I \|S\|_O \\ \left|\operatorname{tr}(AS)\right| &\leq \|A\|_O \|S\|_N. \end{aligned} \tag{3}$$

*Partially supported by a grant from the National Science Foundation.

We denote by F_1 the space $F(B)$ taken with the norm $\|\cdot\|_O$. Similarly, we let $F_2 = (F(B), \|\cdot\|_N)$ and $F_3 = (F(B), \|\cdot\|_I)$. The norm completions of these spaces are denoted by $\bar{F}_1$, $\bar{F}_2$ and $\bar{F}_3$, respectively.

Finally, for a given vector $x \in B$ we let $L_x = \{x \otimes y : \ y \in B^*\}$ and for a vector $y \in B^*$ we let $L_y = \{x \otimes y : \ x \in B\}$. We omit the proofs of the next two trivial lemmas.

1. Lemma. *For each $t = x \otimes y \in T(B)$ one has*

$$\|t\|_O = \|t\|_N = \|t\|_I = \|x\| \cdot \|y\|.$$

2. Lemma. *Let $\varphi(x, y)$ be a continuous bilinear form on $B \times B^*$. Then there exists a continuous linear operator $A : B \to B^{**}$ such that for $(x, y) \in B \times B^*$*

$$\varphi(x, y) = (Ax, y).$$

3. Theorem. *For a Banach space B the following properties are equivalent:*

(a) *B is reflexive.*

(b) $(\bar{F}_1)^* = I(B)$.

(c) $(\bar{F}_2)^* = L(B)$.

Proof. Suppose that (a) is true. Let $\varphi_i \in (\bar{F}_i)^*$, $i = 1, 2$. The restriction of φ_i to the set $T(B) \subset F_i$ defines, by Lemma 1, a continuous bilinear form on $B \times B^*$ and so, by Lemma 2, there exists a continuous linear operator K_i from B to B^{**} such that $(K_i x, y) = \varphi_i(x \otimes y)$. Since B is reflexive, each of these operators belongs to $L(B)$. For each $S \in F(B)$ we have

$$\varphi_i(S) = \mathrm{tr}(K_i S). \tag{4}$$

We claim that

$$\|\varphi_1\| = \|K_1\|_I \qquad \text{and} \qquad \|\varphi_2\| = \|K_1\|_O.$$

The first equality follows easily from the definition of the integral norm (1):

$$\begin{aligned} \|K_1\|_I &= \sup\left\{ \left|\mathrm{tr}(K_1 S)\right| : \quad S \in F(B), \ \|S\|_O \le 1 \right\} \\ &= \sup\left\{ \left|\varphi_1(S)\right| : \quad S \in F(B), \ \|S\|_O \le 1 \right\} \\ &= \|\varphi_1\|. \end{aligned}$$

Using the definition of the norm of the functional φ_2 and the second inequality in (3), we obtain

$$\begin{aligned}\|\varphi_2\| &= \sup\{\varphi_2(S): \quad S \in F(B),\ \|S\|_N \le 1\}\\ &= \sup\left\{|\mathrm{tr}(K_2 S)|: \quad S \in F(B),\ \|S\|_N \le 1\right\}\\ &\le \|K_2\|_O.\end{aligned}$$

To get the converse of the last inequality fix $\varepsilon > 0$ and pick two normed vectors $x_0 \in B$ and $y_0 \in B^*$ such that $|(K_2x_0, y_0)| \ge \|K_2\|_O - \varepsilon$. Notice that $(K_2x_0, y_0) = tr(K_2S_0)$, where $S_0 = x_0 \otimes y_0$, and so we obtain

$$\begin{aligned}\|K_2\|_O &\le |tr(K_2S_0)| + \varepsilon\\ &\le \sup\left\{|\mathrm{tr}(K_2 S)|: \quad S \in F(B),\ \|S\|_N \le 1\right\} + \varepsilon\\ &= \|\varphi_2\| + \varepsilon.\end{aligned}$$

Thus, we have proved that property (a) implies the isometric embeddings $(\bar{F}_1)^* \subset I(B)$ and $(\bar{F}_2)^* \subset L(B)$.

To show that these embeddings are onto let us take $K_1 \in I(B)$ and $K_2 \in L(B)$. In view of inequalities (3) each operator K_i defines, by formula (4), a continuous linear functional on the space F_i which can be continuously extended to the space $\bar{F}_i$. It is clear that $\|\varphi_1\| = \|K_1\|_I$ and $\|\varphi_2\| = \|K_2\|_O$. So we have a one-to-one isometric correspondence between the spaces $(\bar{F}_1)^*$ and $I(B)$ and between $(\bar{F}_2)^*$ and $L(B)$.

Suppose now that either (b) or (c) is true and let us prove that B is reflexive. For any two fixed vectors $x \in B$ and $x_1 \in B^{**}$ the formula

$$\varphi(x \otimes y) = (x_1, y)$$

defines a continuous linear functional φ on the subspace $L_x = \{x \otimes y: \ y \in B^*\}$ of the space $\bar{F}_i$. Let φ_i be a continuous extension of φ to the whole space $\bar{F}_i$, $i = 1, 2$. That is, $\varphi_i \in (\bar{F}_i)^*$, and so by (b) (respectively, by (c)) there exists an operator $K_1 \in I(B)$ (respectively, $K_2 \in L(B)$) such that for each $y \in B^*$ we have $(K_ix, y) = \varphi_i(x \otimes y) = (x_1, y)$. This implies that $K_ix = x_1$ and so $x_1 \in B$, that is, B is reflexive. ∎

4. Theorem. *Given a Banach space B, suppose that the nuclear and integral norms are equivalent on the space $F(B)$. Then the space B is reflexive if and only if the space $\bar{F}_1^{**}$ is isomorphic to the space $L(B)$. Moreover, if the two norms coincide, then the space B is reflexive if and only if*

$$\bar{F}_1^{**} = L(B).$$

Proof. If B is reflexive then, by [1], $\bar{F_3} = I(B)$, so the spaces $I(B)$ and $\bar{F_2}$ are isomorphic and we can apply Theorem 3.

Conversely, suppose that $L(B) = E^*$, where E denotes the Banach space $(\bar{F_1})^*$ equipped with some equivalent norm. It is clear that E contains the space $\bar{F_3}$ as a subspace. Repeating the arguments from the last part of the previous proof, we get that that the space B is reflexive.

The second part of the statement is just a particular case of the theorem. ■

5. Corollary. *Suppose that a Banach space B has an approximation property. Then B is reflexive if and only if the space $L(B)$ is the second dual of the space of compact operators $K(B)$.*

Proof. By Grothendieck's Theorem [1] subspaces $\bar{F_2}$ and $\bar{F_3}$ are the same in this case. So we can apply Theorems 3 and 4. ■

6. Corollary. *Let H be a Hilbert space. Then $K(H)^{**} = L(H)$.*

Note in conclusion that similar problems were considered in [2] and [3].

References

1. J. Diestel and J. J. Uhl Jr., Vector Measures, *Amer. Math. Soc. Math. Surveys* 15 (1977).
2. M. Feder and P. Saphar, Spaces of compact operators and their dual spaces, *Israel J. Math.* 21 (1975), 38–49.
3. G. Godefroy and P. Saphar, Duality in spaces of operators and smooth norms on Banach Spaces, *Illinois J. of Math.* 32 (1988), 672–695.
4. A. Pietsch, *Operator Ideals*, VEB, Deutscher Verlag der Wissenschaften, Berlin, 1978.

Victor Lomonosov
Department of Mathematics and Computer Science
Kent State University
Kent, OH 44242
USA
email: lomonoso@mcs.kent.edu

TOPOLOGICAL TRANSITIVITY AND RECURRENCE AS A SOURCE OF CHAOS

CONSTANTIN P. NICULESCU*

Abstract. This paper builds on previous work due to Glasner–Weiss [7], concerning the connection between topological transitivity, recurrence, and sensitive dependence on initial conditions.

Introduction. The basic feature of the phenomenon of deterministic chaos is sensitive dependence on initial conditions, a property opposite to stability. We shall discuss it in the context of topological dynamical systems acting on metric spaces. In what follows, M will denote a metric space (with a metric d) and $\mathcal{S}$ will denote any one of the following semigroups: $\mathbb{N}$, $\mathbb{Z}$, $\mathbb{R}_+$, or $\mathbb{R}$.

1. Definition (J. Guckenheimer, [8]). *A topological dynamical system $\Phi : \mathcal{S} \times M \to M$ shows sensitive dependence on initial conditions (equivalently, Φ is sensitive) if there exists $\delta > 0$ such that for every $x \in M$ and every neighborhood V of x, one can find a point $y \in V$ and a number $t \in \mathcal{S}$, $t > 0$, for which*

$$d(\Phi_t x, \Phi_t y) > \delta.$$

The terminology above extends to the case of a single continuous mapping $T : M \to M$, by referring to the behavior of the discrete dynamical systems $(T^n)_{n\in\mathbb{N}}$ the mapping T generates. As is well known, the behavior of many dynamical systems can be determined by observing appropriate mappings.

A simple example of a mapping which shows a sensitive dependence on initial conditions is that of doubling angles on the unit circle S^1,

$$T : S^1 \to S^1, \qquad T(z) = z^2.$$

In many cases (including this one), sensitive dependence on initial conditions is a result of topological transitivity and the abundance of nice orbits. To make this assertion precise, we need some preparation, which is mostly due to E. Glasner and B. Weiss [7].

Given a topological dynamical system $\Phi : \mathcal{S} \times M \to M$, a point a of M is said to be *transitive* if its ω–limit set is M. The systems which admit transitive points are called *topologically transitive.* Topological transitivity indicates the

*Research supported in part by CNCSU–Grant 447/1996.

existence of complicated (dense) orbits. It could lead eventually to sensitivity, as it does is the case where M is an interval, see [3].

If the phase space M has an isolated point, then every topological dynamical system acting on M is clearly nonsensitive. So only perfect metric spaces should be considered, in which case the restriction of the given system to the closure of each orbit is topologically transitive.

The following result says that topological transitivity is a stroboscopic property.

2. Lemma (H. Onishi; cf. [1, page 104]). *Let $(t_n)_n$ be a real sequence tending to ∞. Then there exists a residual subset A of $\mathbb{R}_+$ with the following property: for each $t \in A$ one can find a subsequence $(t_{k(n)})_n$ of $(t_n)_n$ and a sequence $(m_{k(n)})_n$ of natural numbers such that $t_{k(n)} - t \cdot m_{k(n)} \longrightarrow 0$ as $n \to \infty$.*

The basic fact relating the topological transitivity and the chaotic behavior is stated next.

3. Lemma. *Let $\Phi : \mathcal{S} \times M \to M$ be a topologically transitive nonsensitive dynamical system. Then for every $\varepsilon > 0$ there exist a transitive point $a \in M$ and a neighborhood U of a such that*

$$\sup_{x\in U} \sup_{t\in\mathcal{S}_+} d(\Phi_t x, \Phi_t a) \leq \varepsilon.$$

Proof. Because Φ is nonsensitive, there must exist a point z and an open neighborhood V of z such that

$$\sup_{y\in V} \sup_{t\in\mathcal{S}_+} d(\Phi_t z, \Phi_t y) < \frac{\varepsilon}{2}.$$

Let b be a transitive point of Φ, so it follows that $a = \Phi_s b \in V$ for some $s \in \mathcal{S}_+$, and thus $U = B_\delta(a) \subset V$ for some $\delta > 0$. Clearly, a is also a transitive point for Φ, so for every $x \in U$ and every $t \in \mathcal{S}_+$ we have

$$d(\Phi_t a, \Phi_t x) \leq d(\Phi_t a, \Phi_t z) + d(\Phi_t z, \Phi_t x) < \varepsilon.$$

∎

4. Corollary. *Let $\Phi : \mathcal{S} \times M \to M$ be a topologically transitive nonsensitive dynamical system. Then there exist $t > 0$ in $\mathcal{S}$ and a strictly increasing sequence $(k(n))_n$ of natural numbers such that*

$$\Phi_{k(n)\cdot t} \to id_M$$

uniformly as $n \to \infty$. Moreover, in the continuous time case, the set of all such t is residual.

Proof. Notice first the existence of $t > 0$ in $\mathcal{S}$ such that Φ_t is topologically transitive; see Lemma 2 for the continuous time case. Because Φ is nonsensitive, Φ_t is nonsensitive too. According to Lemma 3, for each $n \in \mathbf{N}^*$ there exists a transitive point a_n and a neighborhood U_n of a_n such that

$$\sup_{k\in\mathbf{N}} \sup_{x\in U_n} d(\Phi_{kt}x, \Phi_{kt}a_n) < 1/n,$$

which yields an element $k(n) \in \mathcal{S}_+$ such that $k(n) \geq n$ and $\Phi_{k(n)\cdot t}a_n \in U_n$. Then

$$\sup_{m\in\mathbf{N}} d(\Phi_{k(n)\cdot t}(\Phi_{m\cdot t}a_n), \Phi_{m\cdot t}a_n) < 1/n,$$

which implies

$$d(\Phi_{k(n)\cdot t}x, x) < 1/n$$

because of the transitivity of a_n. Consequently, $\Phi_{k(n)\cdot t} \to id_M$ uniformly as $n \to \infty$. ∎

The property outlined in Corollary 4 reflects a certain kind of rigidity of nonchaotic systems. The trajectories of different points visit at the same moment of time all ε–neighborhoods, so the different patterns in the phase space tend to be recovered during the process of iteration.

Quite naturally, transitivity and frequency with which neighborhoods are visited are of a significance in related chaotic situations. We shall need the following definition introducing a class of fast recurrent points.

5. Definition. *A point a of M is said to be algebraically recurrent (for a topological dynamical system $\Phi : \mathcal{S} \times M \to M$) if for every neighborhood U of a there exists a sequence $(k(n))_n$ of elements of $\mathcal{S}_+$ such that $k(n) \to \infty$, $\Phi_{k(n)}a \in U$ for every $n \in \mathbf{N}$, and one of the difference sets*

$$\begin{aligned} A_0 &= \{k(n): \quad n \in \mathbf{N}\} \\ A_n &= \{s - t: \quad t, s \in A_{n-1}, \quad t < s\}, \quad n \geq 1 \end{aligned}$$

has bounded gaps (i.e., for a suitable $L > 0$, every interval $[\alpha, \beta] \subset \mathbb{R}_+$ with $\beta - \alpha > L$ contains an element of that set).

Roughly speaking, algebraic recurrence means that each neighborhood (of the point in question) is visited with polynomial frequency. There are two particular cases already noted in the literature. In the case of uniformly recurrent (equivalently, almost periodic) points, A_0 has bounded gaps. In the case of regular points (i.e., the generic points a for which there exist invariant probability measures μ such that $\mu(U) > 0$ for every neighborhood U of a), A_1 has bounded gaps; see [6, page 75] for details.

If a point a is algebraically recurrent, then so is each point in the orbit of a. Consequently, we can speak of *algebraically recurrent orbits.*

6. Theorem. *Suppose that $\Phi : \mathcal{S} \times M \to M$ is a topological dynamical system satisfying the following two conditions:*

(T) *Φ is topologically transitive.*

(AR) *The union of all algebraically recurrent orbits is dense.*

Then either Φ shows sensitive dependence on initial conditions or Φ is nonsensitive and minimal.

Proof. Suppose that Φ is nonsensitive. We shall show that every point z of M is topologically transitive. Notice that the function Φ has the following equicontinuity–like property:

$$\begin{aligned}&\text{For every } \varepsilon > 0 \text{ there exists } \delta > 0 \text{ such that}\\ &d(y,z) < \delta \quad \text{implies} \quad \sup_{t\in\mathcal{S}_+} d(\Phi_t y, \Phi_t z) < \varepsilon.\end{aligned} \tag{1}$$

In fact, given $\varepsilon > 0$, we can choose (via Lemma 3) a transitive point a and a neighborhood U of a such that

$$\sup_{x\in U}\sup_{t\in\mathcal{S}_+} d(\Phi_t x, \Phi_t a) < \varepsilon.$$

By (AR), there also exists an algebraically recurrent point $p \in U$ and a strictly increasing sequence $(k(n))_n$ of elements of $\mathcal{S}_+$, with the properties stated in Definition 5. By Lemma 2, we can also assume (replacing, if necessary, the sequence $(k(n))_n$ by its translate) that all the mappings $\Phi_{k(n)}$ are topologically transitive. Then for every $t \in \mathcal{S}_+$ and every $n \in \mathbb{N}$ we have

$$d(\Phi_{k(n)}\Phi_t a, \Phi_t a) \le d(\Phi_{k(n)+t} a, \Phi_{k(n)+t} p) + d(\Phi_{k(n)+t} p, \Phi_t a) < 2\varepsilon,$$

and thus

$$\sup_{n\in\mathbb{N}}\sup_{x\in M} d(\Phi_{k(n)} x, x) \le 2\varepsilon$$

because the positive orbit of a is dense.

Let $0 \leq m \leq n$ in $\mathbb{N}$. Then

$$d(\Phi_{k(n)-k(m)}\Phi_{k(m)}x, \Phi_{k(m)}x) = d(\Phi_{k(n)}x, \Phi_{k(m)}x) \leq 4\varepsilon$$

for every $x \in M$. Because the mappings $\Phi_{k(n)}$ have dense images, we get

$$\sup_{x\in M} d(\Phi_s x, x) \leq 4\varepsilon$$

for every s in the difference set $A_1 = \{k(n) - k(m) : 0 \leq m \leq n \in \mathbb{N}\}$. Letting

$$\begin{aligned} A_0 &= \{k(n) : \quad n \in \mathbb{N}\} \\ A_j &= \{s - t : \quad s, t \in A_{j-1}, \quad 0 \leq t \leq s\}, \quad j \geq 1, \end{aligned}$$

we clearly get that

$$\sup_{x\in M} d(\Phi_s x, x) \leq 2^{j+1}\varepsilon \tag{2}$$

for every s in A_j. Since p is uniformly recurrent, one of the above sets, say A_N, has bounded gaps. Thus, writing A_N as an increasing sequence $\{s(n) : n \in \mathbb{N}\}$, we know that $s(n+1) - s(n) \leq L$ for some $L > 0$ and all $n \in \mathbb{N}$.

A simple compactness argument shows that the family $(\Phi_t)_{t\in[0,L]}$ is equicontinuous at z. (Otherwise there would exist an $\varepsilon > 0$, a sequence $(t_n) \subset [0, L]$, and a sequence (y_n) converging to z such that $d(\Phi_{t_n}y_n, \Phi_{t_n}z) \geq \varepsilon$ for each n. As $[0, L]$ is compact, we can assume that (t_n) is also convergent. Letting $n \to \infty$ we get a contradiction.)

Therefore, to derive (1) from (2) (with $n = N$) it suffices to notice that the equality

$$d(\Phi_t y, \Phi_t z) = d(\Phi_{t-s(n)}\Phi_{s(n)}y, \Phi_{t-s(n)}\Phi_{s(n)}z)$$

holds for all $t \in [s(n), s(n) + L]$.

Now, to conclude the proof of the transitivity of z, let $x \in M$ and $\varepsilon > 0$. Then choose $\delta > 0$ as in (1) and note that by the transitivity of a there exists a pair t', t'' in $\mathcal{S}$ with $t'' > t' + 1/\varepsilon > t' > 0$, such that

$$d(\Phi_{t'}a, z) < \delta \qquad \text{and} \qquad d(\Phi_{t''}a, x) < \varepsilon.$$

Then $t'' - t' > 1/\varepsilon$ and

$$d(\Phi_{t''-t'}z, x) \leq d(\Phi_{t''-t'}z, \Phi_{t''-t'}\Phi_{t'}a) + d(\Phi_{t''}a, x) < 2\varepsilon,$$

which assures that $\omega(z) = M$. ■

If a system is minimal, then all points of its state space are uniformly recurrent. The case of irrational rotations shows that minimality alone is not strong enough to imply the sensitivity. On the other hand, there exist sensitive systems which do not satisfy either (T) or (AR).

7. Example (Sensitivity without recurrence and topological transitivity). Consider the mapping

$$T(x) = \begin{cases} 3x/2 & \text{if } 0 \leq x \leq 2/3 \\ -3x/2 + 2 & \text{if } 2/3 \leq x \leq 1. \end{cases}$$

Clearly, T is expansive and thus chaotic. Because the ω–limit set of $(0,1)$ is included in $[1/2, 1]$, no point in $(0, 1/2)$ is recurrent. Because the interval $(1/2, 1)$ is positively invariant, T cannot be topologically transitive. ∎

R. Devaney [4] made the first attempt to define the term *chaotic dynamical system.* Except for a redundancy noticed by J. Banks, J. Brooks, G. Cairns, G. Davis, and P. Stacey [2], his definition is as follows.

8. Definition. *A topological dynamical system is said to be* ***Devaney chaotic*** *if it is topologically transitive, nonminimal, and the union of all periodic orbits is dense.*

In the case of Devaney chaotic behavior, nonminimality is equivalent to the fact that the phase space contains an infinite number of points. Devaney's original definition required also sensitivity, but it was noticed in [2] that sensitivity follows from the other hypotheses. Of course, Theorem 6 also implies this conclusion.

Theorem 6 has counterparts for attractors (i.e., bounded, closed, invariant, and attracting sets). See Haraux [9] for details. We note here only a particular case, strong enough to explain why the chaotic behavior is a more common phenomenon than an exotic one.

9. Theorem. *Let Φ be a topological dynamical system acting on a metric space. Suppose that Φ has a compact attractor A such that*

(P) *The set of periodic orbits of Φ is dense in A.*

(T) $\Phi|_A$ *is topologically transitive.*

Then either A is a periodic attractor, or A is a strange attractor (i.e., the dynamic on it is sensitive).

Conditions (P) and (T) in Theorem 9 above are fulfilled for example by the so called *Axiom A attractors.* See [11] for details. Due to their hyperbolic structure, all these attractors show sensitive dependence on initial conditions. In particular, this is so for the horseshoe, the solenoid, Anosov's tori etc.

Conditions (T) and (AR) both pass to quotients, so in principle we can exhibit new examples of sensitive dynamical systems (or of strange attractors) by passing to appropriate quotients.

References

1. E. Akin, *The General Topology of Dynamical Systems*, Amer. Math. Soc., Providence, R.I., 1993.
2. J. Banks, J. Brooks, G. Cairns, G. Davis, and P. Stacey, On Devaney's definition of chaos, *Amer. Math. Monthly* **99** (1992), 332–334.
3. L. S. Block and W. A. Coppel, Dynamics in one dimension, *Lecture Notes in Math.* **1513**, Springer (1992).
4. R. L. Devaney, *An Introduction to Chaotic Dynamical Systems*, 2nd edition, Addison–Wesley, 1989.
5. H. Furstenberg, Poincaré recurrence and number theory, *Bull. Amer. Math. Soc.* **5** (1981), 211–234.
6. H. Furstenberg, *Recurrence in Ergodic Theory and Combinatorial Number Theory*, Princeton Univ. Press, Princeton, N.J., 1981.
7. E. Glasner and B. Weiss, Sensitive dependence on initial conditions, *Nonlinearity* **6** (1993), 1067–1075.
8. J. Guckenheimer, Sensitive dependence on initial conditions for one–dimensional maps, *Commun. Math. Phys.* **70** (1979), 133–160.
9. A. Haraux, *Systèmes dynamiques dissipatifs et applications*, in: *Collection Recherches en Math. Appl.*, no. 17, Masson, Paris, 1991.
10. C. P. Niculescu, *Chaotic Dynamical Systems*, in: *Lecture Notes*, *University of Craiova* (1995/1996), Craiova University Press, 1996.
11. D. Ruelle, *Elements of Differentiable Dynamics and Bifurcation Theory*, Academic Press, 1989.

Constantin P. Niculescu
Department of Mathematics
University of Craiova
Craiova 1100
Romania
email: niculesc@udjmath1.sfos.ro

ORDER BOUNDED OPERATORS MAY BE FAR FROM REGULAR*

ANTONY W. WICKSTEAD

Abstract. In the main result of this paper, Example 2.2, we construct an order bounded operator $T : E \to F$ between two Banach lattices, such that $\|T - S\| \geq 1$ for each regular operator $S : E \to F$. This shows a drastic difference between the regular and order bounded operators. On the space of the latter operators we introduce a new natural norm, termed the order bound norm and denoted by $\|\cdot\|_b$. In Example 4.1, we show that there exists a compact order bounded operator K such that $\|K - S\|_b \geq 1$ for each regular operator S. Finally, in Section 5, we pose some open problems.

1. Introduction

The most obvious operators to consider between two vector lattices are the *regular operators*, which are the linear span of the positive operators (see [8] or [14] for fundamental terminology). Unfortunately it is not always easy to determine whether or not a given operator is regular. It is rather easier to determine whether or not an operators is *order bounded* (that is, it maps order bounded sets to order bounded sets) if only because this is a local property of the operator which can be tested by considering the restrictions to principal ideals while regularity is very much a global property. In some important special cases, e.g. when the range space is Dedekind complete (see [8, Theorem 1.13] or [14, Theorem 1.3.2]) or when the domain is a separable Banach lattice and the range is a Dedekind σ–complete Banach lattice (see [15, Theorem 5.2]) the two notions coincide. There are however many examples in the literature to show that the two notions differ in general, see for instance [12], [2], [15], [5] and the example due to Lotz in Example 1.11 of [8].

When attention is restricted to Banach lattices, it has long been known that order bounded operators (which include the regular operators) must be norm bounded but that the converse is false even when the range space is Dedekind complete (see [1] and [7]). It is much more likely that an order bounded operator is regular. Arendt and Voigt, in [9], introduced the notion of a *strongly non-regular* operator as being a bounded operator which is not in the norm closure of the regular operators. They showed there that every L_p–space $(1 < p < \infty)$ has strongly non-regular operators defined on it. Since

*This work was executed during a visit by Y. Abramovich and A. Wickstead to Oberwolfach in Spring 1996, which was supported by the Volkswagen–Stiftung (RiP–program at Oberwolfach). The author would like to thank Y. Abramovich for many helpful suggestions in connection with this paper.

such L_p–spaces are Dedekind complete, all order bounded operators on them are regular. In Section 2 of this paper we give an example to show that it is possible for order bounded operators to be strongly non-regular.

Much weaker conditions suffice to force a compact operator to be regular (see [13] and [16]). The combination of the two properties of being compact and order bounded might seem so strong that it would force regularity. After the disappointing properties of compact operators in an order theoretic setting that were shown in [4] and [6], it should not come as a surprise that this is not so. Under the relatively mild assumption that the range space has the approximation property, every compact operator is in the operator norm closure of the finite rank operators, all of which are regular, so there can be no strongly non-regular compact operators. However there is a fairly natural norm, the *order bound norm*, which may be imposed on the order bounded operators between two Banach lattices. In Section 4 we show that there are compact order bounded operators which do not even lie in the order bound norm closure of the regular operators. In the example given in Section 2 the operator norm and the order bound norm coincide and although there are strongly non-regular operators then, every compact operator is regular.

We conclude by posing some open problems.

2. A strongly non-regular order bounded operator

In several places in the paper we make use of an finite Cartesian products of spaces of sequences (either of scalars or vectors) converging to zero. We fix now several pieces of notation. If $X = \mathbb{R}$ then $\mathbf{1}$ will denote the constantly one sequence. If $x \in X$ then $x_{(j)} \in X^n$ will denote $(0, \ldots, 0, x, 0, \ldots, 0)$ where x is the only nonzero element which is in the j–th position. If X is any normed space (possibly $X = \mathbb{R}$) then we define $\mathcal{I} : c_0(X)^n \to c_0(X)$ by requiring that the $(kn + j)$–th entry in $\mathcal{I}(x_1, x_2, \ldots, x_n)$ is the $(k + 1)$–st entry in x_j, when $0 \leq j \leq n$ and $k \in \mathbb{N}$. I.e. we produce the image by taking in turn the first elements of $x_1, x_2, \ldots, x_n$, then their second elements in turn etc.

Recall that the *regular* norm of a regular operator $T : E \to F$ between two Banach lattices is defined by

$$\|T\|_r = \inf\{\|S\| : \quad S : E \to F, \quad S \geq \pm T\}.$$

One good way to construct non-regular operators is by means of a sequence of operators which are regular but with the sequence of regular norms being unbounded.

The example given in this section accomplishes two things at once. It gives an example of an order bounded operator which is strongly non-regular (unlike the examples given by [9], where order boundedness and regularity coincide because of Dedekind completeness) and is at the same time rather more elementary than the examples given in [9]. Notice that because the range space is an M–space, every compact operator will actually be regular because of [13] and that because the range space has a strong order unit every norm bounded operator is order bounded.

As a first step in the construction, which we will re-use later, let us note the following example.

2.1. Example. *For each $n \in \mathbb{N}$ there is an operator $T : c^n \to c$ with $\|T\| = 2$ and $\|T\|_r \geq n$. Furthermore, if $0 < \alpha < 1$ and $T' : c^n \to c$ with $\|T - T'\| < \alpha$, then $\|T'\|_r \geq n(1 - \alpha)$.*

Proof. By Theorem 2.13 of [17], every bounded operator from c into itself is regular. It follows very easily that the same is true for operators from c^n into c, where c^n denotes the produce of n copies of c normed by

$$\|(x_1, x_2, \dots, x_n)\| = \max\{\|x_1\|_\infty, \|x_2\|_\infty, \dots, \|x_n\|_\infty\}.$$

If $x \in c$ let $\ell(x)$ denote its limit. Define a linear operator $Q : c^n \to c_0^n$ by $Q(x_1, x_2, \dots, x_n) = \big(x_1 - \ell(x_1),\ x_2 - \ell(x_2), \dots,\ x_n - \ell(x_n)\big)$. Now define $T = \mathcal{I} \circ Q$ to obtain an operator from $c^n \to c$. The norm of T is 2.

Suppose that $S \geq T, -T$ and consider $S(\mathbf{1}_j)$. It is routine to see that the $(kn+j)$–th entry in $S(\mathbf{1}_j)$ must be at least 1 and hence that the limit the entries in $S(\mathbf{1}_j)$ must be at least 1. It follows that $S(\mathbf{1}, \mathbf{1}, \dots, \mathbf{1}) = \sum_{j=1}^n S(\mathbf{1}_j)$ must have limit at least n and hence that $\|S(\mathbf{1}, \mathbf{1}, \dots, \mathbf{1})\|_\infty \geq n$. It follows that $\|S\| \geq n$ and hence that $\|T\|_r \geq n$.

Now suppose that $0 < \alpha < 1$, that $T' : c^n \to c$ and that $\|T - T'\| \leq \alpha$. Consider, for the moment, T and T' as operators into ℓ_∞ rather than into c, so that we may work in a lattice of operators. It is elementary to check that in this setting, $|Q|$ is precisely the identity embedding and that $|T| = \mathcal{I} \circ |Q|$. We then see that for $1 \leq j \leq n$ we have

$$\begin{aligned} |T'|(\mathbf{1}_j) &\geq |T|(\mathbf{1}_j) - |T' - T|(\mathbf{1}_j) \\ &= \mathcal{I}(\mathbf{1}_j) - x_j \end{aligned}$$

where $x_j \in \ell_\infty$ and $\|x_j\| \leq \alpha$. Returning now to the viewpoint of these operators taking values in c we see that if $S' \geq \pm T'$ then every $(nj + k)$–th entry in $S'(\mathbf{1}_j)$ will be at least $1 - \alpha$ so that the limit of $S'(\mathbf{1}_j)$ will be at least $1 - \alpha$. It follows that the limit of $S'(\mathbf{1}, \mathbf{1}, \dots, \mathbf{1}) = \sum_{j=1}^n S'(\mathbf{1}_j)$ will be at least $n(1 - \alpha)$ and hence $\|S'\| \geq n(1 - \alpha)$. It certainly follows that $\|T_n'\|_r \geq n(1 - \alpha)$. ∎

2.2. Example. *There are unital M–spaces E and F and an order bounded operator $V : E \to F$ which is strongly non-regular.*

Proof. For each $n \in \mathbb{N}$ let T_n be an operator from c^n into c as constructed in Example 2.1. Take $E = \ell_\infty(c^n)$ and $F = \ell_\infty(c)$, both with the supremum norm, so that both are clearly unital M–spaces. Define $V : \ell_\infty(c^n) \to \ell_\infty(c)$ by $V(x_n) = (T_n x_n)$, so that $\|V\| = \|V\|_b = 2$ while V is not regular, for else we would have $n \le \|T_n\|_r \le \|V\|$ for all $n \in \mathbb{N}$. I claim that the distance from V to the regular operators is at least 1.

Suppose that $0 < \alpha < 1$ and that $V' : \ell_\infty(c^n) \to \ell_\infty(c)$ with $\|V - V'\| \le \alpha$. Let T'_n denote the restriction of V' to c^n and let P_n denote the projection of $\ell_\infty(c)$ onto its n–th component. Now $P_n \circ T'_n : c^n \to c$ and certainly $\|P_n \circ T'_n - T_n\| \le \|P_n \circ (V' - V)\| \le \|V' - V\| < \alpha$. The previous example shows that $\|T'_n\|_r \ge n(1-\alpha)$. If $\alpha < 1$ then the fact that $\|V'\|_r \ge \|T'_n\|_r$ for all $n \in \mathbb{N}$ will contradict the regularity of V'. The only way out of this is to admit that $\|V' - V\| \ge 1$ so that V is not in the closure of the regular operators. ∎

3. The order bound norm

In this section we will introduce the order bound norm for an order bounded operator between two Banach lattices and point out a few of its fundamental properties. We will use the notation $[a, b]$ to denote an order interval with end points a, b, respectively, without reference to the space in which it lies. We will denote by $\mathcal{L}^r(E, F)$ the space of regular operators from E into F and by $\mathcal{L}^b(E, F)$ the corresponding space of order bounded operators. We will only use these in setting where all operators involved are necessarily norm bounded. The proof of the next proposition is essentially the classical proof that order bounded operators on Banach lattices must be norm bounded.

3.1. Proposition. *Let E be a Banach lattice and F a normed lattice. If $T : E \to F$ is an order bounded linear operator then there is a real number M such that for all $x \in E_+$ there is $y \in F_+$ with $T([-x, x]) \subseteq [-y, y]$ and $\|y\| \le M\|x\|$.*

Proof. Suppose that this fails then for all $n \in \mathbb{N}$ there is $x_n \in E_+$ such that whenever $T(-x_n, x_n]) \subseteq [-y, y]$ we must have $\|y\| \ge n^3$. Let $x = \sum_{n=1}^\infty x_n/n^2$, which is absolutely convergent and hence convergent then, as T is order bounded, there is $y \in F_+$ with $T([-x, x]) \subseteq [-y, y]$. As $T([-x_n/n^2, x_n/n^2]) \subseteq T([-x, x])$ for all $n \in \mathbb{N}$ we must have $\|y\| \ge n$ for all $n \in \mathbb{N}$, which is impossible. ∎

3.2. Definition. *If E is a Banach lattice, F a normed lattice, and $T : E \to F$ is an order bounded operator, define*

$$\|T\|_b = \inf\left\{ M \in \mathbb{R} : \quad \forall x \in E_+ \ \exists y \in F_+ \ \textit{with} \ \begin{array}{c} T([-x,x]) \subseteq [-y,y], \\ \|y\| \leq M\|x\| \end{array} \right\}.$$

We will refer to this as the ***order bound norm*** *of T.*

It is clear from the proof of Example 2.2 that, for regular operators from $\ell_\infty(c^n)$ into $\ell_\infty(c)$, the regular and order bound norms are not even equivalent.

The proof of the following result is routine and is omitted.

3.3. Proposition. *If E is a Banach lattice and F a normed lattice then $\|\cdot\|_b$ is a norm on $\mathcal{L}^b(E,F)$ which is complete if F is a Banach lattice.*

It is clear that for a regular operator T we certainly have $\|T\|_b \leq \|T\|_r$. If E and F are Banach lattices and $\mathcal{L}^r(E,F) = \mathcal{L}^b(E,F)$ then the two norms will certainly be equivalent by the open mapping theorem. However even in this setting the two norms need not be equal as is shown by Example 2.1 above.

We conclude this short section by recording without proof one further simple property of the order bound norm.

3.4. Proposition. *Let X_n and Y_n be Banach lattices and let $T_n : X_n \to Y_n$ for each $n \in \mathbb{N}$. Define $T(x_n) = (T_n x_n)$ for each $(x_n) \in \ell_\infty(X_n)$. Then T is an order bounded operator from $\ell_\infty(X_n)$ into $\ell_\infty(Y_n)$ if and only if each T_n is order bounded and $\sup\{\|T_n\|_b : n \in \mathbb{N}\} < \infty$. Furthermore, we then have $\|T\|_b = \sup\{\|T_n\|_b : n \in \mathbb{N}\}$.*

4. A compact order bounded operator which is not regular

In fact we prove even more, namely that the following example exists.

4.1. Example. *There are Banach lattices E and F and a compact order bounded operator $V : E \to F$ which is not in the order bound norm closure of the regular operators.*

Proof. Let us first recall from [6] that for all $n \in \mathbb{N}$ there is a finite rank operator K_n on $L_2[0,1]$, with norm $2^{-n/2}$ such that $|K_n|(f) = (\int_0^1 f\,d\mu)\mathbf{1}$ and $\big\||K_n|\big\| = 1$, where μ denotes Lebesgue measure on $[0,1]$ and $\mathbf{1}$ denotes the constantly one function. Let $c(L_2)$ denote the space of all convergent sequence in $L_2[0,1]$ normed by $\|(f_n)\| = \sup\{\|f_n\|_2 : n \in \mathbb{N}\}$ and ordered by the coordinate-wise ordering. These definitions make $c(L_2)$ into a Banach lattice. We will define a linear operator $S : c(L_2) \to c(L_2)$ by $S(f_n) = (K_n f_n)$. Note that S actually takes values in $c_0(L_2)$ and that it is a sum of compact operators with the n–th term having norm $2^{-n/2}$, so is compact.

The operator S is order bounded because if we fix $(f_n) \in c(L_2)_+$ and $|(g_n)| \leq (f_n)$ then for each $n \in \mathbb{N}$ we have $|g_n| \leq f_n$ and hence $|K_n g_n| \leq |K_n||g_n| \leq |K_n| f_n = (\int_0^1 f_n \, d\mu)\mathbf{1}$. Since there is $f \in L_2$ with $f_n \to f$ for the $\|\cdot\|_2$–norm we certainly have $\int_0^1 f_n \, d\mu \to \int_0^1 f \, d\mu$ so that the sequence $\left((\int_0^1 f_n \, d\mu)\mathbf{1}\right)_{(n)}$ certainly lies in $c(L_2)$. This shows that S is order bounded. Note that if $\|(f_n)\| \leq 1$ then this upper bound for $S([-(f_n),(f_n)])$ also has norm at most 1.

If $T \geq \pm S$, then restricting T to the n–th coordinate in $c(L_2)$ we see that for $0 \leq f \in L_2$

$$Tf_{(n)} \geq (|K_n| f)_{(n)} = \left(\left(\int_0^1 f \, d\mu\right)\mathbf{1}\right)_{(n)}.$$

In particular, $T\mathbf{1}_{(n)} \geq \mathbf{1}_{(n)}$, so that $T\mathbf{1} \geq \mathbf{1}$.

Now let us introduce an operator $Q_n : c(L_2)^n \to c(L_2)$, where $c(L_2)^n$ is equipped with the M–product norm and the pointwise ordering. We define $Q_n(a_1, a_2, \ldots, a_n) = \mathcal{I}(Sa_1, Sa_2, \ldots, Sa_n)$. Since each $Sa_j \in c_0(L_2)$ we see that Q_n also takes values in $c_0(L_2)$ and hence in $c(L_2)$. It is clear that Q_n is compact and that Q_n has the same norm as S, namely $2^{-1/2}$.

Our next step is to show that Q_n is order bounded.
Let $0 \leq \mathbf{a} = (a_1, a_2, \ldots, a_n) \in c(L_2)^n$. We know that there are $b_j \in c(L_2)$ with $S([-a_j, a_j]) \subseteq [-b_j, b_j]$ and that we may assume that $\|b_j\| \leq \|a_j\| \leq \|\mathbf{a}\|$. Replacing b_j by c_j, where the $(n(k-1)+j)$–th entry in c_j is the k–th entry in b_j and all others are equal to the limit of b_j, we obtain an element of $c(L_2)$ with the same norm as b_j. It is clear that if $d = \bigvee_{j=1}^n c_j$ then $Q_n([-\mathbf{a}, \mathbf{a}]) \subseteq [-d, d]$ so that Q_n is order bounded. Note also that $\|d\| \leq \sqrt{n}\|\mathbf{a}\|$. To see this it suffices to prove that each entry in d has $\|\cdot\|_2$ at most $\sqrt{n}\|\mathbf{a}\|$. But each entry in d is simply the supremum of the corresponding entries in all the c_j, each of which has norm at most $\|\mathbf{a}\|$. It remains only to note that if $0 \leq f_1, f_2, \ldots, f_n \in L_2$ then

$$\left(\bigvee_{j=1}^n f_j\right)^2 \leq \sum_{j=1}^n f_j^2,$$

so that

$$\left\|\bigvee_{j=1}^n f_j\right\|_2^2 \leq \sum_{j=1}^n \|f_j\|_2^2,$$

and now we need only take square roots to obtain the desired inequality. I.e. we have shown that $\|Q_n\|_n \leq \sqrt{n}$.

If $T : c(L_2)^n \to c(L_2)$ and $T \geq \pm Q_n$ then consider the image of $\mathbf{1}_{(j)}$, that member of $c(L_2)^n$ with j–th entry being the constantly one sequence and all other entries being the zero sequence. Composing T with the projection onto the $\big(n(k-1)+j\big)$–th entries for $k \in \mathbf{N}$ will give us an upper bound for S and we know that under this the image of $(\mathbf{1})$ is at least $(\mathbf{1})$. Thus the $\big(n(k-1)+j\big)$–th entries in $T\mathbf{1}_{(j)}$ will all be at least $\mathbf{1}$ and hence the limit of the entries in $T\mathbf{1}_{(j)}$ will be at least $\mathbf{1}$. Thus if we apply T to $\mathbf{1}_{(1)}+\mathbf{1}_{(2)}+\cdots+\mathbf{1}_{(n)}$, which has norm 1, then its image $T\mathbf{1}_{(1)}+T\mathbf{1}_{(2)}+\cdots+T\mathbf{1}_{(n)}$ will have limit at least $n\mathbf{1}$ and hence have norm at least n. Thus $\|Q_n\|_r \geq n$.

Now let us define an operator $V : \ell_\infty(c(L_2)^{n^3}) \to \ell_\infty(c(L_2))$ by $V(a_n) = (Q_{n^3}a_n/n^2)$. The fact that all the Q_n's have the same operator norm shows that V may be described by an operator norm convergent sum of compact operators, so is compact. For each $n \in \mathbf{N}$ we have $\|Q_{n^3}/n^2\|_b \leq \sqrt{n^3}/n^2 = n^{-1/2}$ so that by Proposition 3.4 V is order bounded. By considering the restriction of V to its n–th coordinate we see that if V were regular then its regular norm would be at least $n^3/n^2 = n$, for all $n \in \mathbf{N}$, so that V cannot be regular.

In fact we show that if V' is an order bounded operator with $\|V - V'\|_b = \alpha < 1$ then V' cannot be regular. The proof of this will proceed from the same basic building blocks that we used in the construction of V. Let us first suppose that $K_n' : L_2 \to L_2$ is regular and with $\|K_n - K_n'\|_b = \|K_n - K_n'\|_r \leq \alpha < 1$. Because $|K_n'| \geq |K_n| - |K_n' - K_n|$, we see that $|K_n'|\mathbf{1} \geq |K_n|\mathbf{1} - |K_n' - K_n|\mathbf{1} = \mathbf{1} - x_n$ (say) where $\|x_n\|_2 \leq \alpha$. Now suppose that $S' \in \mathcal{L}^r(c(L_2))$ and that $\|S' - S\|_b \leq \alpha$. If S_n' (respectively, S_n) denotes the restriction of S' (respectively, S) to the n–th coordinate of $c(L_2)$ then clearly $\|S_n' - S_n\|_b \leq \alpha$ and $S_n', S_n : L_2 \to c(L_2)$. If P_n denotes the projection of $c(L_2)$ onto its n–th coordinate then we similarly have $\|P_n \circ S_n' - P_n \circ S_n\|_b = \|P_n \circ S_n' - K_n\|_b \leq \alpha$. It follows that if $T' \geq \pm S'$ then we have $T_n' \geq \pm S_n'$ (where T_n' denotes the restriction of T' to the n–th coordinate) so that $T_n' \geq P_n \circ T_n' \geq \pm P_n \circ S_n'$ and hence $T_n' \geq |P_n \circ S_n'|$. Hence $T_n'\mathbf{1} \geq \mathbf{1} - x_n$ where $\|x_n\| \leq \alpha$. If follows that $T(\mathbf{1}) \geq (\mathbf{1}) - (x_n)$ where $\|(x_n)\| \leq \alpha$.

Now suppose that $Q_n' \in \mathcal{L}^r(c(L_2)^n, c(L_2))$ with $\|Q_n' - Q_n\|_b \leq \alpha$. The restriction to the k–th element of the domain will possess a similar relationship and hence we see that if $T \geq \pm Q_n'$ then there is $x^k = (x_m^k)_{m=1}^\infty$ with $\|(x_m^k)_{m=1}^\infty\| \leq \alpha$ and $T\mathbf{1}_{(k)} \geq (\mathbf{1}) - (x_m^k)_{m=1}^\infty$. In particular if $y_k = \lim_{m\to\infty} x_m^k$, then $\|y_k\| \leq \alpha$ for each $k = 1, 2, \ldots, n$. It follows that the limit of $T(\sum_{k=1}^n \mathbf{1}_{(k)})$ will be at least $n\mathbf{1} - \sum_{k=1}^n y_k$ so that $\|T(\sum_{k=1}^n \mathbf{1}_{(k)})\| \geq n(1-\alpha)$. Hence $\|Q_n'\|_r \geq n(1-\alpha)$.

Finally, if $V' \in \mathcal{L}\big(\ell_\infty(c(L_2)^n),\ \ell_\infty(c(L_2))\big)$ and $\|V' - V\|_b \leq \alpha$ then considering the projection onto the n–th coordinate in the range space of the restriction to the n–th coordinate of the domain, we see that $\|V'\|_r \geq (n^3(1-\alpha)/n^2) = n(1-\alpha)$ for all $n \in \mathbb{N}$. As $\alpha < 1$ this shows that V' cannot be regular after all. ∎

5. Some open problems

When one considers the space of regular operators as a subspace of the space of all bounded operators there are *a priori* four possibilities. The subspace may be

(i) equal to the whole space,

(ii) a dense proper subspace,

(iii) a proper closed subspace, or

(iv) a non-closed non-dense subspace.

All four possibilities do in fact occur. Possibility (i) happens, for example, when the range space is Dedekind complete with a strong order unit while Example 2.2 shows that (iv) can occur as in that example the operator and regular norms are not equivalent so that the regular operators cannot be a closed subspace of the bounded operators else the Open Mapping Theorem would prove their equivalence.

To see that (ii) can occur, suppose that $p > q$ and recall that from a result of Cartwright and Lotz [10] (see also Abramovich and Janovsky [3]) that $\mathcal{L}^r(\ell_p, \ell_q) \neq \mathcal{L}(\ell_p, \ell_q)$. It follows from Pitts theorem that $\mathcal{L}(\ell_p, \ell_q) = \mathcal{K}(\ell_p, \ell_q)$, so that certainly every bounded operator is in the closure of the finite rank operators, all of which are regular.

To see that (iii) can occur, recall from Theorem 3.1 of [7] that not every bounded operator from $L_1[0,1]$ into c_0 is regular. As c_0 has a Fatou norm it follows from the Riesz–Kantorovich formula that $\|T\|_r = \|j \circ T\|_{r^{**}}$ for every regular operator from $L_1[0,1]$ into c_0, where $j : c_0 \to c_0^{**} = \ell_\infty$ is the natural embedding and $\|\cdot\|_{r^{**}}$ denotes the regular norm in $\mathcal{L}^r(L_1[0,1], c_0^{**})$. But in $\mathcal{L}^r(L_1[0,1], \ell_\infty)$ the regular and operator norms coincide, so the fact that $\|T\| = \|j \circ T\|$ shows that $\|T\|_r = \|T\|$ for all $T \in \mathcal{L}^r(L, c_0)$. As $\mathcal{L}^r(L, c_0)$ is complete under the regular norm it must be complete under the operator norm and hence closed in $\mathcal{L}(L, c_0)$.

It is easy to see that taking $E = \ell_p \oplus L_1[0,1]$ and $F = \ell_q \oplus c_0$, for $p > q$, gives another example of possibility (iv).

The same four possibilities exist for the relationship between $\mathcal{L}^r(E,F)$ and $\mathcal{L}^b(E,F)$. Possibility (i) occurs whenever F is Dedekind complete (and in some other cases), while Example 2.2 (where the order bound norm coincides with the operator norm) shows that possibility (iv) can occur.

5.1. Problem. *Are there Banach lattices E and F such that the regular operators from E into F form a proper closed subspace of the space of order bounded operators from E into F for the order bound norm?*

5.2. Problem. *Are there Banach lattices E and F such that the regular operators from E into F forms a proper dense subspace of the space of order bounded operators from E into F for the order bound norm?*

There are several possible formulations of a similar problem for compact operators, depending on what space one takes as the analogue of the space of regular operators. One could take the space of compact operators which are also regular, or the linear span of the positive compact operators, or the regular norm closure of the finite rank operators or (in view of the example in [11]) the closure of the finite rank operators for the k–norm introduced in [11]. We leave examination of the possible relationships to the interested reader, while warning that the myriad of answers may not be all that interesting when found.

References

1. Y. A. Abramovich, *When each regular operator is continuous*, Functional analysis, optimization, and mathematical economics, Oxford University Press, New York, 1990, 133–140.
2. Y. A. Abramovich and V. A. Gejler, On a question of Fremlin concerning order bounded and regular operators, *Coll. Math.* 46 (1982), 15–17.
3. Y. A. Abramovich and L. P. Janovsky, Applications of the Rademacher systems to operator characterizations of Banach lattices, *Colloq. Math.* 46 (1982), 75–78.
4. Y. A. Abramovich and A. W. Wickstead, A compact regular operator without modulus, *Proc. Amer. Math. Soc.* 116 (1992), 721–726.
5. Y. A. Abramovich and A. W. Wickstead, The regularity of order bounded operators into $C(K)$, II, *Quart. J. Math. Oxford* 44 (1993), 257–270.
6. Y. A. Abramovich and A. W. Wickstead, Solutions of several problems in the theory of compact positive operators, *Proc. Amer. Math. Soc.* 123 (1995), 3021–3026.
7. Y. A. Abramovich and A. W. Wickstead, When each continuous operator is regular, II, *Indag. Math.* 8 (1997), 281–294.
8. C. D. Aliprantis and O. Burkinshaw, *Positive operators*, Academic Press, New York & London, 1985.

9. W. Arendt and J. Voigt, Approximation of multipliers by regular operators, *Indag. Math.* 2 (1991), 159–170.

10. D. I. Cartwright and H. P. Lotz, Some characterizations of AM– and AL–spaces, *Math. Z.* 142 (1975), 97–103.

11. Z. L. Chen and A. W. Wickstead, Incompleteness of the linear span of the positive compact operators, *Proc. Amer. Math. Soc.* 125 (1997), 3381–3389.

12. S. Kaplan, An example in the space of bounded operators from $C(X)$ to $C(Y)$, *Proc. Amer. Math. Soc.* 38 (1973), 595–597.

13. U. Krengel, Über den Absolutbetrag stetiger linearer Operatoren und seine Anweddung auf ergodische Zerlegungen, *Math. Scand.* 13 (1963), 151–187.

14. P. Meyer–Nieberg, *Banach lattices*, Springer–Verlag, Berlin Heidelberg New York, 1991.

15. A. W. Wickstead, The regularity of order bounded operators into $C(K)$, *Quart. J. Math. Oxford* (2) 41 (1990), 359–368.

16. A. W. Wickstead, Dedekind completeness of some lattices of compact operators, *Bull. Pol. Acad. Sci.* 43 (1995), 297–304.

17. H.–Y. Xiong, On whether or not $\mathcal{L}(E,F) = \mathcal{L}^r(E,F)$ for some classical Banach lattices E and F, *Indag. Math.* 47 (1984), 267–282.

Antony W. Wickstead
Department of Pure Mathematics
The Queen's University of Belfast
Belfast, BT7 1NN
Northern Ireland
email: a.wickstead@qub.ac.uk

IMPLICIT PROGRAMMING AND THE INVARIANT MANIFOLD FOR RAMSEY EQUILIBRIA

ROBERT A. BECKER and CIPRIAN FOIAS

Abstract. We present an economically motivated construction of the global invariant manifold for a class of Ramsey equilibria. This invariant manifold is found by solving a functional equation for an implicit programming problem based on the most patient household's necessary condition for an optimum. An iterative process is given which yields the first agent's optimal policy function in the limit and thereby the invariant manifold.

1. Introduction

An important problem in economic dynamics concerns the evolution of individual agents' consumption and capital in a market economy. The model discussed in this paper originates in the seminal paper of Ramsey [11]. He indicated that in a stationary equilibrium profile the most patient of the economy's households would hold all the capital. However, he did not provide a detailed analysis. Becker [1] reconsidered Ramsey's problem and demonstrated Ramsey's steady state conjecture for the type of model considered in this paper.

The underlying economic model is a dynamic general equilibrium model of an economy populated by a finite number of infinitely lived agents with differing rates of time preference or discount factors. A single commodity is produced in the production sector using capital and labor inputs supplied by the household sector. That good can be either consumed or a portion can be converted into capital which is carried over to the next time period. Individual households receive income at each time from supplying their labor services and renting their capital stocks to the production sector. At each time they decide how to divide their income between consumption and capital savings that will yield the next period's rental income. Households are not allowed to borrow against their future labor income. The economic problem is to study the properties of dynamic equilibria with perfect foresight. These are called *Ramsey equilibria.*

Becker, Boyd and Foias [3] gave general conditions for equilibria (including nonstationary ones) to exist from arbitrary initial conditions. The basic dynamics of this model were worked out in a series of papers by Becker and Foias [5], [6], and [7]. Those papers considered conditions on the production technology under which the dynamics exhibited eventual monotone convergence to the steady state. They also constructed an example of an economy with cycles of period–2 as well as showed how those cycles can emerge by a flip bifurcation of the steady state. Sorger ([12] and [13]) showed more complicated equilibria existed including chaotic ones. He also demonstrated multiple equilibria could exist.

The purpose of this paper is to develop an economic approach to the construction of the invariant manifold for a class of Ramsey equilibrium problems. The economic situation is one where it is known from [5] that all equilibria are eventually monotonic and only the most patient agent eventually holds all of the economy's capital. We focus in this paper on equilibria where the only agent with capital is the most patient. In this case we can analyze a dynamical system in the first quadrant of the plane. The evolution of this agent's capital stocks can be found by constructing the global invariant manifold for the two dimensional system describing the equilibrium dynamics. We construct this manifold by a new method which emphasizes the economic structure of our dynamical system. This approach leads us to the formulation of a new type of functional equation whose solution yields the invariant manifold. This equation defines a nonlinear operator which exhibits a monotonicity property on the class of lower semi-continuous strictly increasing functions defined on a particular half–open interval. The monotonicity of this operator can be exploited via a successive approximation argument to find the functional equation's solution.

The functional equation considered in this paper is derived from the no arbitrage or Euler equation of the most patient household. This equation represents the first–order necessary condition for that agent's optimal action given the profile of market determined prices. The way we solve this equation leads to an *implicit programming problem* — a mathematical programming problem where the constraint equation depends on the solution to the optimization problem. The solution to this equation is the most patient agent's *policy function* — it tells that individual how much to consume given the agent's current capital holdings. It is a remarkable feature of our problem that along equilibrium paths this agent need only know the size of his current capital stock to make a consumption and capital accumulation decision consistent with that equilibrium configuration once we have the policy function.

Previous work by economists on related optimal growth models and their equilibrium counterparts has focused on dynamic programming approaches to solving Bellman's equation.[1] The standard dynamic programming approach solves Bellman's equation by showing the corresponding Bellman operator

[1] A comprehensive treatment of dynamic programming and its applications to economic dynamics can be found in Stokey and Lucas [14].

is a contraction on a suitable function space. This yields the existence and uniqueness of the value function, from which the decision maker's optimal policy function can be found. Bellman [8] also suggested a scheme to solve his equation by approximation in policy space. The idea is to choose the initial approximation of the policy function and use it to calculate an approximate value function. By iterating this process a sequence of approximate policy and value functions can be found. Bellman showed that this process produced a sequence of approximate value functions which monotonically converged to the solution of the Bellman equation.[2] We bypass Bellman's equation and focus directly on the most patient individual's Euler equation. Our approximation procedure produces a sequence of policy functions. But unlike the dynamic programming approximation in policy space approach, we obtain monotonic convergence of the policy functions. In this manner, we have a policy improvement algorithm that produces the invariant manifold for the Ramsey dynamical system.

The monotone operator approach to solving our functional equation is basically an application of Amann's fixed point theorem. It says an increasing operator taking an ordered set X, for which every chain in X has a supremum, into itself has a smallest fixed point.[3] This fixed point theorem can also be used to show Bellman's equation has at least one solution found by approximation in policy space. The contraction property of the Bellman operator implies this is the only solution. We obtain the uniqueness of the solution to our implicit programming functional equation by a separate argument that does not rely on showing the operator associated to our equation is a contraction map.

2. The Ramsey equilibrium model

The Ramsey equilibrium model is briefly summarized together with the reduction to the dynamical system governing the equilibrium profiles explored in this paper.

---

[2]See Bellman [8, Theorem 3, p. 18] for an example of this value function improvement algorithm.

[3]Ziedler [16, pp. 503-508] presents a detailed discussion of Amann's result and its variants. Amann's result is utilized since it does not require the ordered set X to be a lattice as is the case with Tarski's Theorem [15]. The set of functions appearing in our construction do not form a lattice. Coleman [9] shows how to apply Tarski's result to a monotone operator problem using the Euler equations for a stochastic equilibrium problem. Our model is deterministic and our operator applies to a different functional equation than the one he examines, although we derive our equation from the Euler equation for the most patient consumer.

2.1. The basic model

There are $H \geq 1$ households indexed by $h = 1, \ldots, H$. There is a single commodity available for consumption or investment at each time. At time zero, households are endowed with capital stocks $k^h \geq 0$. Put $K = \sum_h k^h$ and assume $K > 0$. Time is taken in discrete intervals, $t = 1, 2, \ldots$. Let c_t^h, x_t^h denote the consumption and capital stock of household h at time t. Household h has temporal utility function u_h; c_t^h is the argument of u_h. Household h discounts future utilities by the factor δ_h with $0 < \delta_h < 1$. Let $\mathbb{R}_+ = [0, \infty)$, $\mathbb{R}_{++} = (0, \infty)$, and assume the following.

2.1.1. Assumption. *For each* h, $u_h : \mathbb{R}_+ \to \mathbb{R}$ *is* $C^{(2)}$ *on* $\mathbb{R}_{++}$ *with* $u_h' > 0$, $u_h'' < 0$, $\lim_{c\to 0} u'(c) = \infty$, *and* $\lim_{c\to\infty} u'(c) = 0$.

2.1.2. Assumption. $1 > \delta_1 > \delta_2 > \cdots > \delta_H > 0$.

Assumption 2.1.1 is standard; Assumption 2.1.2 defines the first household as the most patient. Every household supplies one unit of labor at each time and labor services are assumed homogeneous.

The production sector is characterized by the one sector neoclassical production function f; inputs precede outputs by one period. Capital is the only variable factor. The technology is described by the standard assumption:

2.1.3. Assumption. $f : \mathbb{R}_+ \to \mathbb{R}_+$, $f(0) = 0$, f *is* $C^{(2)}$ *on* $\mathbb{R}_{++}$, $f' > 0$, $\lim_{x\to 0} f'(x) = \infty$, $\lim_{x\to\infty} f'(x) = 0$, *and* $f'' < 0$.

This assumption implies there is a maximum sustainable stock, a, with $0 < f(a) = a < \infty$. We regard f as derived from a neoclassical constant returns to scale production function with labor input fixed at H.

Let $\{1 + r_t,\ w_t\}$ be a sequence of one period rental factors and wage rates, respectively. The sequences $\{1 + r_t,\ w_t\}$ are always taken to be nonnegative and nonzero. Households are competitive agents and perfectly anticipate the profile of factor returns $\{1 + r_t,\ w_t\}$. Given $\{1 + r_t,\ w_t\}$, h solves

$$P(h) : \ \sup \sum_{t=1}^{\infty} \delta_h^{t-1} u_h(c_t^h)$$

by choice of nonnegative sequences $\{c_t^h,\ x_t^h\}$ satisfying $x_0^h = k^h$ and

$$c_t^h + x_t^h = w_t + (1 + r_t) x_{t-1}^h \quad (t = 1, 2, \ldots). \tag{2.1}$$

The *no arbitrage* necessary conditions for $\{c_t^h,\ x_t^h\}$ to solve $P(h)$ are $c_t^h > 0$ and

$$\delta_h (1 + r_{t+1}) u_h'(c_{t+1}^h) \leq u_h'(c_t^h) \tag{2.2}$$

with equality whenever $x_t^h > 0$.

The production sector is modeled by the myopic profit maximization problem $P(F)$ at each t, where

$$P(F):\ \sup[f(x_{t-1}) - (1+r_t)x_{t-1}]$$

by choice of $x_{t-1} \geq 0$. The assumption of a point–input point–output production function and a competitive rental market implies the maximization of discounted profits is equivalent to $P(F)$.

2.1.4. Definition. *Sequences* $\{1+r_t,\ w_t,\ K_{t-1},\ c_t^h,\ x_{t-1}^h\}$ *constitute a* ***Ramsey Equilibrium*** *provided*:

(E1) *For each* h, $\{c_t^h,\ x_{t-1}^h\}$ *solves* $P(h)$ *given* $\{1+r_t,\ w_t\}$.

(E2) *For each* t, K_{t-1} *solves* $P(F)$ *given* $1+r_t$.

(E3) $Hw_t = f(K_{t-1}) - (1+r_t)K_{t-1} \quad (t = 1, 2, \ldots)$.

(E4) $\sum_h x_{t-1}^h = K_{t-1} \quad (t = 1, 2, \ldots), \quad 0 < K = K_0 \leq a$.

Thus, consumers maximize utility (E1) and producers maximize profits (E2). The labor market clearing condition is expressed in (E3). The capital market clearing condition is (E4). The output market balance follows by combining (E1)–(E4). Hence

$$\sum_h (c_t^h + x_t^h) = f(K_{t-1}). \tag{2.3}$$

Note that equilibrium consumption and capital sequences are bounded from above by the maximum sustainable stock.

If $0 < 1+r_t < \infty$, then Assumption 2.1.3 implies there is a unique positive stock K_{t-1} which solves $P(F)$ at each t; clearly

$$f'(K_{t-1}) = 1 + r_t; \tag{2.4}$$

furthermore, the corresponding $\{w_t\}$ defined by

$$Hw_t = f(K_{t-1}) - (1+r_t)K_{t-1} \tag{2.5}$$

is positive.

2.2. The stationary equilibrium and the turnpike property

An important example of a Ramsey equilibrium is the *stationary Ramsey equilibrium* defined as a time invariant Ramsey equilibrium. Becker [1] showed there is a unique stationary Ramsey equilibrium. Moreover, in that equilibrium configuration only the most patient household had a positive capital stock; all other agents consumed only wages. This steady state equilibrium level of the first agent's capital stock (which is also the aggregate capital stock) is denoted by $\bar{x}$ and satisfies the equation $\delta_1 f'(\bar{x}) = 1$. Becker and Foias [5] found equilibria converged to this steady state when *capital income*, $f'(K)K$, is a strictly increasing function of K. This condition implies every equilibrium exhibits the *turnpike property* whereby all households but the most patient eventually achieve and maintain a zero capital position. If the production function is specified by the Cobb–Douglas form $f(K) = AK^\alpha$, where $A > 0$ and $0 < \alpha < 1$, then capital income equals αAK^α and is increasing in K.

We will consider the cases where the equilibrium profiles with the turnpike property exist and explore the dynamics resulting from that specification of the model. To this end we first state the necessary and sufficient conditions for the existence of such equilibria.[4] A Ramsey equilibrium for an economy evolves in the case where the turnpike property holds according to the equations

$$x_{t+1} + c^1_{t+1} = g(x_t), \qquad c^h_{t+1} = w_{t+1} \quad (h \geq 2), \tag{2.6}$$
$$\delta_1(1 + r_{t+1})u_1'(c^1_{t+1}) = u_1'(c^1_t), \tag{2.7}$$
$$1 + r_{t+1} = f'(x_t), \tag{2.8}$$

where

$$g(x_t) = (1 + r_{t+1})x_t + w_{t+1} = (1 - \frac{1}{H})f'(x_t)x_t + \frac{1}{H}f(x_t). \tag{2.9}$$

We call g the first household's *income function.* Clearly $g(x) < f(x)$ for $0 < x < \infty$. Equation (2.6) expresses the budget constraint for each agent; equation (2.7) is the first agent's no arbitrage condition. The equilibrium condition (E3) may be used to track agents $h \geq 2$ consumption. The no arbitrage condition for those agents is expressed by the inequality given below.

$$\delta_h(1 + r_{t+1})u_h'(c^h_{t+1}) \leq u_h'(c^h_t).$$

Equilibria characterized by the turnpike property satisfy the necessary and sufficient condition reported next.

2.2.1. Proposition. *Suppose* $\{1 + r_t,\ w_t,\ K_{t-1},\ c^h_t,\ x^h_{t-1}\}$ *is a Ramsey equilibrium with the properties* $\liminf_{t\to\infty} x_t > 0$ *and* $\liminf_{t\to\infty} w_t > 0$ *where* $\Sigma^H_{h=1} x^h_t = x_t$. *Suppose further that* $\limsup c^h_t \leq a$ *and* $\limsup_{t\to\infty} x_t \leq a$.

[4] Becker and Foias [5] give a special version of these conditions for the case of a cycle of period 2.

Then a necessary and sufficient condition for $x^h_{t-1} = 0$ and $c^h_t = w_t$ for $h \geq 2$ and all t is that

$$\begin{aligned} c^1_{t+1} + x_{t+1} &= g(x_t) \\ \delta_1 f'(x_t) u'_1(c^1_{t+1}) &= u'_1(c^1_t) \\ \delta_h f'(x_t) u'_h(w_{t+1}) &\leq u'_h(w_t) \quad (h \geq 2), \end{aligned}$$

where $x_t = x^1_t$.

The proof of this proposition is omitted since it is trivial: the transversality and no arbitrage conditions are sufficient for each agent's optimization problem. We observe the equilibrium level of aggregate capital depends only upon the magnitude of δ_1 but not on the values of the other agents' discount factors. Whenever we have a sequence of aggregate capital stocks $\{K_t\}$ with $K_t = x_t$ satisfying the first two conditions of the Turnpike Property Characterization Theorem, then just taking δ_h small enough for $h \geq 2$ will insure the third condition holds and that sequence is a Ramsey equilibrium.

Becker and Foias [5] showed that the turnpike property held if the income function (2.9) was increasing in the capital stock. They also showed that this condition implied every Ramsey equilibrium capital stock converged (eventually monotonically) to the steady state capital $\bar{x}$. We will maintain this monotonicity condition on the income function as a basic assumption throughout this paper. Note that this condition is a relaxation of the capital income strict monotonicity.

2.2.2. Assumption. *The income function g satisfies $g'(x) > 0$ for every $x \in (0, a]$.*

The monotonicity of the income function and the properties of the production function imply there is a capital stock a_1 such that $\bar{x} < a_1 < a$ and $f'(a_1) = 1$. Clearly $g(x) > x$ for $0 < x < a_1$, and $g(a) < f(a) = a$. Moreover, there is a unique $b \in (a_1, a)$ such that $g(b) = b$, and $g(x) > x$ for $0 < x < b$ and $g(x) < x$ for $b < x \leq a$.

3. Ramsey trajectories

We describe Ramsey trajectories in this section. These are sequences of capital stocks that correspond to equilibrium capital sequences. They are found by considering a dynamical system based on the no arbitrage equation for the first household.

3.1. Definitions and basic concepts

The no arbitrage equation for the first household can be written (due to $u_1'' < 0$) as

$$u_1'(y) = \delta_1 f'(x) u_1'(F(x,y)) \quad \text{for} \quad 0 < x \leq a \quad \text{and} \quad 0 < y,$$

for some differentiable function $F(x,y)$, where $F_y(x,y) > 0$ and $F_x(x,y) < 0$ are the partial derivatives of F. The function F tells the first household how much to consume in the next period given the current period's capital and consumption. That is, $F(x_t, c_t^1) = c_{t+1}^1$ when the first household owns the economy's capital stocks at every time. We can use this formulation of the no arbitrage equation to define a pair of second order difference equations — the first's variables are consumption and capital while the second is defined entirely in terms of capital stocks.

Let

$$\begin{aligned} c_{t+1}^1 &= F(x_t, c_t^1) \\ x_{t+1} &= g(x_t) - c_t^1 \end{aligned}$$

define a map from (x_t, c_t^1) to (x_{t+1}, c_{t+1}^1). We use the compact notation $\Phi(x_t, c_t^1) = (x_{t+1}, c_{t+1}^1)$ to describe this system for $t \geq 1$. We can also write this as the mapping

$$\begin{aligned} y_{new} &= F(x,y) \\ x_{new} &= g(x) - F(x,y), \end{aligned}$$

for $0 < x \leq a$ and $F(x,y) < g(x)$. Now rewrite this system one more time. Let

$$\begin{aligned} x_{0new} &= x_1 \\ x_{1new} &= G(x_0, x_1), \end{aligned} \tag{3.1}$$

where
$G(x_0, x_1) \equiv g(x_1) - F(x_1, g(x_0) - x_1)$, $G_{x_0} = -F_y(x_1, g(x_0) - x_1) g'(x_0) < 0$, $G_{x_1} = g'(x_1) - F_x(x_1, g(x_0) - x_1) + F_y(x_1, g(x_0) - x_1) > 0$, and $(x_0, x_1) \in \Omega$, where

$$\Omega = \{(x_0, x_1) : \quad 0 < x_0 < a; \; 0 < x_1 < g(x_0)\}.$$

Note that $G_{x_0} < 0$ on Ω. This map gives rise to the second order difference equation

$$(x_{t+1}, x_{t+2}) = \Psi(x_t, x_{t+1}) \quad \text{for all } t \geq 0, \tag{3.2}$$

where $\Psi(x_0, x_1) = (x_{0new}, x_{1new})$.

A *Ramsey trajectory* is a sequence $\{x_t\}_{t=1}^{\infty}$ such that $(x_t, x_{t+1}) \in \Omega$ for all $t \geq 1$ and $\{(x_{t+1}, x_{t+2})\}_{t=0}^{\infty} = \{\Psi^t(x_0, x_1)\}_{t=0}^{\infty}$, where Ψ^t is the t–th iterate of Ψ.

3.1.1. Remark. *If a Ramsey equilibrium sequence satisfies the turnpike property, then its associated capital sequence is a Ramsey trajectory. In particular, the steady state Ramsey equilibrium capital sequence $\{\bar{x}, \bar{x}, \dots\}$ is a Ramsey trajectory.*

The next are elementary observations about the mapping Ψ.

3.1.2. Remark. *If $\Psi(x_0, x_1) = (x_0, x_1)$, then $x_0 = x_1 = \bar{x}$. That is, a fixed point of Ψ is the steady state Ramsey equilibrium capital stock.*

3.1.3. Remark. *If $\Psi(x_0, x_1) = \Psi(x_0', x_1')$, then $(x_0, x_1) = (x_0', x_1')$. That is, Ψ is a one-to-one map.*

3.2. Analysis of Ramsey trajectories

We will show here that a Ramsey trajectory converges to the steady state capital stock under our basic assumptions. This result is a slight improvement of the one reported in Becker and Foias [6].

3.2.1. Theorem. *If a Ramsey trajectory $\{x_t\}_{t=0}^{\infty}$ is generated by a Ramsey equilibrium having the turnpike property, then $\lim_{t\to\infty} x_t = \bar{x}$.*

The proof is developed in a series of lemmas.

We begin by noticing that if $\theta = \{x_t\}_{t=0}^{\infty}$ is a Ramsey trajectory, then $0 < x_{t+1} < g(x_t)$ for all $t \geq 0$. Define the set Ω_{11} as follows:

$$\Omega_{11} = \{(x_0, x_1) : \quad x_1 \leq \bar{x} < x_0 \quad \text{or} \quad x_1 \leq x_0 < \bar{x} \quad \text{or} \quad x_1 < x_0 = \bar{x}\}.$$

3.2.2. Lemma. $\{(x_t, x_{t+1}) : x_t \in \theta\} \cap \Omega_{11} = \emptyset$.

Proof. If $(x_t, x_{t+1}) \in \Omega_{11}$ for some $x_t \in \theta$ then either

$$\text{(i) } x_{t+1} \leq \bar{x} < x_t \quad \text{or} \quad \text{(ii) } x_{t+1} \leq x_t < \bar{x} \quad \text{or} \quad \text{(iii) } x_{t+1} < x_t = \bar{x}.$$

Since

$$\begin{aligned} x_{t+2} &= g(x_t) - F(x_{t+1}, g(x_t) - x_{t+1}) \leq g(x_{t+1}) - F(\bar{x}, g(x_t) - x_{t+1}) \\ &= x_{t+1} - [g(x_t) - g(x_{t+1})] \leq x_{t+1}, \end{aligned}$$

where the first (respectively, last) inequality is strict in the case (ii) (respectively, in the cases (i) and (ii)). Thus

$$x_{t+2} < x_{t+1} \leq \bar{x}.$$

It follows that $(x_{t+1}, x_{t+2}) \in \Omega_{11}$ too and therefore $\{(x_{t+s+1}, x_{t+s})\}_{s\geq 0} \in \Omega_{11}$, and

$$\bar{x} \geq x_{t+1} > x_{t+2} > \cdots > x_{t+s} > \cdots .$$

Hence

$$x_\infty = \lim_{s\to\infty} x_s$$

exists and $x_\infty \in [0, \bar{x})$. If $x_\infty > 0$, then $\Psi(x_\infty, x_\infty) = (x_\infty, x_\infty)$, hence $x_\infty = \bar{x}$, a contradiction. So $x_\infty = 0$. Since, for $k \geq 1$,

$$x_{t+k+1} \leq x_{t+k-1} - [g(x_{t+k-1}) - g(x_{t+k})],$$

we obtain (by summing the above for $k = 1, 2, \ldots, n$ and then letting $n \to \infty$)

$$0 \leftarrow x_{t+n+1} \leq x_{t+1} - [g(x_{t1}) - g(x_{t+n})] \to x_{t+1} - g(x_t),$$

that is, $(x_t, x_{t+1}) \notin \Omega$, a contradiction. ■

3.2.3. Lemma. *If* $\{(x_t, x_{t+1}) : x_t \in \theta\} \subset \Omega_{12} = \{(y_0, y_1) \in \Omega : y_0 \leq \bar{x}\}$, *then* $x_0 \leq x_1 \leq \cdots \to \bar{x}$.

Proof. In the present situation, due to Lemma 3.2.2, we have $x_t \leq x_{t+1}$ for all $t \geq 0$, thus $\lim_{s\to\infty} x_s = x_\infty$ exists and $x_\infty \in (0, \bar{x}]$. Since $\Psi(x_\infty, x_\infty) = (x_\infty, x_\infty)$ we must have $x_\infty = \bar{x}$. ■

For the next lemma we define a set Ω_{13} as follows:

$$\Omega_{13} = \{(x_0, x_1) : \quad (x_0, x_1) \in \Omega,\ x_1 \geq x_0 \geq \bar{x},\ (x_0, x_1) \neq (\bar{x}, \bar{x})\}.$$

3.2.4. Lemma.

(i) $\Psi(\Omega_{13}) \subset \Omega_{13}$ *and* $\Psi^n(x_0, x_1) = (\Psi \circ \cdots \circ \Psi)(x_0, x_1) \to (b, b)$.

(ii) *If* $\{(x_t, x_{t+1}) : x_t \in \theta\} \cap \Omega_{13} \neq \emptyset$, *then eventually* $x_t < x_{t+1} < \cdots \to b$.

Proof. If $(x_0, x_1) \in \Omega_{13}$, then $(x_1, x_2) = \Psi(x_0, x_1)$ satisfies

$$\begin{aligned} x_2 &= g(x_1) - F(x_1, g(x_0) - x_1) \geq g(x_1) - F(\bar{x}, g(x_0) - x_1) \\ &= x_1 + g(x_1) - g(x_2) \geq x_1 \end{aligned}$$

and therefore $\bar{x} \leq x_1 \leq x_2 < g(x_1)$. So $(x_1, x_2) \in \Omega$, $x_1 \leq x_2$, and $(x_1, x_2) \neq (\bar{x}, \bar{x})$, hence $(x_1, x_2) \in \Omega_{13}$. Note that $x_2 > \bar{x}$. This establishes statement (i).

Concerning (ii), note that if $(x_t, x_{t+1}) \in \Omega_{13}$ for some $x_t \in \theta$, then $(x_{t+k}, x_{t+k+1}) \in \Omega_{13}$ for all $k \geq 0$, by virtue of (i). Thus $x_{t+1} > \bar{x}$ and

$$\bar{x} \leq x_t \leq x_{t+1} \leq \cdots \leq x_{t+k} \leq \cdots \leq b.$$

Therefore, $x_\infty = \lim_{s\to\infty} x_s$ exists and $x_\infty \leq b$. If $x_\infty < b$, then $(x_\infty, x_\infty) \in \Omega$ and $\Psi(x_\infty, x_\infty) = (x_\infty, x_\infty)$, which implies $x_\infty = \bar{x}$, a contradiction, since $x_\infty \geq x_{t+1} > \bar{x}$. So $x_\infty = b$. ∎

3.2.5. Lemma. *Let $\Omega_{14} = \Omega \setminus (\{(\bar{x}, \bar{x})\} \cup \Omega_{11} \cup \Omega_{12} \cup \Omega_{13})$ and assume $\{(x_t, x_{t+1}) : x_t \in \theta\} \subset \Omega_{14}$. Then*

$$x_0 > x_1 > x_2 \cdots \geq \bar{x},$$

so there exists $x_\infty = \lim_{t\to\infty} x_t$. Moreover, either $x_\infty = \bar{x}$, or $x_\infty = b$.

Proof. It is obvious that

$$\Omega_{14} = \{(x_0, x_1) \in \Omega : \bar{x} < x_1 < x_0\}.$$

Therefore, in the present situation

$$x_t > x_{t+1} > \bar{x} \quad \text{for all } t \geq 0.$$

Let $x_\infty = \lim_{s\to\infty} x_s$. If $(x_\infty, x_\infty) \in \Omega$ then, again since $\Psi(x_\infty, x_\infty) = (x_\infty, x_\infty)$, we have $x_\infty = \bar{x}$. If $(x_\infty, x_\infty) \notin \Omega$, then $x_\infty = g(x_\infty)$, and thus $x_\infty = b$. ∎

Taken together the last three lemmas imply the following.

3.2.6. Proposition. *If $\theta = \{x_t\}_{t=0}^\infty$ is a Ramsey trajectory, then either $x_t \to \bar{x}$ or $x_t \to b$. In both cases the convergence is eventually monotonic.*

The proof of Theorem 3.2.1 follows once we show the case where $x_t \to b$ is not a Ramsey equilibrium.

3.2.7. Lemma. *If $\theta = \{x_t\}_{t=0}^\infty$ is a Ramsey trajectory and $x_t \to b$, then $c_t^1 = g(x_{t-1}) - x_t \to 0$. This trajectory does not correspond to a Ramsey equilibrium.*

Proof. The assumption that $x_t \to b$ implies $g(x_{t-1}) - x_t \to g(b) - b = 0$. Hence, $\lim_{t\to\infty} c_t^1 = 0$. But according to Corollary 1 [5], in a Ramsey equilibrium $\limsup_{t\to\infty} c_t^h > 0$ for each household. Therefore θ cannot come from a Ramsey equilibrium. ∎

Proposition 3.2.6 and Lemma 3.2.7 prove Theorem 3.2.1.

3.2.8. Remark. *The Ramsey trajectories $\theta = \{x_t\}_{t=0}^\infty$ which come from Ramsey equilibria are exactly those for which $x_t \to \bar{x}$.*

4. Existence of the invariant manifold

The steady state capital stock is a saddle point in the phase space determined by the domain of Ψ. This implies the existence of a local stable manifold defined on a neighborhood of the point $(\bar{x}, \bar{x})$. This local stable manifold is the graph of a function defined on an interval $(\bar{x}-\epsilon,\ \bar{x}+\epsilon)$. We also show that this function can be extended to $(0, a)$ and its graph is the global invariant manifold for the dynamical system defined by Ψ.

4.1. The local stable manifold

The construction of the local stable manifold follows from an analysis of the linear approximation system defined on a neighborhood of the steady state $(\bar{x}, \bar{x})$. This part of the analysis reworks results in [6].

The derivative of Ψ at (x_0, x_1) can be written as the matrix

$$D\Psi(x_0, x_1) = \begin{bmatrix} 0 & 1 \\ G_{x_0} & G_{x_1} \end{bmatrix} = \begin{bmatrix} 0 & 1 \\ -F_y g'(x_0) & g'(x_1) - F_{x_0} + F_y \end{bmatrix}, \tag{4.1}$$

where $x = x_0$, $y = g(x_0) - x_1$, $G_{x_i} = \partial G(x_0, x_1)/\partial x_i$ $(i = 0, 1)$, and $F_x = \partial F(x,y)/\partial x$, $F_y = \partial F(x,y)/\partial y$. The determinant of (4.1) is positive on Ω, hence Ψ is locally invertible. In fact, it is a diffeomorphism from Ω onto $\Psi(\Omega)$. In particular, at the steady state $(\bar{x}, \bar{x})$ we see

$$D\Psi(\bar{x}, \bar{x}) = \begin{bmatrix} 0 & 1 \\ -\gamma & \gamma+\eta+1 \end{bmatrix}, \tag{4.2}$$

where $\gamma = g'(\bar{x})$, $\eta = -F_{x_0}(\bar{x}, \bar{y})$ for $\bar{y} = g(\bar{x}) - \bar{x}$.

The eigenvalues λ of (4.2) are given by

$$\lambda^2 - (\gamma+\eta+1)\lambda + \gamma = 0$$

and consequently they are real. We let λ_1 denote the smaller one and λ_2 the other. In [6] we showed that

$$0 < \lambda_1 < 1 < \lambda_2, \qquad \lambda_1 + \lambda_2 > 1. \tag{4.3}$$

The corresponding eigenvectors are $(1,\ \lambda_1)$ and $(1,\ \lambda_2)$. Thus $(\bar{x}, \bar{x})$ is a saddle point with two locally invariant manifolds given by the graphs of the functions $\psi_i : (\bar{x}-\epsilon,\ \bar{x}+\epsilon) \to \mathbb{R}$ $(i = 1, 2)$ for $\epsilon > 0$ small enough. The functions ψ_i are C^1 and

$$\psi_i(\bar{x}) = \bar{x}, \qquad \psi_i'(\bar{x}) = \lambda_i.$$

If ϵ is small enough, then the only points $(x_0, x_1) \in (\bar{x} - \epsilon,\ \bar{x} + \epsilon) \times (\bar{x} - \epsilon,\ \bar{x} + \epsilon)$ for which $\Psi^t(x_0, x_1) \to (\bar{x}, \bar{x})$ as $t \to \infty$ are those on the graph of ψ_1. By decreasing ϵ further if necessary we have

$$0 < \psi_1'(x) < 1 \qquad \text{and} \qquad 1 < \psi_2'(x) < \infty$$

for each $x \in (\bar{x} - \epsilon,\ \bar{x} + \epsilon)$. It follows that

$$\begin{aligned} \psi_1(x) > x \quad & \text{for } x \in (\bar{x} - \epsilon,\ \bar{x}) \\ \psi_1(x) < x \quad & \text{for } x \in (\bar{x},\ \bar{x} + \epsilon), \end{aligned}$$

and

$$\begin{aligned} \psi_2(x) < x \quad & \text{for } x \in (\bar{x} - \epsilon,\ \bar{x}) \\ \psi_2(x) > x \quad & \text{for } x \in (\bar{x},\ \bar{x} + \epsilon). \end{aligned}$$

Let $\psi_i^2(x) = \psi_i(\psi_i(x))$. The invariance properties mean that

$$\begin{aligned} \Psi(x, \psi_1(x)) = (\psi_1(x), \psi_1^2(x)) \quad & \text{for all } x \in (\bar{x} - \epsilon,\ \bar{x} + \epsilon) \\ \Psi(x, \psi_2(x)) = (\psi_2(x), \psi_2^2(x)) \quad & \text{for all } x \in (\bar{x} - \epsilon,\ \bar{x} + \epsilon), \end{aligned}$$

such that $\psi_2(x) \in (\bar{x} - \epsilon,\ \bar{x} + \epsilon)$. Thus, if the initial capital stock $x_0 \in (\bar{x} - \epsilon,\ \bar{x} + \epsilon)$ and x_1 is chosen so that $\psi_1(x_0) = x_1$, then $x_1 \in (\bar{x} - \epsilon,\ \bar{x} + \epsilon)$. Continuing this, we see that $x_{t+1} = \psi_1(x_t) \in (\bar{x} - \epsilon,\ \bar{x} + \epsilon)$ for all $t \geq 0$. It follows that the iterates of ψ_1 form a Ramsey trajectory which converges to the steady state capital stock. The properties of ψ_1 imply the convergence is monotonic.

4.1.1. Proposition. *$\{\psi_1^t(x_0)\}_{t=0}^{\infty}$ is a Ramsey trajectory and $\lim_{t\to\infty} \psi_1^t(x_0) = \bar{x}$. Moreover, the convergence is monotonic: $\psi_1^t(x_0) \uparrow \bar{x}$ if $x_0 \in (\bar{x} - \epsilon,\ \bar{x})$ and $\psi_1^t(x_0) \downarrow \bar{x}$ if $x_0 \in (\bar{x},\ \bar{x} + \epsilon)$.*

4.2. The global invariant manifold

We extend the domain of ψ_1 to a maximal interval in $(0, a)$ and show the graph of this function is the invariant manifold for the dynamical system defined by Ψ.

Let $I = (\bar{x} - \epsilon,\ \bar{x} + \epsilon) \subset (0, a)$ be the *largest* interval such that there exists $\psi : I \to \mathbb{R}$ with the properties

(S1) $\psi(x) < \psi(x')$ for $x, x' \in I$, $x < x'$;

(S2) ψ is continuous;

(S3) $\psi(x) = \psi_1(x)$ for all $x \in (\bar{x} - \epsilon,\ \bar{x} + \epsilon)$;

(S4) $\psi(x) < x$ for $x > \bar{x}$ and $\psi(x) > x$ for $x < \bar{x}$;

(S5) $\psi(\psi(x)) = G(x, \psi(x))$ for all $x \in I$.

4.2.1. Lemma. *A maximal interval satisfying* (S1)–(S5) *exists.*

Proof. For intervals I and $\widetilde{I}$ satisfying properties (S1)–(S5) (except possibly being maximal) define a binary relation $\prec$ by letting $(I, \psi) \prec (\widetilde{I}, \widetilde{\psi})$ if $I \subset \widetilde{I}$ and $\psi = \widetilde{\psi}\big|_I$ where $\widetilde{\psi}$ is the function associated with $\widetilde{I}$. Apply Zorn's Lemma to obtain a maximal pair (I, ψ). ■

4.2.2. Lemma. *The maximal pair* (I, ψ) *is uniquely determined.*

Proof. Denote by $I = (\bar{x} - \epsilon,\ \bar{x} + \epsilon)$. Let I_u be the maximal interval containing $(\bar{x} - \epsilon,\ \bar{x} + \epsilon)$ where $\psi_u = \psi\big|_{I_u}$ is uniquely determined. Let $I_u = (\bar{x} - \eta_-,\ \bar{x} + \eta_+)$ for $\eta_- \geq \epsilon_-$ and $\eta_+ \leq \epsilon_+$. If one of the inequalities is strict, say $\eta_- < \epsilon_-$, then for $x \leq \bar{x} - \eta_-$ but near enough to $\bar{x} - \eta_-$, we will have $\bar{x} - \eta_- < \psi(x) < \psi_u(\psi(x))$,

$$(\psi(x), \psi_u(\psi(x))) = \Psi(x, \psi(x)).$$

By Remark 3.1.3 the inverse ψ^{-1} of ψ on a small interval of $[\bar{x} - \eta_-,\ \bar{x} + \eta_+]$ is uniquely determined and therefore ψ on $[\psi^{-1}(\bar{x} - \eta_-),\ \psi^{-1}(\bar{x} + \eta_+)]$ is also uniquely determined. This contradicts the assumed maximality of I_u. ■

The map ψ defined on the maximal interval I has several interesting properties.

4.2.3. Remark. *If* $(x_0, x_1) \in \operatorname{graph} \psi$, *then* $\lim_{t \to \infty} \Psi^t(x_0, x_1) = (\bar{x}, \bar{x})$.

This remark, the observation that Ψ is a diffeomorphism from Ω onto $\Psi(\Omega)$ and the fact that $\psi_1 \in C^1$ exists implies the next result.

4.2.4. Remark. *The function* ψ *defined on the maximal interval* I *is* C^1.

In fact we can say a little more, namely

$$\psi'(x) > 0 \quad \text{for all } x \in I. \tag{4.4}$$

Indeed from $G(x, y) - \psi(y) = 0$ for $y = \psi(x)$, $x \in I$, we infer by differentiation that

$$(G_y(x, y) - \psi'(y))\big|_{y = \psi(x)} \psi'(x) = F_\eta(\xi, \eta)\big|_{\substack{\xi = \psi(x), \\ \eta = g(x) - \psi(x)}} \qquad g'(x) > 0. \tag{4.5}$$

So $\psi'(x) \neq 0$; since $\psi'(x) \geq 0$, we obtain (4.4).

Suppose that (ψ, I) is the maximal pair and $I = (x_-, x_+)$. The following limits clearly exist:

$$y_- = \lim_{x \searrow x_-} \psi(x), \qquad y_+ = \lim_{x \nearrow x_+} \psi(x).$$

Let $\partial\Omega$ denote the boundary of Ω.

4.2.5. Lemma. *$(x_\pm, y_\pm) \in \partial\Omega$, that is, $g(x_-) = y_-$, and either $g(x_+) = y_+$ or $x_+ = a$.*

Proof. We notice first that the equation $G(x,y) - \psi(y) = 0$ has a solution $y = \psi(x)$ for each $x \in I$.

If (any of) $(x_\pm, y_\pm)$ is in Ω, then from (4.5) we infer, by continuity, that $G_y(x,y) - \psi'(y) > 0$ at (that) $(x_\pm, y_\pm)$.

The Implicit Function Theorem implies that if $(x_\pm, y_\pm) \in \Omega$, the equation $G(x,y) - \psi(y) = 0$ has one and only one solution $y = \theta(x)$ for each x near $x_\pm$ such that $\theta(x_\pm) = y_\pm$. Clearly $\theta(x) = \psi(x)$ when that x is in I. We can now extend the definition of ψ by putting $\theta(x) = \psi(x)$ for x near $x_\pm$ outside I. This contradicts the maximality of I, so $(x_\pm, y_\pm) \in \partial\Omega$. ∎

4.2.6. Lemma. *$x_- = 0$, hence*

$$\Psi(\text{graph}(\psi|_{(0,\bar{x}]})) = \text{graph}(\psi|_{(0,\bar{x}]}). \tag{4.6}$$

Proof. For $x \downarrow x_-$ we have $\psi(x) \downarrow y_-$. If $x_- > 0$, then

$$\psi(y_-) \leftarrow G(x, \psi(x)) = g(\psi(x)) - F(\psi(x), g(x) - \psi(x)) \to g(y_-)$$

since $\psi(x) \to y_- > 0$ and $g(x) - \psi(x) \to g(x_-) - y_- = 0$, by Lemma 4.2.5. But $y_- = g(x_-) > x_-$ so $(y_-, \psi(y_-)) \in \text{graph}\,\psi$ in contradiction to the fact that $(y_-, \psi(y_-)) = (y_1, g(y_-)) \notin \Omega$. This establishes that $x_- = 0$. The equality (4.6) follows now from the fact that $\psi((0,\bar{x}]) = (0,\bar{x}]$ and thus

$$\begin{aligned}\Psi(\text{graph}(\psi|_{(0,\bar{x}]})) &= \{(y, \psi(y)) : y \in \psi((0,\bar{x}])\} \\ &= \{(y, \psi(y)) : y \in (0,\bar{x}]\} = \text{graph}(\psi|_{(0,\bar{x}]}).\end{aligned}$$

∎

4.2.7. Lemma. *If $\theta = \{x_t\}_{t=0}^{\infty}$ is a Ramsey trajectory such that $x_t \nearrow \bar{x}$, then $(x_t, x_{t+1}) \in \text{graph}\,\psi$ for all $t \geq 0$. Hence, $\psi(x_t) = x_{t+1}$ for all $t \geq 0$.*

Proof. Let $(x_0, x_1) \in \Omega$ and $x_0 < x_1$ be such that $\Psi^t(x_0, x_1) = (x_t, x_{t+1}) \to (\bar{x}, \bar{x})$. Then eventually $(x_t, x_{t+1}) \in \text{graph}\,\psi_1 \subset \text{graph}\,\psi$ (where ψ is as above). We can take $t = 1$ without loss of generality, that is

$$\Psi(x_0, x_1) = (x_1, \psi(x_1)), \qquad \text{and} \qquad x_0 < x_1 < \cdots \nearrow \bar{x}.$$

By Lemma 4.2.6, $(x_0, x_1) = (\psi^{-1}(x_1), x_1) \in \text{graph}\,\psi$. ∎

Next we define the region Ω_{24} as the set:

$$\Omega_{24} = \{(x, x') \in \Omega : \quad x > \bar{x}, \quad x' < g(x), \quad x' > g(x) - [g(\bar{x}) - \bar{x}]\}.$$

4.2.8. Lemma. *If* $(x_0, x_1) \in \Omega_{24}$, *then* $\Psi(x_0, x_1) \in \Omega_{24}$.

Proof. Note that $x_1 > \bar{x}$, and if $\Psi(x_0, x_1) = (x_1, x_2)$, then $x_2 < g(x_1)$ by the definition of Ψ. But

$$\begin{aligned} x_2 &= g(x_1) - F(x_1, g(x_0) - x_1) \\ &\geq g(x_1) - F(\bar{x}, g(x_0) - x_1) \\ &= g(x_1) - [g(x_0) - x_1], \end{aligned}$$

so $x_2 - g(x_1) > x_1 - g(x_0) > -[g(\bar{x}) - \bar{x}]$. Thus $(x_1, x_2) \in \Omega_{24}$. ∎

The next result shows that $x_- = 0$ and $x_+ = a$, hence $I = (0, a)$. We will use the following fact to show this.

4.2.9. Remark. *A Ramsey trajectory* $\theta = \{x_t\}_{t=1}^{\infty}$ *such that* (x_0, x_1) *is in* Ω_{24} *converges to* (b, b). *In particular*, $(\operatorname{graph} \psi) \cap \Omega_{24} = \emptyset$.

Proof. A simple computation shows that $\lambda_1 < \gamma = g'(\bar{x})$. If $x_t \to \bar{x}$, then $x_{t+1} = \psi(x_t)$ for t large enough and so $(x_t, x_{t+1}) \notin \Omega_{24}$, contradicting Lemma 4.2.8. The conclusion now follows from Proposition 3.2.6. ∎

4.2.10. Lemma. *The maximal interval* $I = (0, a]$, *and* $G(x, \psi(x)) = \psi(\psi(x))$ *for all* $x \in (0, a]$.

Proof. Due to Lemma 4.2.6, it remains to prove only that $x_+ = a$. If $y_+ = g(x_+)$ then as in the proof of Lemma 4.2.6, we obtain $g(y_+) = \psi(x_+)$. But, $(y_+, \psi(y_+)) \in \operatorname{graph} \psi$, and graph ψ is disjoint from the closure of Ω_{24} containing $(y_+, g(y_+)) = (y_+, \psi(y_+))$, a contradiction. By virtue of Lemma 4.2.5, y_+ equals a. It is clear that we can define $\psi(a)$ by continuity. Therefore from now on we will consider ψ defined on $(0, a]$. Finally, the equality in the lemma plainly follows from the invariance under Ψ of the graph ψ. ∎

Now let $(x_0, x_1) \in \Omega$ with $x_1 < x_0$ such that $\Psi^t(x_0, x_1) = (x_t, x_{t+1}) \to (\bar{x}, \bar{x})$. Then eventually $(x_t, x_{t+1}) \in \operatorname{graph} \psi$. Without loss of generality we may take $t = 1$ so that $\Psi(x_0, x_1) = (x_1, \psi(x_1))$ and $x_0 > x_1 > \cdots \searrow \bar{x}$. If $x_1 \leq \psi(a)$, then $(x_0, x_1) = (\psi^{-1}(x_1), x_1) \in \operatorname{graph} \psi$, so if $(x_0, x_1) \in \Omega$, $x_1 \leq \psi(a)$, $x_1 < x_0$, and $\Psi^t(x_0, x_1) = (x_t, x_{t+1}) \to (\bar{x}, \bar{x})$ as $t \to \infty$, then $(x_0, x_1) \in \operatorname{graph} \Psi$.

Consider now the case $\psi(a) < x_1 < x_0 \leq a$, $\Psi(x_0, x_1) = (x_1, \psi(x_1))$. Note that

$$G(x_0, \psi(x_0)) = \psi(\psi(x_0)) \leq \psi(\psi(a)) < \psi(x_1) = G(x_0, x_1)$$

and that $G_y > 0$. So, $x_1 > \psi(x_0)$. So, $x_1 > \psi(x_0)$. This leads to the following.

4.2.11. Remark. *If for any $(x_0, x_1) \in \Omega$ such that $\Psi^t(x_0, x_1) = (x_t, x_{t+1}) \to (\bar{x}, \bar{x})$ as $t \to \infty$, then $(x_0, x_1) \in \operatorname{graph} \psi$, or there is a t_0 such that*

$$\begin{aligned} &\psi(a) < x_{t_0+1} < x_{t_0} \le a, \\ &x_{t+1} = \psi(x_t) \quad \text{for all } t > t_0, \\ &x_{t+1} > \psi(x_t) \quad \text{for all } t = 0, 1, \ldots, t_0. \end{aligned}$$

We conclude this section by noting that whenever $y < \psi(x)$, then the decision maker's capital in one period will be smaller than it should be since the policy function tells the planner to choose $\psi(x)$. Thus, the decision maker underaccumulates capital. This observation is recorded as the following proposition.

4.2.12. Proposition. *If $0 < x \le a$, $0 < y < \psi(x)$, then $G(x, y) < \psi(y)$.*

Proof. From $G(x, y) - \psi(\psi(x)) = 0$ for all $x \in (0, a]$, it follows that (by taking derivatives)

$$G_y(x, y) - \psi'(y) > 0 \quad \text{for} \quad y = \psi(x) \quad \text{and for all } x \in (0, a].$$

Thus there exists an $\epsilon(x) > 0$ such that

$$G(x, y) - \psi(y) < 0 \quad \text{for} \quad y \in (\psi(x) - \epsilon(x),\ \psi(x)) \quad \text{and for all } x \in (0, a].$$

If for some $x_0 \in (0, a]$ and $0 < y < \psi(x_0)$ we have $G(x_0, y) > \psi(y)$, then there exists y_0 with $y < y_0 < \psi(x_0)$ such that $G(x_0, y_0) = \psi(y_0)$. Then

$$\Psi(x_0, y_0) = (y_0, G(x_0, y_0)) = (y, \psi(y_0)) \in \operatorname{graph} \psi.$$

If $x_0 \in (0, \psi(a)]$, then this is impossible by Lemma 4.2.6 and Remark 4.2.9. If $x_0 \in (\psi(a), a]$, then this is also impossible by Remark 4.2.9. ∎

This implies the following result.

4.2.13. Theorem. *The policy function $\psi(x)$ solves the following programming problem:*

$$\begin{aligned} \psi(x) &= \inf \{y : G(x, y) \ge \psi(y)\} \\ &= \min \{y : G(x, y) \ge \psi(y)\}. \end{aligned} \tag{4.7}$$

We will show in the next section how this problem leads to a new construction of the global invariant manifold based on economic considerations.

We also emphasize that this invariant manifold is not the stable manifold for iterations of the map Ψ may converge to other points in Ω.[5] However, this invariant manifold does coincide with the local stable manifold associated to the steady state $(\bar{x}, \bar{x})$ in a neighborhood of that point.

[5] Lemma 4.2.8 and Remark 4.2.9 imply that (b, b) is the limit of a sequence found by iterating a point in Ω_{24}.

5. An implicit programming construction of the global stable manifold

The functional equation (4.7) describes the Euler or no arbitrage condition for the most patient household in terms of the total capital stocks when that person always has all the capital. The map ψ gives the next period's capital stocks given the current stocks that are consistent with the no arbitrage condition. It should be noted that ψ is similar to the policy function found by dynamic programming although the most patient household is *not* solving a first best optimum problem for the economy. The equilibrium calculated by iterating ψ is stationary in the sense that what ψ tells the first agent to do is independent of calendar time.

Equation (4.7) is a new type of functional equation that represents the global invariant manifold of the Ramsey dynamical system. In fact, this equation represents a new form of the *implicit programming problem.*[6] The programming problem defined by the right–hand side of (4.7) has the choice of y constrained by the value of the program, ψ, which is in turn clearly determined by solving the problem. The solution to this functional equation is an economic based method for finding the invariant manifold of the Ramsey dynamical system as a type of variational principle.[7]

The implicit program defined in (4.7) can be interpreted as follows. The constraint $G(x, y) \geq \psi(y)$ expresses the requirement that the stocks realized over two periods following the Euler equations with the given initial stocks when y is chosen in the first period are not smaller than what the "policy" function ψ would select for the second period stocks given the input y. This is a consistency requirement on ψ; the programming problem on the right–hand side of (4.7) will force the constraint to bind as an equality at the optimum value. The appearance of ψ as both the value function and component of the constraint reflects the second order character of the Euler equation. Some further intuition for the use of the min operator in (4.7) is that the agent chooses the smallest capital stock for tomorrow consistent with the Euler equation in order to provide maximal consumption within the period. Iteration of this improvement procedure yields a nondecreasing sequence of policy functions

[6]See Becker and Foias [4] and Becker and Boyd [2] for a discussion of the implicit programming problem that describes the steady states of discounted optimal growth models. Also see Kehoe, Levine and Romer [10] for a related family of problems with side constraints.

[7]Coleman [9] gives a construction of an equilibrium consumption function using an operator approach that is similar in spirit to our problem. He exploits both the no arbitrage equation and the envelope properties of the value function for an underlying dynamic programming problem in his construction. It is of some interest to determine the specific relationship between our approach and his operator methods.

which converges to ψ.

The process of iteration and the proof that it yields ψ in the limit is really a proof that the minimum gain operator introduced below has a fixed point. The argument proceeds in two steps. First, we show the minimum gain operator is increasing. The policy improvement monotonicity argument for convergence is basically showing that Amann's fixed point theorem applies to the prescribed class of functions. Since we carry out all the necessary calculations directly we do not explicitly draw on Amann's result. The final step is to show the uniqueness of the fixed point within our class of functions — a fact that cannot be proven from the monotonicity properties of the minimum gain operator alone.

5.1. The minimum gain operator

In order to show how the function ψ can be obtained directly by an approximation procedure with natural economic undertones, we introduce the class of functions from which a solution to (4.7) is sought. We are interested in those functions $\theta : (0, a] \to (0, a]$ that satisfy the following properties:

(C1) θ is continuous;

(C2) $\theta(x) \gtrless x$ according to $x \lessgtr \bar{x}$;

(C3) $\theta(x) < g(x)$ for all x.

The basic idea is to construct ψ as the limit of an increasing sequence of trial policy functions found by iterating the functional equation for our implicit programming problem. We will carry out this iteration for a particular initial seed function satisfying (C1)–(C3). To this end we introduce the *minimum gain operator*. It is the operator T defined according to the equation:

$$T\xi(x) = \inf \{y : G(x, y) \geq \xi(x)\}, \tag{5.1}$$

where $\xi : (0, a] \to (0, a]$ satisfies $\xi \leq \psi$. This operator's name is suggested by the fact that it is associated with the smallest consumption level consistent with continuing iteration of the Euler equation for the first agent assuming that person's consumption decisions are following a prescribed trial policy function.

We record an elementary monotonicity property of the function $\eta(x) = \inf \{y : G(x, y) \geq \xi(y)\}$ for a given function ξ that does not necessarily satisfy conditions (C1)–(C2).

5.1.1. Remark. *If* $\eta(x) = \inf \{y : G(x, y) \geq \xi(y)\}$, *then* $x_1 \leq x_2$ *implies* $\eta(x_1) \leq \eta(x_2)$. *If* ξ *is continuous, then* $x_1 < x_2$ *implies* $\eta(x_1) < \eta(x_2)$.

Proof. The first part follows since $G(x_1, y) \geq G(x_2, y)$ for all $0 < y < g(x)$. If ξ is continuous, then

$$\xi(T\xi(x)) = G(x, T\xi(x))$$

for all $x \in (0, a]$. In particular, if $x_1 < x_2$ and $T\xi(x_1) = T\xi(x_2) = y_0$ for $x_1 \leq x \leq x_2$ and $\xi(y_0) = G(x, y_0)$ which contradicts $G_x < 0$. ∎

The next result tells us that the only function satisfying conditions (C1)–(C3) and solving the functional equation (4.7), i.e. $T\xi = \xi$, is $\psi = \xi$.

5.1.2. Lemma. *If θ satisfies properties* (C1)–(C3) *and satisfies*

$$\theta(x) = \inf\{y : G(x, y) \geq \theta(y)\} \quad \textit{for all } x \in (0, a], \tag{5.2}$$

then $\theta \equiv \psi$.

Proof. By continuity $G(x, \theta(x)) = \theta(\theta(x))$ and

$$\Psi(x, \theta(x)) = (\theta(x), \theta(\theta(x))),$$

and letting $\Psi^t(x, \theta(x)) = (x_t, x_{t+1})$ we have

$$x = x_0 < x_1 < \cdots < \bar{x} \quad \text{if } x_0 < \bar{x};$$
$$x = x_0 > x_1 > \cdots > \bar{x} \quad \text{if } x_0 > \bar{x}.$$

The limit point x_∞ exists in either case and satisfies $\theta(x_\infty) = x_\infty = \bar{x}$ by (C2).

By Remark 4.2.11, $(x, \theta(x)) = (x, \psi(x))$ if $\theta(x) = x_1 \leq \psi(a)$. In particular,

$$\psi(x) = \theta(x) \quad \text{for } x \in (0, \psi(a)].$$

Let $c_0 \in [\psi(a), a]$ be the largest c such that

$$\psi(x) = \theta(x) \quad \text{for } x \leq c.$$

If $c_0 < a$, then for $\epsilon > 0$ small enough $\theta((c_0-\epsilon,\ c_0+\epsilon))$ contains a neighborhood of $\psi(c_0)$ by Remark 5.1.1. Then for $x \in (c_0 - \epsilon,\ c_0 + \epsilon)$

$$\Psi(x, \theta(x)) = (\theta(x), \psi(\theta(x))) = \Psi(\psi^{-1}(\theta(x)), \theta(x)),$$

so $x = \psi^{-1}(\theta(x))$, hence $\theta(x) = \psi(x)$ for all $x \in (c_0 - \epsilon,\ c_0 + \epsilon)$, contradicting the maximality of c_0. We conclude $c_0 = a$. ∎

The construction of ψ as the limit of an increasing sequence of approximate policy functions is analogous to policy iteration in dynamic programming, but uses our Euler equation instead of Bellman's equation.[8] We do not compute approximate value functions as found in the dynamic programming approach, but instead focus on constructing monotonic improvements of the policy function. To this end we establish a limited form of monotonicity for the minimum gain operator defining the minimization operator (5.1).

5.1.3. Remark. *If* $\xi, \eta \leq \psi : (0,a] \to (0,a]$ *satisfy* $\xi, \eta \leq \psi$ *and* $\xi \leq \eta$, *then* $T\xi \leq T\eta$.

Indeed in this case (for any $x \in (0,a]$)

$$\{y : G(x,y) \geq \eta(y)\} \subset \{y : G(x,y) \geq \xi(y)\}$$

and therefore the infima $T\eta(x)$ and $T\xi(x)$ of these sets satisfy $T\eta(x) \geq T\xi(x)$.

5.1.4. Remark. *If* $\xi(\bar{x}) = \bar{x}$, *then* $T\xi(\bar{x}) = \bar{x}$.

First note that $G(\bar{x},\bar{x}) = \bar{x}$, so $T\xi(\bar{x}) \leq \bar{x}$. If $T\xi(\bar{x}) < \bar{x}$, there exists $y < \bar{x}$ such that $G(\bar{x}, y) \geq y$. But

$$\begin{aligned} G(\bar{x},y) &= g(y) - F(y, g(\bar{x}) - y) \leq g(y) - F(\bar{x}, g(\bar{x}) - y) \\ &= g(y) - [g(\bar{x}) - y] < y, \end{aligned}$$

a contradiction.

5.1.5. Remark. T *is increasing and* $T\xi \leq \psi$.

The operator T is increasing by virtue of Remark 5.1.3 and $T\xi \leq \psi$ since $y = \psi(x)$ satisfies $G(x,y) = \psi(y) \geq \xi(y)$.

5.1.6. Lemma. *If* $\xi_0(x) = x$ *for* $x \in (0,\bar{x}]$ *and* $\xi_0(x) = \bar{x}$ *for* $x \in [\bar{x}, a]$, *then* $\xi_0 \leq T\xi_0 \leq \psi$.

Proof. Clearly $\xi_0 \leq \psi$, so $T\xi_0 \leq \psi$ by Remarks 5.1.3 and 5.1.5. Remarks 5.1.3 and 5.1.4 imply $T\xi_0(x) \geq \bar{x} = \xi_0(x)$ for $x \in [\bar{x}, a]$. If $x \in (0, \bar{x})$, then

$$T\xi_0(x) = \inf\{y : G(x,y) \geq \xi_0(y)\} = \inf\{y : G(x,y) \geq y\}.$$

But if $y < x < \bar{x}$, then

$$\begin{aligned} G(x,y) &= g(y) - F(y, g(x) - y) < g(y) - F(\bar{x}, g(x) - y) \\ &= y - [g(x) - g(y)] < y. \end{aligned}$$

Hence, $T\xi_0(x) \geq x = \xi_0(x)$. ■

Recall that an increasing function $\varphi : (0,a] \to (0,a]$ is *lower semi-continuous* at x if $\varphi(x-0) = \varphi(x)$ where $\varphi(x-0) = \liminf_{x' \to x} \varphi(x')$.

5.1.7. Lemma. *If* ξ *is continuous, then*

[8] Policy iteration solutions to dynamic programming problems are discussed in [8].

(i) *$T\xi$ is strictly increasing;*

(ii) *$T\xi$ is lower semi-continuous, that is, $(T\xi)(x-0) = (T\xi)(x)$ for all $x \in (0,a]$.*

Proof. (i) is immediate from Remark 5.1.1. To prove (ii) let $x_j \to x$; we have $G(x_j, (T\xi)(x_j)) = \xi((T\xi)(x_j))$. Let $(T\xi)(x_j) \to y$ along a subsequence (which we continue to index by j). Then $G(x,y) = \xi(y)$ and hence $(T\xi)(x) \le y$. It follows that $\liminf_{x' \to x}(T\xi)(x') \ge (T\xi)(x)$. Hence, $T\xi$ is lower semi-continuous. ∎

Our intention is to prove that if we start with the ξ_0 defined above and recurrently define $\xi_{n+1} = T\xi_n$ $(t = 0,1,2,\ldots)$, then $\xi_n \uparrow \psi$. To this aim we need several supplementary properties of the minimum gain operator. They are summed up in the following statements.

5.1.8. Proposition. *Let $\xi : (0,a] \to (0,a]$ be an increasing, lower semi-continuous function such that $\xi \le \psi$. Then:*

(i) *$G(x, T\xi(x)) = \xi(T\xi(x))$ for all $x \in (0,a]$;*

(ii) *$T\xi$ is strictly increasing;*

(iii) *$T\xi$ is lower semi-continuous.*

Proof. We start by noting that $T\xi$ is lower semi-continuous by Remark 5.1.1. Let $y_j \to T\xi(x)$ such that $G(x,y_j) \ge \xi(y_j)$, $y_j > T\xi(x)$. Then

$$G(x, T\xi(x)) \leftarrow G(x,y_j) \ge \xi(y_j) \to \xi(T\xi(x+0)).$$

Here and throughout, for a monotone function $\eta(x)$ we set

$$\eta(x \pm 0) = \lim_{\varepsilon \downarrow 0} \eta(x \pm \varepsilon).$$

For $y_j \to T\xi(x)$ such that $y_j < T\xi(x)$, we have

$$G(x, T\xi(x)) \leftarrow G(x,y_j) < \xi(y_j) \to \xi(T\xi(x-0)).$$

Since ξ is increasing and lower semi-continuous we obtain

$$G(x,T\xi(x)) \ge \xi(T\xi(x)+0) \ge \xi(T\xi(x)) = \xi(T\xi(x-0)) \ge G(x,T\xi(x)),$$

which establishes (i). The statement (ii) now follows by the same argument as in the proof of Remark 5.1.1, by using the relation (i).

It remains to prove (iii). For this let $x_j \to x$, $x_j < x$. Then

$$\xi(T\xi(x-0)) \leftarrow \xi(T\xi(x_j)) = G(x_j, T\xi(x_j)) \to G(x, T\xi(x-0)).$$

So $\xi(y) = G(x,y)$ for $y = T\xi(x-0)$. Therefore $y \geq T\xi(x)$. Together with $y \leq T\xi(x)$, this yields $y = T\xi(x)$. ■

The next is a continuity property of the nonlinear operator T.

5.1.9. Proposition. *Let $\xi_j : (0,a] \to (0,a]$, $j = 1,2,\dots$ be a sequence of increasing lower semi-continuous functions such that*

$$\xi_1 \leq \xi_2 \leq \cdots \leq \psi.$$

Let $\xi = \lim_{j\to\infty} \xi_j$. Then $T\xi = \lim_{j\to\infty} T\xi_j$.

Proof. The existence of the limit ξ is obvious. It is also evident that ξ is increasing, lower semi-continuous and that $\xi \leq \psi$. From

$$\xi_j \leq \xi_{j+1} \leq \xi \leq \psi$$

we infer (see Remark 5.1.3)

$$T\xi_j \leq T\xi_{j+1} \leq T\xi \leq T\psi = \psi,$$

so that $\eta = \lim_{j\to\infty} T\xi_j$ exists and $\eta \leq T\xi$. If $\eta \neq T\xi$, then there exists x_0 such that $\eta(x_0) < T\xi(x_0)$. So

$$G(x_0,y) < \xi(y) \quad \text{for} \quad 0 < y < T\xi(x_0),$$

and (due to Proposition 5.1.8)

$$G(x_0, T\xi_j(x_0)) = \xi_j(T\xi_j(x_0)) \quad \text{for } j = 1,2,\dots .$$

Clearly for j_0 fixed we have (with $j \geq j_0$, $j \to \infty$)

$$\xi_{j_0}(\eta(x_0)) \leftarrow \xi_{j_0}(T\xi_j(x_0)) \leq \xi_j(T\xi_j(x_0)) = G(x_0, T\xi_j(x_0)) \to G(x_0, \eta(x_0))$$

and therefore by letting $j_0 \to \infty$ we obtain

$$\xi(\eta(x_0)) \leq G(x_0, \eta(x_0)) < \xi(\eta(x_0)),$$

since $\eta(x_0) < T\xi(x_0)$, a contradiction. Thus $\eta = T\xi$. ■

5.2. The iterative construction of ψ

Let ξ_0 be the trial policy function defined in Lemma 5.1.6. It is the seed for our iterative process. Define $\xi_{j+1} = T\xi_j$ $(j = 0,1,2,\dots)$. By virtue of

Lemma 5.1.6, Remark 5.1.3, and Proposition 5.1.8, each ξ_j is, for $j = 1, 2, \ldots$, a lower semi-continuous strictly increasing function such that

$$\xi_1 \leq \xi_2 \leq \cdots \leq \psi,$$
$$\xi_{j-1}(\xi_j(x)) = G(x, \xi_j(x)), \quad x \in (0, a], \quad j = 1, 2, \ldots .$$

Thus

$$\xi_\infty(x) = \lim_{j\to\infty} \xi_j(x) \quad \text{for } x \in (0, a]$$

exists and is a lower semi-continuous increasing function of x such that $\xi \leq \psi$.

This sequence of approximate policy functions computed from the particular seed ξ_0 produces a larger consumption at each stage than the previous iteration for a given starting capital stock. The use of the minimization by the operator T allows us to find successive approximations of the true policy function by pushing consumption upwards while still being able to continue the iteration at the next stage. At each stage the smallest consumption is found that allows us to continue the calculation one more time. This leaves room for us to find a larger consumption that might also be consistent with continuing the iterative process. The iteration of T and the approximation by policy improvement is a way to find larger and larger consumption levels that permit continued iteration without violating the Euler equations. The true policy function is found when it is possible to continue this iterative process for every period.

With these preliminaries, we can now state our main result.

5.2.1. Theorem. $\xi_\infty = \psi$.

Proof. By virtue of Proposition 5.1.9, we have

$$T\xi_\infty = \lim_{j\to\infty} T\xi_j = \lim_{j\to\infty} \xi_{j+1} = \xi_\infty$$

and then by Proposition 5.1.8 (i),

$$\xi_\infty(\xi_\infty(x)) = \xi_\infty(T\xi_\infty(x)) = G(x, T\xi_\infty(x)) = G(x, \xi_\infty(x)), \qquad 0 < x \leq a.$$

Therefore

$$\Psi^t(x, \xi_\infty(x)) = (\xi_\infty^t(x), \xi_\infty^{t+1}(x))$$

is valid for all $t \geq 0$ and (since $\xi_0 \leq \xi_\infty \leq \psi$) $\xi_\infty^t(x)$ increases (respectively, decreases) in t if $x < \bar{x}$ (respectively, $x > \bar{x}$). Clearly $\xi_\infty^t(x) \to \bar{x}$. Referring to Remark 4.2.11 and the fact that $\xi_\infty \leq \psi$ we conclude $\xi_\infty = \psi$. ∎

This theorem shows how to construct by an economically appealing iterative process the function ψ whose graph gives the invariant manifold for this economy. This policy function improvement routine focuses entirely on the first agent's optimality conditions and ignores the no arbitrage equations for the remaining households without capital. The resulting Ramsey trajectory gives rise to a Ramsey equilibrium only in those cases where *all* the households no arbitrage conditions obtain. Since the more impatient households' utility functions are fixed, it is possible that for an arbitrary initial capital stock in $(0, a]$ that the turnpike property only holds after a finite number of periods even though the more impatient people start without capital. For this reason, one can produce examples with as few as two households where the neighborhood of the steady state for which our construction yields the economy's invariant manifold is a proper subset of $(0, a]$. However, the main lesson remains — an economically motivated construction of the invariant manifold is possible once the turnpike property has taken hold.

6. Conclusion

We have shown how an economically motivated construction of the stable manifold for a particular dynamical system can be carried out. Our analysis assumed that the income function is strictly increasing. However, there are interesting specifications of the underlying one–sector technology for which this condition does not hold. In those cases equilibrium profiles with complicated dynamics are possible. One problem for future work is to see if our construction can be carried through for models without income monotonicity.

A second open question concerns determination of the part of the invariant manifold actually realized along an equilibrium program when particular functional forms for utility functions are assumed. For example, we would be interested in knowing which utility functions yield a connected invariant manifold.

References

1. Robert A. Becker, On the long–run steady state in a simple dynamic model of equilibrium with heterogeneous households, *Quarterly Journal of Economics* **95** (1980), 375–382.

2. Robert A. Becker and John H. Boyd III, *Capital Theory, Equilibrium Analysis and Recursive Utility*, Basil Blackwell, Cambridge, 1997.

3. Robert A. Becker, John H. Boyd III, and Ciprian Foias, The existence of Ramsey equilibrium, *Econometrica* **59** (1991), 441–460.

4. Robert A. Becker and Ciprian Foias, A minimax approach to the implicit programming problem, *Economic Letters* 20 (1986), 171–175.
5. Robert A. Becker and Ciprian Foias, A characterization of Ramsey equilibrium, *Journal of Economic Theory* 41 (1987), 173–184.
6. Robert A. Becker and Ciprian Foias, Convergent Ramsey equilibrium, *Libertas Mathematica* 10 (1990), 41–52.
7. Robert A. Becker and Ciprian Foias, The local bifurcation of Ramsey equilibrium, *Economic Theory* 4 (1994), 719–744.
8. Richard Bellman, *Dynamic Programming*, Princeton University Press, 1957.
9. Wilbur John Coleman II, Equilibrium in a production economy with an income tax, *Econometrica* 59 (1991), 1091–1104.
10. Timothy J. Kehoe, David K. Levine, and Paul M. Romer, On characterizing equilibria in economies with externalities and taxes as solutions to optimization problems, *Economic Theory* 2 (1992), 43–68.
11. Frank Ramsey, A mathematical theory of saving, *Economic Journal* 38 (1928), 453–559.
12. Gerhard Sorger, On the structure of Ramsey equilibrium: cycles, indeterminacy, and sunspots, *Economic Theory* 4 (1994), 745–764.
13. Gerhard Sorger, Chaotic Ramsey equilibrium, *International Journal of Bifurcation and Chaos* 5 (1995), 373–380.
14. Nancy Stokey and Robert E. Lucas with Edward Prescott, *Recursive Methods in Economic Dynamics*, Harvard University Press, 1989.
15. Alfred Tarski, A lattice–theoretical fixpoint theorem and Its applications, *Pacific Journal of Mathematics* 5 (1955), 285–309.
16. Eberhard Ziedler, *Nonlinear Functional Analysis and Its Applications* I: *Fixed Point Theorems*, Springer, 1991.

Robert A. Becker
Department of Economics
Indiana University
Bloomington, IN 47405
USA
email: becker@indiana.edu

Ciprian Foias
Department of Mathematics
Indiana University
Bloomington, IN 47405
USA
email: foias@ucs.indiana.edu

AN APPROACH TO BARGAINING FOR GENERAL PAYOFFS REGIONS*

D. GLYCOPANTIS and A. MUIR

Abstract. This paper discusses an approach to bargaining which provides a solution to a demand game even when the payoffs region is not convex. The game is played repeatedly but the demands of the two players at each play are only used as information in subsequent calculations. This process goes on until convergence to a feasible vector is established. This will be the negotiated outcome of the game. The properties of such solutions are considered and the relation with the axiomatically justified Nash bargaining solution is discussed.

1. Introduction

We discuss here a simple approach to bargaining which provides a solution to a demand game even when the payoffs region is not convex. Indeed there is no reason to expect that we shall always have a concave boundary of Pareto efficient pairs of utility payoffs as the Nash theory assumes. For example, the utility possibilities frontier of an exchange economy is not in general concave and it is desirable to have an approach to bargaining which will apply to such cases as well.

Of course, an arbitrary payoffs boundary can be convexified by accepting the possibility of joint mixed strategies by the players. However resort to such a device might not always be possible, either because it is excluded by the rules of the game or because, as we shall see below, it is not in the interest of one of the players. The discussion here concerns a generalized demand game, where the generalization is in terms of the payoffs region. Its boundary will originally be assumed to be described by a strictly decreasing function and the case when it contains flat segments is discussed next. The separation of the two cases leads to a convenient presentation of the arguments.

For both types of games we shall obtain a unique bargaining solution through an algorithm which employs the general idea introduced in [4]. We assume that the game is repeated at discrete points in time but the publicly known demands of the two players at each play are only used as information in subsequent calculations. It is assumed that each player's demand at stage n depends continuously on and lies strictly between what they demanded and the maximum utility their opponent was offering at stage $n-1$. This pro-

*A comment made by a referee of "Economic Notes" on an earlier version of [4] also turned out to be useful in the initial development of our approach to generalized demand games as discussed in this paper.

cess goes on until convergence to a feasible vector of demands, which is then handed out, is established. This will be the negotiated outcome of the game.

We show that the discrete dynamical process described by such an algorithm for adjusting the players' demands is stable in the sense that it implies convergence to a unique Pareto efficient point, which of course depends on the initially declared demands of the players. These initial demands will be identified with the status–quo payoffs. It is also pointed out that when the boundary of the payoffs region is described by a strictly decreasing function all Pareto efficient points are fixed points of the algorithm.

Obviously the solution obtained here satisfies Nash's axioms of individual rationality and efficiency. We shall impose later a further condition on the functions characterizing the algorithm to guarantee that the solution satisfies also his axiom of linear invariance which we consider to be a reasonable requirement on any outcome of bargaining. We shall distinguish between invariance referring to the solution of the game and the stronger step by step invariance implying that as the sequence of demands unfolds we make exactly the same steps in terms of the original variables as before the linear transformation. In particular we satisfy step by step invariance when we employ the simple and intuitively appealing algorithm which states that the demand of a player at stage n is the weighted sum of what he demanded and the maximum utility that their opponent was willing to concede at stage $n-1$, with the weights remaining constant in time.

We then show that algorithms of the above type do not imply that the solution satisfies necessarily the controversial axiom of independence of irrelevant alternatives and therefore, for the case when the payoffs region is convex and comparisons are meaningful, the solution obtained is in general different from the Nash bargaining solution. A simple example supports the McDonald and Solow [7] alternative intuition to that brought forward in support of this axiom.

Finally we look at more general algorithms for adjusting the demands of the players, as for example when the adjustment functions depend also on the surplus payoffs set available for further negotiations before x_n and y_n are announced, possibly contracted by some factor, and explicitly on the status–quo payoffs. Furthermore the adjustment mechanism might not necessarily be the same in each iteration. It is then discussed how in the context of such algorithms the Nash bargaining solution could be obtained for convex games by a one–step or a many–step algorithm.

2. The basic model

The fundamental principles of the bargaining approach employed here and the nature of the results obtained are best understood by first considering demand games in which the boundary of the utility payoffs region contains no flats.

In the basic model the payoffs region will be assumed to be of the general form

$$X = \{(x,y) \in \mathbb{R}^2_+ : \quad y \leq f(x) \text{ with } x \in [0,c] \text{ and } f(c) = 0\},$$

where $f(x)$ is a continuous and strictly decreasing function. Writing $f(0) = d$, the inverse function $g(y)$ exists and is also continuous and strictly decreasing on $[0,d]$. The utility payoffs x and y refer to the two players P_1 and P_2, respectively. The game consists of the payoffs region X and the status–quo payoffs of the players which denote the minimum utility level they are prepared to accept from any negotiations. The players seek a negotiated outcome and we assume that they are allowed to bargain taking advantage of information as it becomes available at each stage of the negotiations.

Explicitly, at each stage of the game the players get a signal of each other's strength through the record of their previous demands and the assumption is that they adopt a rule whereby their demand at stage n lies strictly between what they demanded and the maximum utility that their opponent was willing to concede at stage $n-1$, i.e.,

$$x_n = F(x_{n-1}, \tilde{x}_{n-1}), \quad \text{where} \quad \tilde{x}_{n-1} = g(y_{n-1}) \tag{2.1}$$

$$y_n = G(y_{n-1}, \tilde{y}_{n-1}), \quad \text{where} \quad \tilde{y}_{n-1} = f(x_{n-1}), \tag{2.2}$$

with the provisos that the functions F and G are continuous, that x_n lies strictly between x_{n-1} and $\tilde{x}_{n-1}$ for $x_{n-1} \neq \tilde{x}_{n-1}$, and y_n lies strictly between y_{n-1} and $\tilde{y}_{n-1}$ for $y_{n-1} \neq \tilde{y}_{n-1}$. It follows that x_n is equal to x_{n-1} or $\tilde{x}_{n-1}$ if and only if $x_{n-1} = \tilde{x}_{n-1}$, and similarly for y_n. The algorithm requires initial conditions (x_1, y_1) and we are interested in whether the sequence of pairs of demands converges to a feasible payoff vector, which will be the negotiated outcome of the game[1]. If such convergence cannot be achieved then the negotiations break down and the players receive their status–quo payoffs.

Throughout, the assumption is that each player knows X and his own rule for adjusting his demand but not the rule employed by his opponent. As we

[1]Of course this an ideal solution. In practice an infinite number of steps is never realized as such and convergence should be taken within a prescribed accuracy, relative to the size of X. The players are willing to accept any feasible payoffs vector within this range of accuracy.

shall see below the continuity of the functions guarantees convergence of the algorithm to a unique feasible vector. The maximum utility that a player is willing to concede will be referred to as the offer made to his opponent.

At the beginning the players inform each other of their status–quo payoffs (threats), as in [9]. At this point, this is the only publicly available information about the intentions of the players and these threats can be considered to be their demands at stage 1 of the game. From the declared status–quo payoff a player can deduce the maximum utility his opponent is prepared to concede and the algorithm is set in motion. The status–quo payoff vector is denoted by (ξ_1, ξ_2). As in [8, 9], throughout the paper it is assumed that $(\xi_1, \xi_2) \in X$.

The relation between the demands in stages $n-1$ and n of the game is shown in Figure 2.1 below.

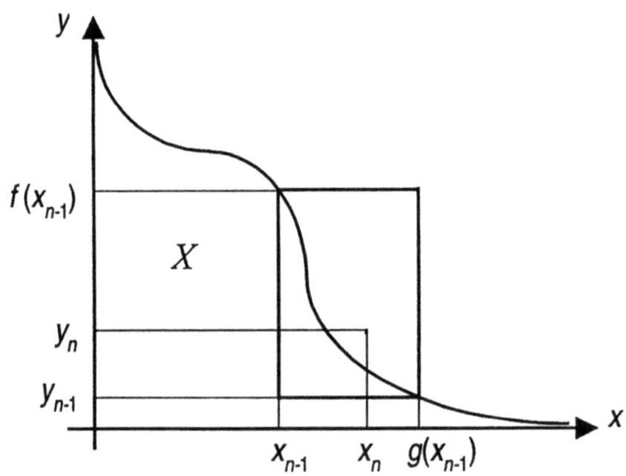

Figure 2.1

At stage $n-1$ of the play of the game P_1 demands for himself utility x_{n-1} and offers to P_2 utility $\tilde{y}_{n-1} = f(x_{n-1})$. On the other hand P_2 demands for himself utility y_{n-1} and offers to P_1 utility $\tilde{x}_{n-1} = g(y_{n-1})$. The figure shows a typical pair of demands of the players, x_n and y_n, at stage n of the game. The conditions on the algorithm imply that this pair will lie in the interior, whenever it is non-empty, of the parallelogram formed here by the heavy horizontal and vertical lines. On intuitive grounds a player will expect, by assigning the same rational behavior to his opponent as the one he adopts, that the demands at stage n will be somewhere in the parallelogram. Namely an opponent will reasonably be expected not to demand more than what he asked for at the previous stage of the game and did not get, and also not to ask for less than what he was offered. The exact position, though, of the demands x_n and y_n is determined by the choices of F and G.

All boundary points, defined to be those for which $y = f(x)$, are Pareto efficient and they are fixed points of the algorithm. This follows from the fact that if at some stage the pair of demands lies on the boundary of the feasible payoffs then what each player offers coincides with what the other player demands and the game terminates. Apart from the boundary points, the algorithm will be seen to have no other fixed points.

We show below that, under the condition of continuity of our functions, an algorithm of the type proposed here implies, for any payoffs region, a unique bargaining solution, which is individually rational and efficient. The solution will depend on the initially declared demands, which will form the starting point of the algorithm. We shall impose later a further condition on the functions to guarantee that the solution satisfies also Nash's axiom of linear invariance which on intuitive grounds is a reasonable requirement on any outcome of bargaining. As we shall see the controversial axiom of independence of irrelevant alternatives will not in general be satisfied. This can be seen to lend support to the intuition of McDonald and Solow.

3. Convergence of the sequential decisions

We now consider the issue of convergence and prove that an algorithm of the type given by relations (2.1) and (2.2), and the stated conditions on the functions, implies that, starting from an arbitrary vector of demands (x_1, y_1), the prescribed dynamical process terminates at an efficient point. Above, we have given a reason why we could take $(x_1, y_1) = (\xi_1, \xi_2)$.

Consider a sequence of demands $\{(x_n, y_n)\}$ generated by the algorithm. As the demands of the players are bounded below by their status–quo payoffs and above by the corresponding maximum utility they could possibly have, $\{(x_n, y_n)\}$ lies in a compact set and therefore it has a convergent subsequence. First, suppose the sequence itself converges with limit (x_0, y_0). We have $\{(x_n, y_n)\} \to (x_0, y_0)$ and therefore $x_0 = \lim x_n = \lim F(x_{n-1},\ g(y_{n-1}))$. The continuity of the functions F and G implies $\lim F(x_{n-1},\ g(y_{n-1})) = F(\lim x_{n-1},\ g(\lim y_{n-1})) = F(x_0,\ g(y_0))$. Therefore we have $x_0 = F(x_0,\ g(y_0))$, and from the property of the algorithm that if x_n is equal to x_{n-1} or $\tilde{x}_{n-1}$ then $x_{n-1} = \tilde{x}_{n-1}$ it follows that $x_0 = g(y_0)$, which means that the limit of the sequence is a boundary point. Arguing similarly we obtain $y_0 = G(y_0,\ f(x_0))$, from which it follows that $y_0 = f(x_0)$, confirming again that the limit of the sequence is a boundary point. Therefore the convergence of the sequence of the pairs of utility demands implies that it tends to an efficient, Pareto optimal point. This is the equilibrium payoff vector that will be handed out to the players. The algorithm implies that this vector is also individually rational.

Next we assume that although it contains a convergent subsequence, the sequence of demands does not itself converge. Since the sequence is bounded, there must exist at least two convergent subsequences with different limits. On the other hand there cannot be more than one of these subsequences on the same side of the boundary, for starting at an element sufficiently far along either of them we would then exclude all but possibly a finite number of elements of the other, as they would fall outside the parallelogram in which the rest of the sequence should lie. Therefore under the hypothesis that the sequence of demands does not converge the only possibility left to be investigated is the one depicted in Figure 3.1.

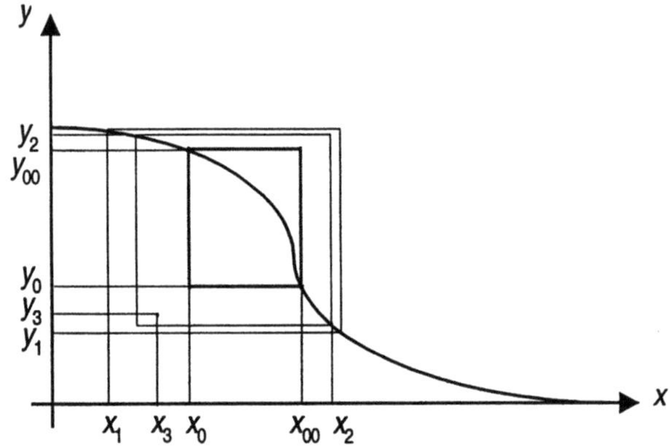

Figure 3.1

Suppose that there are two subsequences with limit points (x_0, y_0) and (x_{00}, y_{00}), inside and outside the feasible set, respectively. They must lie at the corners of a parallelogram as shown on the figure. This follows from the fact that given a pair of demands the next and all subsequent moves must be inside the parallelogram defined by the specific point and the boundary curve.

We want to argue away the situation in Figure 3.1 as leading to a contradiction. In doing so, the continuity of the functions will again be invoked. Notice that the vectors of the subsequence tending to the lower point must have coordinates strictly less than those of (x_0, y_0) and those of the one that tends to the upper point must have coordinates strictly greater than those of (x_{00}, y_{00}). The figure shows a way the two limit points could be approached, in alternating manner, if they existed.

Now consider the subsequence $\{(x_n^s, y_n^s)\}$ tending to (x_0, y_0). There must be an infinite number of its elements which map through the algorithm to a vector with coordinates greater than those of (x_{00}, y_{00}). Otherwise starting the whole sequence from a sufficiently advanced element of $\{(x_n^s, y_n^s)\}$ we could never approach the point (x_{00}, y_{00}) which we know to be the limit of another subsequence.

Consider the subsequence $\{(x_n^{ss}, y_n^{ss})\}$ of $\{(x_n^s, y_n^s)\}$ the elements of which map to a vector with coordinates greater than those of (x_{00}, y_{00}). The subsequence $\{(x_n^{ss}, y_n^{ss})\}$ tends itself to (x_0, y_0) and its images under the algorithm to (x_{00}, y_{00}), where

$$x_{00} = \lim F(x_{n-1}^{ss}, g(y_{n-1}^{ss})) \qquad \text{and} \qquad y_{00} = \lim G(y_{n-1}^{ss}, f(x_{n-1}^{ss})).$$

On the other hand if we were to start the algorithm at (x_0, y_0) we would next obtain the pair of demands $x = F(x_0, g(y_0))$ and $y = G(y_0, f(x_0))$ which lies inside the parallelogram, away from the point (x_{00}, y_{00}). However this is not allowed by the continuity of F, G, f, and g which, given the convergence of $\{(x_n^{ss}, y_n^{ss})\}$, would require $x_{00} = F(x_0, g(y_0))$ and $y_{00} = G(y_0, f(x_0))$. Therefore the assumption that there exist two subsequences with different limit points (x_0, y_0) and (x_{00}, y_{00}) leads to a contradiction. It follows that the sequence itself must converge and, as we have seen above, it must do so to a point on the boundary. Finally notice also that the assumed limit point (x_0, y_0) of the argument above cannot be placed on either of the axes as it would then be easy to see that the sequence would immediately be moving away from this point, which would lead to a contradiction.

The discussion above can also be used to explain why the continuity of F and G cannot be dispensed with. For example, lack of continuity could allow for the possibility which appears in Figure 3.1.

4. Boundaries with flats

Up to this point we have not allowed for either vertical or horizontal flats in the utility payoffs boundary. That is we have been working with functions $y = f(x)$ and $x = g(y)$ and not with correspondences as for example the one in Figure 4.1 (on the next page). We now allow the boundary to consist of horizontal and vertical flats as well as strictly decreasing segments.

In cases with flats the boundary of the utility payoffs frontier consists of vectors such that there is no other feasible utility pair which would make both players better off. Namely it consists of utility allocations which cannot be blocked, in the usual core theoretic sense. This is how the set of Pareto optimal (efficient) vectors is now defined (see, for example, [1, 5, 6]). This definition allows that a reallocation of utilities might exist which makes one player better off without making the other one worse off.

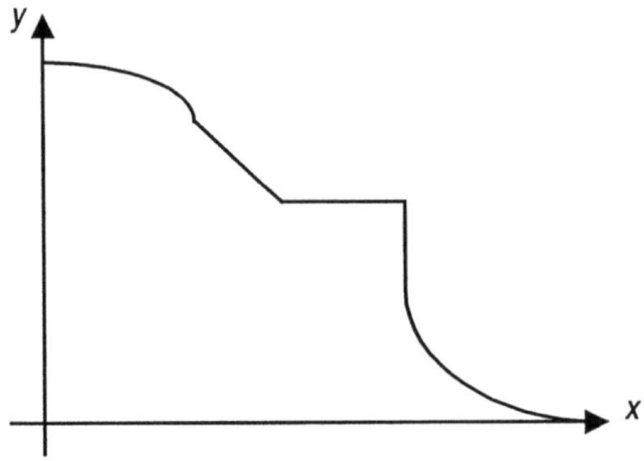

Figure 4.1

In order to consider also such cases, we must adjust the analysis above. We shall denote by $\mathcal{Y} = f(x)$ and $\mathcal{X} = g(y)$, respectively, the correspondence and the inverse correspondence of the boundary. The algorithm is now described by

$$x_n = F(x_{n-1}, \tilde{x}_{n-1}), \quad \text{where} \quad \tilde{x}_{n-1} = \max g(y_{n-1})$$
$$y_n = G(y_{n-1}, \tilde{y}_{n-1}), \quad \text{where} \quad \tilde{y}_{n-1} = \max f(x_{n-1}),$$

with initial condition $(x_1, y_1) = (\xi_1, \xi_2)$. All other properties of the algorithm remain the same. Notice that although $g(y_{n-1})$ and $f(x_{n-1})$ are sets, $\tilde{x}_{n-1} = \max g(y_{n-1})$ and $\tilde{y}_{n-1} = \max f(x_{n-1})$ are unique numbers. The implications of this algorithm for specific cases are seen in Figures 4.2 and 4.3 below.

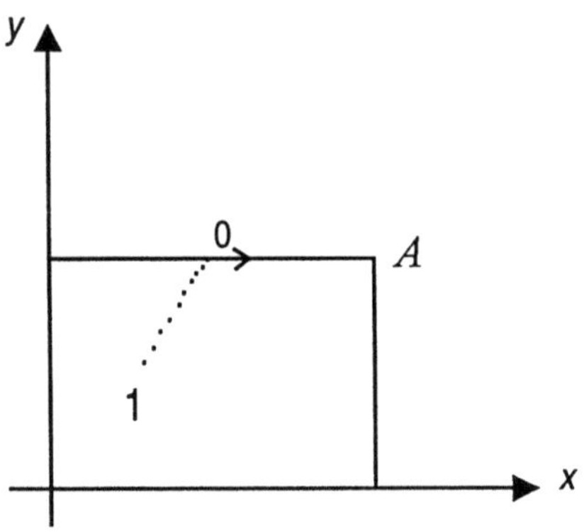

Figure 4.2

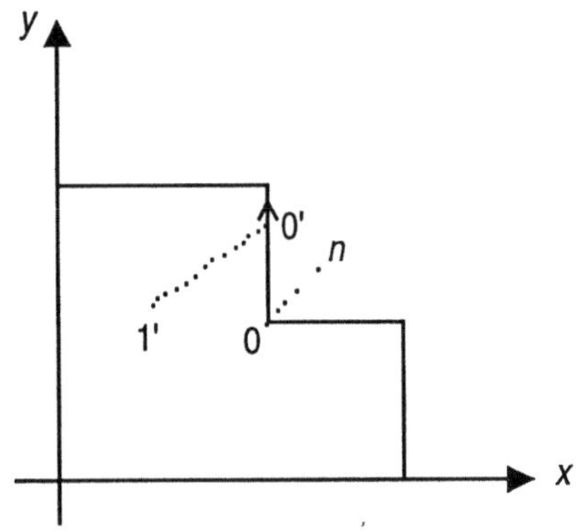

Figure 4.3

Figure 4.2 makes the point that for such a boundary, irrespective of the initial demands, the algorithm will always terminate at the upper corner A. Consider first sequences which start away from the boundary of X. Any such sequence must converge due to the fact that it is bounded above and the coordinates of its vectors increase. However, because F and G are continuous functions, it could not tend to a point (x_0, y_0) in the interior of X for as the sequence approaches the supposed interior limit point the changes of its coordinates tends to zero. On the other hand if the algorithm started at (x_0, y_0) there would be a finite increase in the coordinates of this vector, which is not allowed by the continuity of F and G. This would require a zero change, i.e. that (x_0, y_0) repeats itself, and hence interior points are excluded as limit points.

Next suppose, for example, that the sequence converges to the boundary point 0. Again in the limit the changes in the sequence vectors tend to zero. On the other hand if the algorithm started at the assumed limit point 0 there would be a finite increase in x which cannot be allowed by the continuity of the function F as $\max g(y_{n-1})$ is constant throughout the sequence and at its limit point. Therefore it is not possible to have the sequence as indicated in Figure 4.2. This is shown by the fact that an arrow points away from 0. Similarly the sequence cannot have a limit point on the vertical flat of the boundary. Finally starting from any point on the flats we tend towards the corner point A. This example highlights the fact that when there are flat segments it is no longer the case that any point on the boundary is a fixed point of the algorithm.

Figure 4.3 shows that it is now possible to end up at a point such that a reallocation exists which would make one player better off without reducing the utility of the other one. For example, we can specify explicitly the algorithm so that if at stage n the demands of the players are at point n the sequence traces from then on the positions indicated by the consecutive dots towards 0, which is its limit point and therefore the negotiated outcome of the game. Notice that since $\max f(x_{n-1})$ and $\max g(y_{n-1})$ are discontinuous in the limit, there is no continuity argument that can be used to argue this possibility away. On the other hand if the sequence converges from inside the feasible area then the continuity argument of Figure 4.2 applies and therefore it must end up at a limit point such that there is no feasible allocation which makes one player better off without making the other one worse off. For example, starting at point $1'$ it is not possible to have a sequence in the interior of X which has $0'$ as its limit point. Again an arrow points away from $0'$.

Figure 4.3 can also be used to explain that the negotiated outcome of the game, defined here to be the feasible vector to which the sequence of pairs of demands converges, is not always an equilibrium vector of a dynamical system which would require that starting at that point the system stays there. We saw above that 0 can be approached in such a way so that it becomes the bargaining solution of the game. However this is not an equilibrium point of the algorithm in the mathematical sense, for starting at 0 there would be an increase in the coordinates of this vector. On the other hand a slight alteration of the boundary of the feasible set, obtained, for example, by replacing a small part of the boundary around the point 0 with a segment of a straight line with a negative slope, would eliminate this possibility.

5. General results of the sequential decisions for boundaries with flats

Now we investigate the behavior of the demand vectors implied by the algorithm for general boundaries with flats. As before consider a sequence $\{(x_n, y_n)\}$ generated by the algorithm. It lies in a compact set and therefore it has a convergent subsequence. First, suppose that the sequence itself converges and has as its limit (x_0, y_0). Suppose that the limit point is not on the boundary of X. Typical cases are shown on Figures 5.1 and 5.2 below.

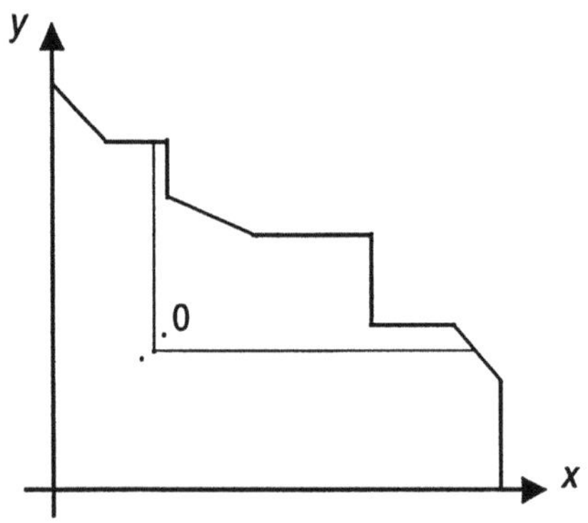

Figure 5.1

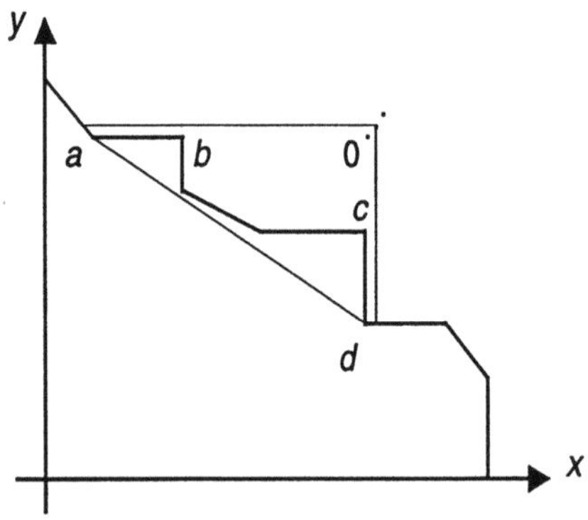

Figure 5.2

The assumed limit point (x_0, y_0) is indicated by 0. Let (x_0, y_0) be a feasible, non-boundary point of X as for example in Figure 5.1. As the sequence progresses the distance of the coordinates of its vectors from those of the assumed limit point tends to zero. On the other hand starting the algorithm from (x_0, y_0) we have a finite increase in both coordinates as the limit point is away from the boundary. However $f(x_{n-1})$, viewed as a function, by ignoring the vertical segments of the boundary, is left–continuous, which implies that such a discrete change in y is not possible. The continuity of $y_n = G(y_{n-1}, f(x_{n-1}))$ does not allow for a change which tends to zero and in the limit jumps to a discrete value. A similar argument applies with respect to $x_n = F(x_{n-1}, \tilde{x}_{n-1})$.

Therefore the limit of the convergent sequence cannot be an interior point of X.

Next let (x_0, y_0) lie outside the boundary of X. Such a possibility cannot be excluded through an argument similar to the one we have invoked for the case when (x_0, y_0) was assumed to be a feasible, non-boundary point of X. The reason is that we might now have a jump in both $\max f(x)$ and $\max g(y)$ and therefore an analogous continuity argument cannot be applied. Figure 5.2 shows the possibility of jumps at the assumed limit point 0 which is placed on the extension of *ab* and *cd*. We exclude the possibility of a limit point outside the boundary of X as follows.

It is sufficient to consider the game in Figure 5.2 as a representative case with jumps in both $\max f(x)$ and $\max g(y)$. One can now design a new game by replacing the segments of the boundary of the original game between a and d with a dashed line as shown on the figure. The new game gives rise to the same sequence of vectors of demands as before and in the limit we have that $\max f(x)$ and $\max g(y)$ attain their limit values.

However the continuity of F and G excludes the possibility that such a sequence of demands is feasible for the new game. As the sequence approaches the supposed limit point the changes of its coordinates, brought about by the application of F and G, tends to zero. On the other hand if the algorithm started at this limit point there would be a finite increase in the coordinates of this vector, which is not allowed by the continuity of F and G which would require a zero change.

Finally, the fact that such a sequence is not feasible for the new game implies that it is not feasible for the original game either as the two games have different boundaries only between points a and d and coincide where the sequence of pairs of demands is formed through the algorithm.

Alternatively, as in the case for functions, we assume that the bounded subsequence of demands does not itself converge and therefore there must exist at least two convergent subsequences with different limits. We show such limits on Figure 5.3 on the next page.

Again, there cannot be more than one of these subsequences on the same side of the boundary. Suppose therefore that there are two subsequences with limit points (x_0, y_0) and (x_{00}, y_{00}), inside the feasible set and outside the boundary, respectively. Then they must lie at the corners of a parallelogram as shown on the figure. We prove that this leads to a contradiction.

Take the subsequence $\{(x_n^s, y_n^s)\}$ in the interior of X tending to (x_0, y_0). As in the case for functions, investigated earlier, there must exist a subsequence $\{(x_n^{ss}, y_n^{ss})\}$ of $\{(x_n^s, y_n^s)\}$ the elements of which map on the other side of the boundary. The subsequence $\{(x_n^{ss}, y_n^{ss})\}$ tends itself to (x_0, y_0) and the coordinates of these images decrease and tend to those of (x_{00}, y_{00}). Invoking, for example, again the left–continuity of $f(x_{n-1})$, viewed as a function by ignoring the vertical segments of the boundary, we can argue that this is not possible because if we were to start the algorithm at (x_0, y_0) we would obtain an image inside the parallelogram and away from (x_{00}, y_{00}). On the other hand the continuity of the function $y_n = G(y_{n-1}, f(x_{n-1}))$ would require that we have $y_{00} = G(y_0, f(x_0))$. Therefore the assumption that there exist two convergent subsequences with different limit points leads to a contradiction. It follows that the sequence itself must converge and it must do so to a point on the boundary. Again the algorithm implies that this equilibrium payoffs vector is also individually rational.

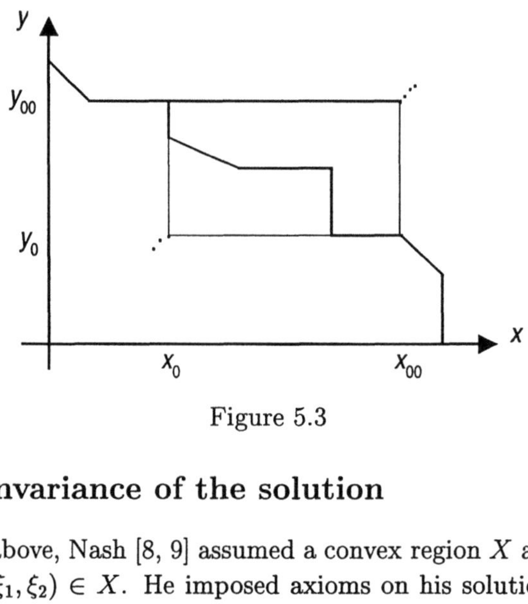

Figure 5.3

6. Linear invariance of the solution

As mentioned above, Nash [8, 9] assumed a convex region X and a status–quo payoff vector $(\xi_1, \xi_2) \in X$. He imposed axioms on his solution and obtained that there is a unique pair of payoffs, depending on X and (ξ_1, ξ_2), which satisfies them. These axioms are discussed also in a number of places in the literature as for example in [5] and [2]. They are briefly that (i) Player i will not agree to a solution which gives him/her less than his/her status–quo payoff ξ_i (individual rationality), (ii) positive linear transformations of the utilities of the players transform the solution in the same manner (linear invariance), (iii) the unique solution must lie on the boundary of the feasible payoffs set (efficiency), (iv) if the payoffs region is reduced while (ξ_1, ξ_2) and the solution stay in the new region this should also be the solution of the new game (independence of irrelevant alternatives), and (v) the solution is independent of which player is called Player 1 and which Player 2 (symmetry). With respect to the symmetry axiom there is agreement that it can be removed and one then obtains what is called in the literature the generalized Nash bargaining solution. In our proposed alternative approach symmetry between the two players would require $F = G$, but in what follows we are not particularly concerned whether our system satisfies this condition.

The axiom of independence of irrelevant alternatives remains controversial, in spite of its intuitive appeal. For example, McDonald and Solow [7] argue that it is also intuitively clear that if the truncated region, which contains the original solution and has the same status–quo payoffs, is such that for a whole range of payoffs to P_i the payoff to the other player is now less, then P_i should now do better. In the next section we shall consider whether the solution obtained here satisfies in general axiom (iv).

We have seen that an algorithm of the type proposed here implies a unique bargaining solution, which is individually rational and efficient. Next we investigate invariance of the solution in the sense of axiom (ii); i.e., suppose the utility payoffs are transformed linearly by $x' = \alpha_1 x + \alpha_2$ and $y' = \beta_1 y + \beta_2$ with $\alpha_1, \beta_1 > 0$. Invariance means that if x^* and y^* solve the original game then the solution of the game with the transformed payoffs is $x'^* = \alpha_1 x^* + \alpha_2$ and $y'^* = \beta_1 y^* + \beta_2$. Here, as we are considering algorithms, invariance can be thought of in two ways. Either the usual one referring to the solution of the game or a step by step invariance, which means that as the sequence of demands unfolds we make exactly the same steps in terms of the original variables as before the transformation. The second type of invariance is stronger as it implies the former. In particular we retain invariance when we employ the simple and intuitively appealing specific algorithm which will be discussed further on.

For the step by step linear invariance we now see that the functions F and G must satisfy the relations

$$F(\alpha_1 x_{n-1} + \alpha_2,\ \alpha_1 \tilde{x}_{n-1} + \alpha_2) = \alpha_1 F(x_{n-1}, \tilde{x}_{n-1}) + \alpha_2 \tag{6.1}$$
$$G(\beta_1 y_{n-1} + \beta_2,\ \beta_1 \tilde{y}_{n-1} + \beta_2) = \beta_1 G(y_{n-1}, \tilde{y}_{n-1}) + \beta_2. \tag{6.2}$$

We consider first the case when the boundary of feasible payoffs is given by a function. Following the linear transformations $x' = \alpha_1 x + \alpha_2$ and $y' = \beta_1 y + \beta_2$ the boundary function will be given by $y' = \beta_1 f((x' - \alpha_2)/\alpha_1) + \beta_2$ and its inverse by $x' = \alpha_1 g((y' - \beta_2)/\beta_1) + \alpha_2$. The algorithm now applies to the new variables and the description of its steps changes from

$$x_n = F(x_{n-1}, \tilde{x}_{n-1}) \qquad \text{and} \qquad y_n = G(y_{n-1}, \tilde{y}_{n-1}),$$

where $\tilde{x}_{n-1} = g(y_{n-1})$ and $\tilde{y}_{n-1} = f(x_{n-1})$, into the relations

$$x'_n = F(x'_{n-1}, \tilde{x}'_{n-1}) \qquad \text{and} \qquad y'_n = G(y'_{n-1}, \tilde{y}'_{n-1}),$$

where $\tilde{x}'_{n-1} = \alpha_1 g((y'_{n-1} - \beta_2)/\beta_1) + \alpha_2$ and $\tilde{y}'_{n-1} = \beta_1 f((x'_{n-1} - \alpha_2)/\alpha_1) + \beta_2$.

Substituting from $x' = \alpha_1 x + \alpha_2$ and $y' = \beta_1 y + \beta_2$ in the new relations we obtain the expressions, in terms of the original variables,

$$\begin{aligned} \alpha_1 x_n + \alpha_2 &= F(\alpha_1 x_{n-1} + \alpha_2,\ \alpha_1 \tilde{x}_{n-1} + \alpha_2) \\ \beta_1 y_n + \beta_2 &= G(\beta_1 y_{n-1} + \beta_2,\ \beta_1 \tilde{y}_{n-1} + \beta_2), \end{aligned}$$

where $\tilde{x}_{n-1} = g(y_{n-1})$ and $\tilde{y}_{n-1} = f(x_{n-1})$.

It is obvious that relations (6.1) and (6.2) are sufficient conditions in order to make exactly the same step in terms of x_n and y_n as before the variables were transformed. In order to prove that they are also necessary we argue as follows.

We consider the function F. Applying to the variables a rescaling by α_1 without a translation, and a translation by α_2 without a rescaling, step by step invariance means that we require that F satisfies the conditions

(i) $F(\alpha_1 x_{n-1},\ \alpha_1 \tilde{x}_{n-1}) = \alpha_1 F(x_{n-1}, \tilde{x}_{n-1})$.

(ii) $F(x_{n-1} + \alpha_2,\ \tilde{x}_{n-1} + \alpha_2) = F(x_{n-1}, \tilde{x}_{n-1}) + \alpha_2$.

The combination of (i) and (ii) implies that we require that F satisfies relation (6.1). An analogous argument applies to the function G and we establish relation (6.2). These conditions together with the continuity property of the functions imply that F and G must be of particular explicit forms.

Condition (i) means that F must be homogeneous of degree one in its arguments. Therefore there exists function h such that we we can write

$$F(x_{n-1}, \tilde{x}_{n-1}) = \tilde{x}_{n-1} h(x_{n-1}/\tilde{x}_{n-1})$$

and condition (ii) can now be cast in the form

$$(\tilde{x}_{n-1} + \alpha_2) h((x_{n-1} + \alpha_2)/(\tilde{x}_{n-1} + \alpha_2)) = \tilde{x}_{n-1} h(x_{n-1}/\tilde{x}_{n-1}) + \alpha_2.$$

We consider $x_{n-1} \neq \tilde{x}_{n-1}$. Setting $\alpha_2 = -x_{n-1}$ we obtain

$$(\tilde{x}_{n-1} - x_{n-1}) h(0) = \tilde{x}_{n-1} h(x_{n-1}/\tilde{x}_{n-1}) - x_{n-1}$$

and rearranging terms we have

$$F(x_{n-1}, \tilde{x}_{n-1}) = \tilde{x}_{n-1} h(x_{n-1}/\tilde{x}_{n-1}) = (1 - \lambda) x_{n-1} + \lambda \tilde{x}_{n-1},$$

where $\lambda = h(0)$. Therefore for step by step invariance F must be of the form $F(x_{n-1}, \tilde{x}_{n-1}) = (1 - \lambda) x_{n-1} + \lambda \tilde{x}_{n-1}$. The condition on the algorithm that x_n lies strictly between x_{n-1} and $\tilde{x}_{n-1}$ for $x_{n-1} \neq \tilde{x}_{n-1}$, implies $0 < \lambda < 1$. If

$x_{n-1} = \tilde{x}_{n-1}$ then x_n is equal to both x_{n-1} and $\tilde{x}_{n-1}$ and therefore F is again of the same form. A similar argument implies that G must be of the form $G(y_{n-1}, \tilde{y}_{n-1}) = (1-\mu)y_{n-1} + \mu\tilde{y}_{n-1}$, where $0 < \mu < 1$.

Nothing that has been said implies that λ must be a fixed constant for all $(x_{n-1}, \tilde{x}_{n-1})$, or μ for all $(y_{n-1}, \tilde{y}_{n-1})$. On the other hand, λ must be constant in $x_{n-1} > \tilde{x}_{n-1}$ and also constant in $x_{n-1} < \tilde{x}_{n-1}$. Similarly μ must be constant in $y_{n-1} > \tilde{y}_{n-1}$ and constant in $y_{n-1} < \tilde{y}_{n-1}$. We show this as follows.

First we notice that any point $(x', \tilde{x}')$ on either side of the 45° line, $\tilde{x}_{n-1} = x_{n-1}$, can be reached, through an appropriate linear transformation, from any point $(x, \tilde{x})$ on the same side but not from a point on the other side. We see this from the fact that following the linear transformation $x' = \alpha_1 x + \alpha_2$ we must have $(x' - \tilde{x}') = \alpha_1(x - \tilde{x})$ and as α_1 is positive the vectors must stay on the same side of the line. Constancy of λ now follows from the step by step linear invariance which requires that we must have $x_n = F(x_{n-1}, \tilde{x}_{n-1}) = (1-\lambda)x_{n-1} + \lambda\tilde{x}_{n-1}$ before the transformation took place and $x'_n = F(x'_{n-1}, \tilde{x}'_{n-1}) = (1-\lambda')x'_{n-1} + \lambda'\tilde{x}'_{n-1}$ after the transformation. Given that $x' = \alpha_1 x + \alpha_2$, the two expressions are consistent if and only if $\lambda = \lambda'$ and therefore λ must be constant on each side of the 45° line. Similarly we can show that μ must be constant in $y_{n-1} > \tilde{y}_{n-1}$ and constant in $y_{n-1} < \tilde{y}_{n-1}$. We give explicitly an algorithm of this form later.

Next we consider the case when the boundary of feasible payoffs contains flats. Recall that $\mathcal{Y} = f(x)$ and $\mathcal{X} = g(y)$ denote, respectively, the correspondence and inverse correspondence of the boundary. Following the transformations $x' = \alpha_1 x + \alpha_2$ and $y' = \beta_1 y + \beta_2$, these correspondences will be given by the sets $\mathcal{Y}' = \beta_1 f((x' - \alpha_2)/\alpha_1) + \beta_2$ and $\mathcal{X}' = \alpha_1 g((y' - \beta_2)/\beta_1) + \alpha_2$, respectively. The algorithm applies now to the new variables and the description of its steps changes from

$$x_n = F(x_{n-1}, \tilde{x}_{n-1}) \qquad \text{and} \qquad y_n = G(y_{n-1}, \tilde{y}_{n-1}),$$

where $\tilde{x}_{n-1} = \max g(y_{n-1})$ and $\tilde{y}_{n-1} = \max f(x_{n-1})$, into the relations

$$x'_n = F(x'_{n-1}, \tilde{x}'_{n-1}) \qquad \text{and} \qquad y'_n = G(y'_{n-1}, \tilde{y}'_{n-1}),$$

where $\tilde{x}'_{n-1} = \max(\alpha_1 g((y'_{n-1} - \beta_2)/\beta_1) + \alpha_2)$
and $\tilde{y}'_{n-1} = \max(\beta_1 f((x'_{n-1} - \alpha_2)/\alpha_1) + \beta_2)$.

Now, substituting from $x' = \alpha_1 x + \alpha_2$ and $y' = \beta_1 y + \beta_2$ in the new relations and taking into account the fact that

$$\max(\alpha_1 g((y'_{n-1} - \beta_2)/\beta_1) + \alpha_2) = \alpha_1 \max g((y'_{n-1} - \beta_2)/\beta_1) + \alpha_2$$

and also that

$$\max(\beta_1 f((x'_{n-1} - \alpha_2)/\alpha_1) + \beta_2) = \beta_1 \max f((x'_{n-1} - \alpha_2)/\alpha_1) + \beta_2,$$

we obtain the expressions, in terms of the original variables,

$$\begin{aligned} \alpha_1 x_n + \alpha_2 &= F(\alpha_1 x_{n-1} + \alpha_2,\ \alpha_1 \tilde{x}_{n-1} + \alpha_2) \\ \beta_1 y_n + \beta_2 &= G(\beta_1 y_{n-1} + \beta_2,\ \beta_1 \tilde{y}_{n-1} + \beta_2), \end{aligned}$$

where $\tilde{x}_{n-1} = \max g(y_{n-1})$ and $\tilde{y}_{n-1} = \max f(x_{n-1})$.

Relations (6.1) and (6.2) above are again necessary and sufficient conditions in order to make exactly the same step in terms of x_n and y_n as before the variables were transformed. The rest of the discussion is the same as for the case where the boundary is described by a function.

In the next section we discuss the implications of the simple and intuitively appealing algorithm with λ and μ constant. The players adopt the same simple rule at each iteration of the game. Explicitly, when the boundary is given by functions the algorithm takes the form

$$x_n = (1-\lambda)x_{n-1} + \lambda g(y_{n-1}) \tag{6.3}$$

$$y_n = (1-\mu)y_{n-1} + \mu f(x_{n-1}), \tag{6.4}$$

where the constants λ and μ satisfy the condition $0 < \lambda < 1$, $0 < \mu < 1$. When the boundary contains flat segments, $g(y_{n-1})$ and $f(x_{n-1})$ are replaced by $\max g(y_{n-1})$ and $\max f(x_{n-1})$.

7. Comparison with the Nash bargaining solution

We now show by means of explicit examples that even when invariance is satisfied the solution obtained through this sequential decisions approach does not in general satisfy the axiom of independence of irrelevant alternatives. We also look at an example which makes it clear that convexification of the payoffs region might cause a player to end up with less utility than otherwise.

Consider first the game of dividing a cake, i.e. $y = 1 - x$. Relations (6.3) and (6.4) above with $\lambda = \mu = 1/2$ describe a one-step algorithm which for initial demands anywhere on the 45° line gives the Nash bargaining solution $(x^*, y^*) = (1/2, 1/2)$. Explicitly, we have

$$x_2 = 1/2\, x_1 + 1/2\,(1 - y_1) \qquad \text{and} \qquad y_2 = 1/2\, y_1 + 1/2\,(1 - x_1)$$

with $y_1 = x_1$ and the result follows.

For the cake game with general λ, μ we have

$$x_n = (1-\lambda)x_{n-1} + \lambda(1 - y_{n-1}) \qquad \text{and} \qquad y_n = (1-\mu)y_{n-1} + \mu(1 - x_{n-1}).$$

This is not a one-step algorithm as we can see by taking $(x_1, y_1) = (0, 0)$ and $\lambda = 1/3$ while $\mu = 1/4$. In the next iteration it implies the vector of demands

$(x_2, y_2) = (1/3, 1/4)$ which is not Pareto efficient. Suppose now that the initial demands are on the line $y_1 = (\mu/\lambda)x_1$. Then multiplying the first relation by μ and the second by $-\lambda$ we obtain that $y_n = (\mu/\lambda)x_n$ and in the limit we obtain the solution vector $(x, y) = (\lambda/(\lambda+\mu), \mu/(\lambda+\mu))$, which we know to be efficient. One could interpret this as the generalized Nash bargaining solution for the cake game, with the bargaining powers of the players given by λ and μ. These can be thought of as expressing the eagerness of the players to adjust their demands given the information available from the last iteration.

On the other hand, relations (6.3) and (6.4) do not in general lead to a Nash solution. For example, if $f(x) = 1 - x^2$ with $(x_1, y_1) = (0, 0)$ and $\lambda = \mu = 1/2$, then the algorithm converges to the point $(0.608, 0.630)$ while the Nash bargaining solution is easily seen to be $(1/\sqrt{3}, 2/3)$. This implies that the solution obtained from the algorithm does not satisfy the axiom of independence of irrelevant alternatives. We see this as follows.

The solution $(x, y) = (0.608, 0.630)$ satisfies the axioms of individual rationality, efficiency and step by step linear invariance and we now investigate whether the axiom of symmetry also holds. For the specific game considered here relations (6.3) and (6.4) take the form

$$\begin{aligned} x_n &= 1/2\, x_{n-1} + 1/2\,(1 - y_{n-1})^{1/2} \\ y_n &= 1/2\, y_{n-1} + 1/2\,(1 - x_{n-1}^2), \end{aligned}$$

and $(x_1, y_1) = (0, 0)$. If we were now to consider the symmetric game by interchanging the roles of the two players we would have

$$\begin{aligned} x_n &= 1/2\, x_{n-1} + 1/2\,(1 - y_{n-1}^2) \\ y_n &= 1/2\, y_{n-1} + 1/2\,(1 - x_{n-1})^{1/2}, \end{aligned}$$

and $(x_1, y_1) = (0, 0)$. The solution of the new game would be $(x, y) = (0.630, 0.608)$ and therefore the axiom of symmetry is satisfied. It follows that the axiom of independence of irrelevant alternatives is not satisfied, as it is one of a set of necessary and sufficient conditions for the Nash bargaining solution and we have established that all other axioms are satisfied. If axiom (iv) was also satisfied then we would not have obtained $(x, y) = (0.608, 0.630)$, as we did, but instead the Nash bargaining solution $(x, y) = (1/\sqrt{3}, 2/3)$ would have emerged as the negotiated outcome.

We also give, through an easy example, a direct demonstration of the fact that algorithms of the above type, even when constructed to guarantee invariance, do not imply that the solution satisfies the axiom of independence of irrelevant alternatives. We show this with the help of the simple games in Figure 7.1 below.

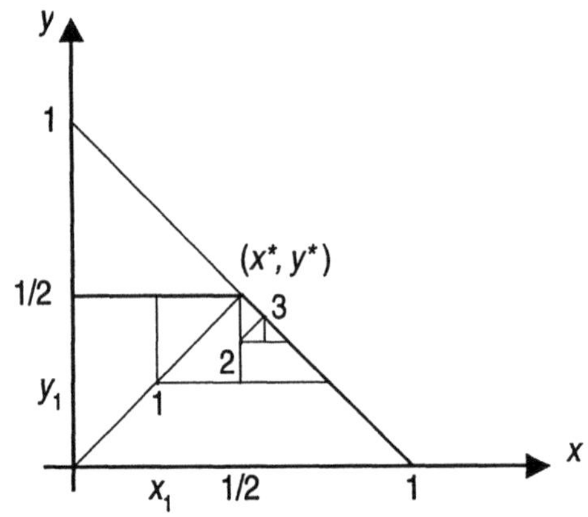

Figure 7.1

Suppose we have again the game of dividing the cake and the algorithm described by (6.3) and (6.4) with $\lambda = \mu = \frac{1}{2}$. Then, as mentioned above, in one step we go from any initial position 1 on the 45° line to the Nash bargaining solution $(x^*, y^*) = (\frac{1}{2}, \frac{1}{2})$. Suppose now the payoff region is truncated so that $y \leq \frac{1}{2}$ when $0 \leq x \leq \frac{1}{2}$, which leaves the Nash bargaining solution and the initial position still available. However the algorithm now implies the movement, always half the distance between what was demanded and what was offered, from the initial point 1 to point 2, and then to the efficient point 3, away from the Nash solution. Hence the axiom of independence of irrelevant alternatives does not hold. The structure of this example and the solution obtained here lend support to the McDonald and Solow [7] intuition.

Next we look briefly at an algorithm where λ and μ are not constant throughout. We suppose that the boundary of the payoffs region is given by a function. Explicitly we assume that

$$x_n = x_{n-1} + \max\left\{\tfrac{2}{3}(\tilde{x}_{n-1} - x_{n-1}),\ \tfrac{1}{3}(\tilde{x}_{n-1} - x_{n-1})\right\} \tag{7.1}$$

$$y_n = y_{n-1} + \max\left\{\tfrac{2}{3}(\tilde{y}_{n-1} - y_{n-1}),\ \tfrac{1}{3}(\tilde{y}_{n-1} - y_{n-1})\right\}. \tag{7.2}$$

The functions have all the required properties, and the algorithm clearly satisfies the axiom of linear invariance. Relation (7.1) implies that for $\tilde{x}_{n-1} \geq x_{n-1}$ we have $x_n = \frac{1}{3}x_{n-1} + \frac{2}{3}\tilde{x}_{n-1}$ and for $\tilde{x}_{n-1} \leq x_{n-1}$ the relation $x_n = \frac{2}{3}x_{n-1} + \frac{1}{3}\tilde{x}_{n-1}$. Hence x_n is continuous in the two areas of its definition and its two respective relations agree on the 45° line and therefore it is overall continuous. Similarly we can show that y_n is continuous in its arguments. The interpretation of this algorithm is that each player acts always in an optimistic manner, leaning in his decisions towards the largest utility between what he demanded and what he was offered.

We now turn to discuss briefly the assumption of convexification of the region of feasible payoffs and why we do not make it here. A justification for assuming that the set of feasible payoffs is convex is provided by allowing that the players take joint action in the form of joint mixed strategies. As mentioned above this might not be possible, if, for example, it is not allowed by the rules of the game, or if it is not in the interest of one of the players to accept such a device. Although the players do not know each other's adjustment function they can still perform calculations under alternative hypotheses and persuade themselves that they might be losing out through convexification.

Figure 7.2 below can be used to make precisely the point that it will not always be to a player's advantage to accept convexification of the payoffs region through joint mixed strategies.

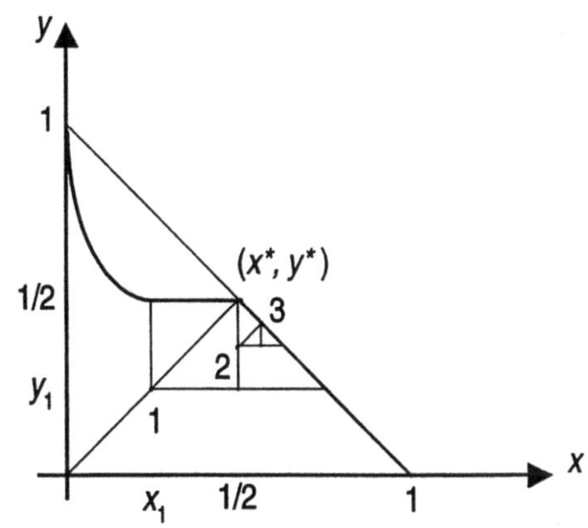

Figure 7.2

Inspection of the graph shows that P_1 could end up with more utility if he stuck to the non-convex region of payoffs rather than accepting convexification which would result in a game of dividing a cake of size one. For example if the algorithm is given by relations (6.3) and (6.4) with constants $\lambda = \mu = 1/2$ and the initial demands are at position 1, then he is better off to stay with the non-convex region. Although he does not know exactly what P_2 will do at each stage of the game, P_1 cannot exclude the possibility that he might end up with more utility from the non-convex region.

8. More general algorithms

More general algorithms for adjusting the demands at stage n of the game can also be considered, as for example those of the form

$$x_n = F(x_{n-1}, \tilde{x}_{n-1}, x_1, y_1, n, \lambda_{n-1}X_{n-1}), \quad \text{where} \quad \tilde{x}_{n-1} = \max g(y_{n-1}) \quad (8.1)$$

$$y_n = G(y_{n-1}, \tilde{y}_{n-1}, x_1, y_1, n, \lambda_{n-1}X_{n-1}), \quad \text{where} \quad \tilde{y}_{n-1} = \max f(x_{n-1}), \quad (8.2)$$

with $(x_1, y_1) = (\xi_1, \xi_2)$ and $0 < \lambda_{n-1} \leq 1$. In this formulation the adjustment functions could depend also on the surplus payoffs set X_{n-1} available for further negotiations before x_n and y_n are announced, possibly contracted by a factor λ_{n-1}, and explicitly on the status–quo payoffs. Furthermore the adjustment mechanism is not necessarily the same in each iteration. All properties concerning the continuity of F and G and the relation between current demands and demands and offers in the previous stage of the play of the game are retained. Of course if the boundary of the payoff region is given by a function then we have $\tilde{x}_{n-1} = g(y_{n-1})$ and $\tilde{y}_{n-1} = f(x_{n-1})$.

A particular example of such a more general algorithm which satisfies the axioms of individual rationality, efficiency and linear invariance is given by

$$x_n = [1 - \tfrac{1}{2}(\tilde{x}_{n-1} - \xi_1)/(\tilde{x}_1 - \xi_1)]x_{n-1} + [\tfrac{1}{2}(\tilde{x}_{n-1} - \xi_1)/(\tilde{x}_1 - \xi_1)]\tilde{x}_{n-1} \quad (8.3)$$

$$y_n = [1 - \tfrac{1}{2}(\tilde{y}_{n-1} - \xi_2)/(\tilde{y}_1 - \xi_2)]y_{n-1} + [\tfrac{1}{2}(\tilde{y}_{n-1} - \xi_2)/(\tilde{y}_1 - \xi_2)]\tilde{y}_{n-1}. \quad (8.4)$$

At each stage of the game and assuming there is further room for negotiations, the players adjust their demands taking into account the fraction of the surplus utility, over their status–quo payoff, they were offered in the previous stage over the surplus utility they were offered at the beginning of the game. The larger this fraction is the more permanent will a player perceive his opponent's offer to be. Consequently they will attach a larger weight to what they have just been offered. Notice that the weights of x_{n-1} and $\tilde{x}_{n-1}$ in (8.3), and of y_{n-1} and $\tilde{y}_{n-1}$ in (8.4) now vary in each iteration.

Next we show that in the context of the more general algorithms described by (8.1) and (8.2) we can also obtain, for convex X, the Nash bargaining solution (x^*, y^*) as the limit of a sequence of demands. As we know this solution can be obtained from the following equivalent problems.

Problem 1	**Problem 2**	**Problem 3**
$\max\ (x - x_1)(y - y_1)$	$\max\ (x - x_1)(f(x) - y_1)$	$\max\ (y - y_1)(g(y) - x_1)$
Subject to $y = f(x)$,	Subject to	Subject to
$x \geq x_1,\ y \geq y_1$	$x_1 \leq x \leq g(y_1)$	$y_1 \leq y \leq f(x_1)$,

where $(x_1, y_1) = (\xi_1, \xi_2)$ is the status–quo payoffs vector and $g(y)$ is the inverse function of $f(x)$.

Problem 1 is in the conventional form and is the outcome of arguing on the basis of the Nash axioms. Problems 2 and 3 formulate the wish of P_1 and P_2, respectively, to maximize their expected utility functions, as in [3, 4]. Each player has no evidence about the behavior of the other one and acts as follows. He attaches a uniform density function to the choices of his opponent, i.e., he considers all of them to be equally likely, and for each one of his own choices he obtains his expected surplus utility payoff. He then announces as his demand the choice that will maximize his expected surplus utility function. These individual demands of the players are the same as those of the Nash bargaining solution which emerges now as the outcome of separate actions taken by the players.

First we see that the solution to Problem 2 and that to Problem 3 describe in effect a one–step algorithm which gives immediately the Nash bargaining solution. This type of adjustment process of the demands of the players belongs to the family of generalized algorithms given by relations (8.1) and (8.2) above and is of the form

$$x_n = F(x_{n-1}, \tilde{x}_{n-1}, X_{n-1}), \quad \text{where} \quad \tilde{x}_{n-1} = g(y_{n-1}) \tag{8.5}$$

$$y_n = G(y_{n-1}, \tilde{y}_{n-1}, X_{n-1}), \quad \text{where} \quad \tilde{y}_{n-1} = f(x_{n-1}), \tag{8.6}$$

with $(x_1, y_1) = (\xi_1, \xi_2)$. The surplus payoffs set X_{n-1} from which x_n and y_n will be drawn is the area of X with $x \geq x_{n-1}$ and $y \geq y_{n-1}$.

In particular F and G are now defined in the following way. Given X_{n-1}, the players P_1 and P_2 calculate their demands of next period by solving the following two problems, respectively.

Problem 2′	**Problem 3′**
$\max\ (x - x_{n-1})(f(x) - y_{n-1})$	$\max\ (y - y_{n-1})(g(y) - x_{n-1})$
Subject to $x_{n-1} \leq x \leq g(y_{n-1})$	Subject to $y_{n-1} \leq y \leq f(x_{n-1})$.

P_1 will announce as x_n the solution to Problem 2′ and P_2 will announce as y_n the solution to Problem 3′. The construction of the algorithm is such that given $(x_1, y_1) = (\xi_1, \xi_2)$ the two players, through their separate calculations, arrive in one step at (x^*, y^*) and the game terminates, as any subsequent calculations will simply confirm the Nash bargaining solution.

On the other hand we can also consider many–step algorithms. The idea is that the players do not negotiate all at once over the whole of the surplus payoff region, but instead, in successive steps, over appropriately defined sub-regions. As each subgame is solved they advance to a new position from which they will carry out the next round of negotiations. Although payoffs will only be handed out if eventually agreement is reached over the whole game, this new position can in effect be considered as their status–quo payoff for the round of bargaining over the next sub-region of surplus payoffs. At each stage demands are generated by the individual decisions of the players.

We can think, for example, of amounts of different currencies which have been translated into a common unit. The eventual aim of the players is to negotiate over the total sum of money. It is possible however that they keep in mind the amounts of the various currencies and they negotiate first over pounds, then they negotiate further by adding the amount of dollars, etc., temporarily retaining by agreement the outcome of each negotiation. This then becomes their status–quo payoff for the next round of negotiations, etc. Or it might be that the size of the payoffs region is very large and that the agents prefer to negotiate starting with a smaller region, reach a temporary agreement and then gradually proceed to bigger regions. There could be an infinite or a finite number of subgames. In the former case X will be approached as a limit.

We can obtain the Nash bargaining solution from a many–step algorithm of the form

$$x_n = F(x_{n-1}, \tilde{x}_{n-1}, \lambda_{n-1}X_{n-1}), \quad \text{where} \quad \tilde{x}_{n-1} = \max g(y_{n-1}) \tag{8.7}$$

$$y_n = G(y_{n-1}, \tilde{y}_{n-1}, \lambda_{n-1}X_{n-1}), \quad \text{where} \quad \tilde{y}_{n-1} = \max f(x_{n-1}), \tag{8.8}$$

with $(x_1, y_1) = (\xi_1, \xi_2)$ and $0 < \lambda_{n-1} \leq 1$. Again, this type of adjustment process belongs to the family of generalized algorithms given by relations (8.1) and (8.2) above. We explain below by means of graphs for a typical game the construction of such a many–step algorithm. It generates a sequence of demands which tends to the Nash bargaining solution (x^*, y^*).

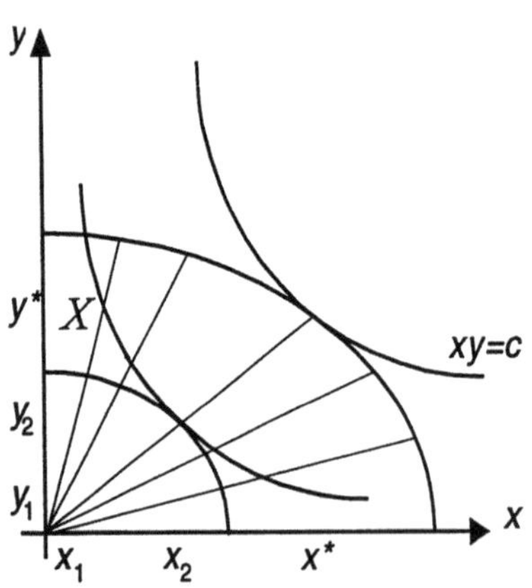

Figure 8.1

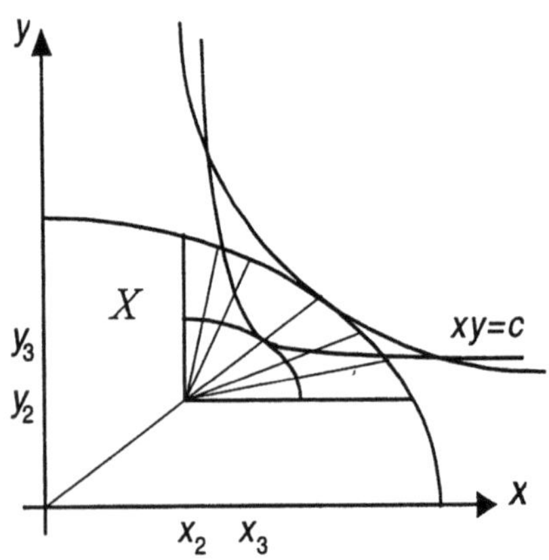

Figure 8.2

Figure 8.1, in which without loss of generality the status–quo payoffs are taken to be equal to zero, shows the first stage and Figure 8.2 the second stage of the game and the role of the contracting factor λ_{n-1} becomes clear.

In the first stage of the game the players negotiate over a sub-region of X which is constructed as follows. The boundary is subjected to a homothetic λ_1–contraction, i.e., it is pulled towards the origin by a factor λ_1. The contracted boundary is shown on the diagram. This contraction means that the slope of the original boundary and of the new one are equal at the points where they meet the same ray through the origin. Now the iso–level curves of the Nash formulation are also homothetic with respect to the status–quo payoffs which is here at the origin. Therefore the solution of the new Nash problem with the contracted boundary will be at the point $(\lambda_1 x^*, \lambda_1 y^*)$ and we know that this will also emerge as the outcome of independent decisions made by the players. They will announce as their demand the choice that will maximize their expected surplus utility function in the context of the new problem. The solution $(\lambda_1 x^*, \lambda_1 y^*)$ will be the starting point for the second round of the negotiations.

Next consider how we arrive at the calculations of the second stage of the algorithm. The Nash and its equivalent problems of the two players with status–quo payoffs $(\lambda_1 x^*, \lambda_1 y^*)$ are now the following.

Problem 1″	**Problem 2″**	**Problem 3″**
$\max\ (x - x_2)(y - y_2)$	$\max\ (x - x_2)(f(x) - y_2)$	$\max\ (y - y_2)(g(y) - x_2)$
Subject to $y = f(x)$,	Subject to	Subject to
$x \geq x_2,\ y \geq y_2$	$x_2 \leq x \leq g(y_2)$	$y_2 \leq y \leq f(x_2)$,

where $(x_2, y_2) = (\lambda_1 x^*, \lambda_1 y^*)$.

However these are not the problems that will be solved at the second stage of the game. Instead the boundary of X is again contracted by λ_2 but now in the direction of (x_2, y_2). The players will now bargain over the sub-region determined by this contraction and (x_2, y_2) will be considered as the status–quo payoffs vector. The relevant segment of the new contracted boundary is shown in Figure 8.2. On the other hand in order to obtain the solution to the new problem we have to investigate the properties of the new iso–level curves $(x - x_2)(y - y_2) = c'$ referring to status–quo payoffs vector (x_2, y_2). They are homothetic with respect to the origin (x_2, y_2) but they do not have the same shape as $(x - x_1)(y - y_1) = c$. It is a matter of routine calculations to show that the slopes of the two types of indifference curves are only the same along the ray through the origin and the point (x^*, y^*), which we call the Nash ray. To the left of this ray the new indifference curves have a sharper slope and to the right they are flatter.

The significant thing is, though, that on the Nash ray the slopes are the same. Therefore the solution of the new reduced problem will also be on the Nash ray and again it will be confirmed by the individual decisions of the players. Continuing in this manner we can obtain the Nash bargaining solution as the limit of a sequence of demands generated by the individual decisions of the players as they solve a number of sub-problems. This will be the negotiated outcome of the game that will be handed out to the two players.

The process above describes an algorithm which belongs to the family of those given by relations (8.1) and (8.2). Of course it is not necessary that contraction of the relevant boundary takes place by the same factor throughout the construction of the various steps of the adjustment process. We still obtain that the sequence of demands tends to the Nash bargaining solution in a number of steps. We could even terminate the process after a finite number of steps by choosing in the last one $\lambda_{n-1} = 1$. We shall again obtain the Nash bargaining solution as the negotiated outcome of the game.

An analogous argument can be applied to obtain the generalized Nash bargaining solution as the limit of a sequence of demands generated as the players solve appropriately constructed sub-problems.

References

1. J. K. Arrow and F. H. Hahn, *General Competitive Analysis*, Holden–Day, Inc., San Francisco, Oliver & Boyd, Edinburgh, 1971.
2. K. Binmore, Nash bargaining theory I, ICERD, Discussion Paper no. 9, London School of Economics, 1980, 1–40. Also published in: *The Economics of Bargaining* (eds. K. Binmore, P. Dasgupta), Basil Blackwell, Oxford, 1987, 27–46.
3. D. Glycopantis and A. Muir, On the Nash bargaining solution: an alternative approach, *British Review of Economic Issues* 16 (1994), 55–71.
4. D. Glycopantis and A. Muir, A probabilistic approach to the Nash bargaining solution, *Economic Notes, Monte dei Paschi di Sena* 24 (1995), 197–218.
5. J. C. Harsanyi, Bargaining, in: *The New Palgrave, A Dictionary of Economics* (eds. J. Eatwell, M. Milgate, and P. Newman), MacMillan, London, 1987, 190–195.
6. W. Hildenbrand and A. P. Kirman, *Introduction to Equilibrium Analysis*, North-Holland, Amsterdam, Oxford; American Elsevier, New York, 1976.
7. I. M. McDonald and R. M. Solow, Wage bargaining and employment, *American Economic Review* 71, no. 5 (1981), 896–908.

8. J. F. Nash, The bargaining problem, *Econometrica* 18 (1950), 155–162.

9. J. F. Nash, Two–person cooperative games, *Econometrica* 21 (1953), 128–140.

D. Glycopantis
Department of Economics
City University
Northampton Square
London EC1V OHB
Great Britain
email: d.glycopantis@city.ac.uk

A. Muir
Department of Mathematics
City University
Northampton Square
London EC1V OHB
Great Britain
email: a.muir@city.ac.uk

SOCIAL STATES OF BELIEF AND THE DETERMINANT OF THE EQUITY RISK PREMIUM IN A RATIONAL BELIEF EQUILIBRIUM

MORDECAI KURZ*

Abstract. We review the issues related to the formulation of endogenous uncertainty in Rational Belief Equilibria (RBE). In all previous models of RBE, individual states of belief were the foundation for the construction of the endogenous state space where individual states of belief were described with the method of assessment variables. This approach leads to a lack of "anonymity" where the belief of each individual agent has an impact on equilibrium prices but as a competitor he ignores it. The solution is to study a replica economy with a finite number of types but with a large number of agents of each type. This gives rise to "type–states" which are distributions of beliefs within each type. The state space for this economy is then constructed as the set of products of the exogenous states and the social states of belief which are vectors of distributions of all the types. Such an economy leads to RBE which do indeed solve the problem of anonymity. We then study via simulations the implications of the model of RBE with social states for market volatility and for the determinants of the equity risk premium in an RBE. Under i.i.d. assessments one uses the law of large numbers to induce a single social state of belief and we show that the RBE of such economies have the same number of prices as in rational expectations equilibrium (REE). However, the RBE may exhibit large fluctuations if agents are allowed to hold extreme beliefs. Establishing 5% boundary restrictions on beliefs we show that the model with a single social state of belief cannot explain all the moments of the observed distribution of returns. We then introduce correlation among beliefs and this leads to the creation of new social states. We next show that under correlation among beliefs the model simulations reproduce the values of four key moments of the empirical distribution of returns. The observed equity premium is then explained by two factors. First, investors demand a higher risk premium to compensate them for the endogenous increase in the volatility of returns. Second, at any moment of time there are both rational optimists as well as rational pessimists in our financial markets and such a distribution leads automatically to a decrease in the riskless rate and to an increase of the risk premium. We show that correlation among beliefs of agents leads to fluctuations over time in the social distribution of beliefs and such fluctuations add to endogenous volatility and lead to a higher equilibrium equity risk premium.

---

*This research was supported by Fondazione ENI Enrico Mattei, Milano, Italy. The author thanks Stanley Black, William Brock, Mark Garmaise, Kenneth Judd, Michael Magill, Maurizio Motolese, Carsten Nielsen, and Martine Quinzii for many helpful suggestions regarding the research reported in this paper. He is grateful to Stanley Black and Maurizio Motolese for dedicated assistance in carrying out the computations. This paper is a drastic revision of [26].

1. The state space and the emergence of endogenous uncertainty

The role of the market mechanism in the optimal allocation of risk bearing has been one of the most extensively studied problems in economics. The theory of general equilibrium, as developed by Arrow and Debreu [3] and by Arrow [4], provided an extremely fertile framework for the examination of the behavior of markets for uncertain prospects in general and the markets for insurance and risky securities in particular. The rapid development of the field of finance is a noteworthy example of the impact of this framework of analysis. Yet, despite these impressive achievements important foundational questions regarding the nature of social uncertainty remain unresolved. This paper explores alternative equilibrium concepts in which a central role is played by Endogenous Uncertainty, a concept which was defined in [31] and explored in a sequence of recent papers (see [29, 30, 22, 23, 25, 28, 27, 24, 36]). In order to explore the emergence of endogenous uncertainty, it would be useful first to review some of the problematics arising out of the treatment of uncertainty in the Arrow–Debreu [3] model, the role played by securities in Arrow's [4] equilibrium and the modifications of the theory by Radner [40, 39, 38].

As is well understood, the full generality of the Arrow–Debreu formulation enables the incorporation of uncertainty merely by a reinterpretation of the symbols employed. In the original Arrow–Debreu [3] paper terms like "risk" or "uncertainty" *are not even mentioned.* In his explicit treatment of uncertainty Arrow [4] defines the exogenous "state space" and explicitly introduces markets for state contingent claims on commodity bundles and the utility of such uncertain commodities. He notes that the treatment of the uncertain case is entirely analogous to the case of certainty except for the enlarged dimension of the commodity space (which equals the number of physical commodities *multiplied* by the number of exogenous "states"). Motivated by markets for insurance, Debreu [14] uses the broader terminology of "events" to identify subsets of states but his formal treatment requires the trading of a complete set of *state* contingent commodities. Apart from the formal interpretation of the concept of a "commodity" the uncertainty interpretation raises only one issue of substance with respect to the assumption of convexity of preferences. Since in the case of uncertainty[1] convexity implies risk aversion, both the existence of competitive equilibrium in the Arrow–Debreu theory as well as the optimality theorem in [4] are proved under the assumption of universal risk aversion.

[1]And assuming expected utility maximization with preferences which are state independent.

It is clear that the crucial step taken in the Arrow–Debreu formulation of uncertainty within general equilibrium theory is the introduction of the concept of "the state" into the theory. This concept, however, is the cornerstone of the theory of *individual* decision theory and subjective probability. In Savage's [42] treatment the concept of "the state of the world" is nothing more than a formal description of what a decision maker is uncertain about. Consequently Savage [42] defines the "world" to be "the object about which the person is concerned" whereas "a state" (of the world) is defined as " ... a description of the world, leaving no relevant aspect undescribed."[2]

Arrow learned mathematical statistics from Hotelling and Wald, and was influenced by Savage's approach to subjective probability. In some early papers he does not even provide a definition of the concept of the "state" and takes it to be both known as well as naturally applicable to the economic problem at hand (e.g. [5, 4]). In later papers (e.g. [1] or [2]) he provides a precise definition of the "state of the world" as " ... a description of the world so complete that, if true and known, the consequences of every action would be known."[3]

In the context of decision theory the concept of the "state" is no more than a tool for the formulation of the individual decision problem. As such, it is entirely satisfactory and indispensable. In fact, it is hard to visualize how one can formulate a stochastic dynamic decision problem without a concept like a "state." Moreover, the formulation of any decision problem as well as Savage's theory of subjective probability neither require the "state" to be observable nor need its description be communicable to or be understood by other decision makers.

The generality of the decision theoretic framework naturally led Arrow and Debreu to adopt this framework for the formulation of the problem of choice under uncertainty of every economic agent in a competitive economy. The important theoretical step which they took was to endow all the agents *with the same state space* and to provide them with the market opportunity of trading the uncertainty defined by the "state." That means that the concept of "the state" became a major tool of general equilibrium analysis. In contrast with the context of the individual decision problem where the "state of the world" is merely an expression of individual uncertainty, in the general equilibrium framework "the state of the world" becomes a description of commodities, it identifies markets and becomes a basis for specifying contracts and property rights. In such a framework the concept must satisfy the same marketability criteria as "navel oranges available in Palo Alto, California, on November 29,

[2]See [42, page 9].
[3]See [1, page 20].

1997": it must be precisely defined, commonly observable and unequivocally comprehended by all economic agents. These requirements clearly raise some difficult practical problems of description.[4] However, the theoretical structure of the exogenous state space enabled Arrow and Debreu to achieve a complete integration of the theory of value.

It is noteworthy that the example of insurance motivated the Arrow–Debreu approach to uncertainty. Indeed, for a description of commodities, the concept of "the state of the world" is extremely useful in characterizing markets for insurance. This is so since an insurance policy is a contract in which the owner receives specified compensations if the state of the world belongs to an event such that the insured object meets a long list of described conditions. In this case "the state of the world" description of the commodity has the precise interpretation of the "sample space" in probability models. Insurance markets function well when the contingency conditions are unambiguous and their probability distributions are truly exogenous and cannot be altered by the behavior of the insured.

Notwithstanding the importance of the integrated vision of the Arrow–Debreu theory, it is evident that the construct of markets for claims which are contingent on the exogenous states constitutes an unsatisfactory solution to the problem of allocating risk in a market economy. Arrow [4] himself observes that outside the insurance framework, markets for commodity claims which are contingent upon the exogenous states do not exist. Moreover, even the insurance markets do not function as visualized in the theory. More specifically, in order to study insurance markets, Malinvaud [33, 32] considers a large economy with individual risks for which a complete set of insurance markets exists in the form of insurance pools that are used for averaging individual risks. In a large economy all risk averse agents clearly purchase fair insurance. It is then shown by Malinvaud that given such pools, in the equilibria of these economies agents trade only in certainty contracts: individual uncertainty disappears from general equilibrium considerations. The implication of the Malinvaud analysis is that in a general equilibrium context the main problems of allocating risk are not associated with the allocation of individual idiosyncratic risks but rather, the allocation of *collective* risk bearing for which the laws of large numbers are not available. We argue in this paper that this conclusion continues to hold when endogenous uncertainty is introduced. Whether exogenous shocks can account for all observed social risks as reflected in the economic fluctuations of quantities and prices is probably the central question at hand. It is evident that the list of *observed* variables which are truly exogenous to the economic

[4]For a description of the exogenous state see [14, page 98].

universe is very short and the range of their variability and impact are much too small to account for the observed variability of economic variables. Thus, one must conclude that if the exogenous shocks are all that matters then the most relevant components of the "state" are not commonly observable and cannot provide a basis for contingent contracts.

The non-existence of markets for contingent claims posed a problem to general equilibrium theory. Arrow's [4] celebrated solution has become the foundation of modern general equilibrium theory of finance. He recognized that without markets for contingent claims one must think of an economy *as a sequence of spot markets linked together by a market for securities* which enable the reallocation of incomes across the different state–date combinations. In Arrow's [4] formulation and in the extension by Radner [39], an equilibrium consists of a set of market clearing spot price functions p_t of commodities associated with each of the finite number of the state–date pairs (s, t), and a set of market clearing prices of securities which pay different dividends in different "states." Since the equilibrium is established at the date $t = 0$ which we can think of as "the present," such an equilibrium requires the agents *to know at* $t = 0$ *all prices* $p_t(s)$ *that would prevail at all future dates and all states* s. This assumption of "Rational Expectations" is the foundation of the optimality theorem of Arrow [4]. It is also the basis for most work in finance which seeks to show that Pareto optimality is obtained whenever the set of securities "spans" the set of exogenous states.

The rational expectations equilibrium concept of Arrow [4] and its extension by Radner [39] elevates the exogenously specified "state" substantially above Arrow's own definition (e.g. [1, page 20]). It is no longer just such a complete description that the consequences of all *individual actions* are known; now the requirement is that the knowledge of the exogenously specified state enables every agent to know the consequences of all collective actions as well and, in particular, to know all future *prices* in the economy. These ideas extend further to the treatment of general equilibrium with private information (e.g. see [38, 37]). The agent's knowledge of the *price maps* $p_t(s)$ plays a crucial role in the public revelation of private information.

The assumption of rational expectations in the Arrow–Radner equilibrium is viewed, almost universally, as placing excessive and unreasonable demands on the agents: since the map $p_t(s)$ is not observable, how could the agents know it at date 0? The term "rational" in connection to the knowledge of this map appears to mean that agents know the structure of the economy so completely (including technology and resources as well as preferences and endowments of other agents) that for each exogenous state s the agents can carry out all general equilibrium calculations needed to deduce the map $p_t(s)$ [5] for all future dates. It is then natural to ask what if the agents do not know the map since they do not have "structural knowledge."[6] The Arrow–Radner equilibrium theory does not apply since agents cannot carry out, at date 0, the kind of intertemporal planning which the theory calls upon them to do. The needed extension of the theory to the case where agents do not have structural knowledge has been recently proposed (see, for example, [30, 23, 25, 28, 27, 36] all of which are included in the volume [21]) by the theory of Rational Belief Equilibrium (in short, RBE). The theory of RBE leads, in a natural way, to the emergence of endogenous uncertainty (see [31]) which is that part of social uncertainty (and hence economic fluctuations) which is propagated within the economy rather than being "caused" by exogenous shocks. We now explore this connection in some detail.

Recall that it was Arrow's [4] and Radner's [39] views that without markets for contingent claims at date $t = 0$ an equilibrium for the economy *is a sequence of market clearing spot prices of the reopened markets at the different dates.* But then at $t = 0$ agents are uncertain about future spot prices at $t = 1, 2, \dots, T$. If we then follow Savage's [42] dictate, then future spot prices are part of the "world" about which all agents are uncertain. This means that the state, which is a description of the world, *should include future spot prices.* Agents are therefore uncertain about their future utilities not only because of the effect of exogenous random variables but also because they are uncertain about those future spot prices that would prevail, at any configuration of the exogenous variables. However, if prices are part of "the state of the world" then agents cannot view prices as a known equilibrium map like $p_t(s)$. Moreover, from the point of view of each agent the state space does not consist of abstract and unknown objects but rather, in the case of M equilibrium prices, the price state space is simply the set of integers $\{1, \dots, M\}$ and in the case of a continuum of prices, the space is the unit interval. With this enlargement of the "state space" we lower the concept of "the state of the world" back to where it is merely *a terminology for the description of what agents are uncertain about.* However, this change of the state space has far reaching implications for the way we need to think about uncertainty in a general equilibrium context and for our perspective on what *social uncertainty* is.

Once agents view prices as random variables they must form probability beliefs about future prices in the same way they form beliefs about exogenous

[5] For this reason the assumption is sometimes called "conditional perfect foresight."

[6] We have introduced this term earlier (see [29]) in order to distinguish knowledge about the *state of the economy* which is considered "information" and knowledge about *the functioning of the economy* which we call "structural knowledge."

variables. Since Savage [42], Arrow [4] and Radner [39] allow agents to have different probability beliefs about what they are uncertain about, it follows that if an equilibrium concept is to permit agents to be uncertain about future prices, then *equilibrium prices at each date must depend upon what agents expect future equilibrium prices to be*! Formally, suppose that in an economy with K agents we denote by $y_t = (y_t^1, \ldots, y_t^K)$ the date t vector of conditional probabilities of the K agents about all equilibrium events after date t conditional *upon the entire past.* y_t is the "state of belief" in the economy and y_t^k is the state of belief of agent k. The decision functions of the agent at each date take the general form

$$x_t^k = F^k(p_{(t)},\ s_{(t)},\ y_t^k),$$

where $z_{(t)} = (z_0, z_1, \ldots, z_t)$ denotes the entire history of a variable z. Market clearing conditions establish equilibrium prices p_t at each date t as

$$p_t = \varphi_t(y_t, s_{(t)}) \tag{1.1}$$

and in the special and useful case of finite memory equilibria, (1.1) takes the simpler form

$$p_t = \varphi(y_t, s_t). \tag{1.2}$$

The map (1.1) which is unknown to the agents in an RBE corresponds to the Arrow–Radner price map $p_t(s)$ which is assumed to be known to the agents. The crucial difference is the emergence of the state of belief which becomes part of the enlarged state space for the economy. In either case (1.1) or case (1.2) the fluctuations of prices over time are in part due to fluctuations of the exogenous shocks s_t and in part to the fluctuations in the state of beliefs y_t.

In a dynamic economy consisting of a sequence of markets, economic risk is an intertemporal phenomenon in the sense that what agents perceive as risk is directly linked to the fluctuations of the economy over time and against such variability they wish to insure themselves. Endogenous uncertainty is then that component of economic fluctuations which is due to the impact of the agent's beliefs on the variability of prices or other endogenous variables. This effect is generated both by the time variability of the states of beliefs of the agents as well as by the structure of the maps (1.1) or (1.2). Since the agents do not know the true equilibrium map between states (y_t, s_t) and prices and since they do not observe states of beliefs, they can learn something from an examination of the data generated by the economy. One of the main conclusions of [29] is that there is no basis to expect that agents will learn the true structure of the maps (1.1) or (1.2) and what is the true probability distribution of exogenous shocks. For this reason the agents form probability beliefs about prices and exogenous states knowing that *the exogenous state space is a partition of the price state space.*

The emergence of endogenous uncertainty in economies where agents do not have structural knowledge points to the observation that in such economies "expectations matter" and have real effects on equilibrium allocations. The theory of Rational Beliefs establishes the limits within which individual conditional probability beliefs may vary if they satisfy the basic rationality principle that such expectations are compatible with the data generated by the economy. An RBE is an equilibrium in which agents do not have structural knowledge and hold rational beliefs.

In what sense should endogenous uncertainty, as defined above, be taken to be "endogenous" and "stochastic"? Observe that endogenous uncertainty is generated by *variations* in the state of beliefs of the agents each of whom selects a rational belief from a set of probability beliefs which satisfy the axioms of rationality. Since the selection of a rational belief is an endogenous phenomenon and their adopted beliefs *cause* aggregate risk and fluctuations, the uncertainty which is induced by these selections is "endogenous" in the sense that it is generated *within* the economy rather than caused by exogenous shocks. In the development below we employ the technique of a Markov model in which each agent uses a *privately generated stochastic assessment variable* (for details on this approach see [27]). On the basis of this realization the agent determines which of a finite number of transition matrices to use on that date. This leads to a tractable modeling of the aggregate states of belief since the individual state of belief is fully described by the realization of his private assessment variable.

The present paper aims to explore alternative ways of defining the expanded state space of an economy with endogenous uncertainty. One may represent the states of belief in the economy either as vectors of the *states of beliefs of the individual agents* or as *distributions* of individual states of belief. Such two descriptions are obviously closely related but we note that the RBE concept used in all the papers cited above defines the states of belief using the first of these two alternatives. We explain below that with a finite number of agents such an RBE lacks a desired property of "anonymity" in the sense that the belief of an agent has an impact on equilibrium prices but, as a competitor, *he is required to ignore it.* Needless to say, lack of anonymity is a universal problem which is common to all competitive models with a finite number of agents. The interest in the second approach is based on the fact that it has two important implications. On the one hand it leads to a concept of an RBE which possesses the anonymity property and thus demonstrates that in a large economy the belief of any one agent does not matter for aggregate behavior.

On the other hand, this view of equilibrium explains how the *distribution* of beliefs affects aggregate behavior and why in applications it is important to focus on the properties of this distribution. Our exploration is carried out both analytically in Section 3 as well as via simulations in Section 4 of this paper.

2. Rational Belief Equilibria (RBE) with individual states of belief

2.1. A family of OLG models with a finite number of equilibrium prices

Some of the papers mentioned earlier (i.e. [25, 28, 27, 24, 36]) use a standard two period OLG model with a single consumption good but vary in the structure of securities which are available. Nevertheless, the construction of the expanded state space which includes the *vector* of individual states of belief is the same in all of them. Since the aim of the present paper is to show how an endogenous state space can be constructed so as to depend only on *social states of belief* and not on *individual states of belief*, we select one of these models and follow its development. This enables us to explain why RBE with individual states of belief lack anonymity. In Section 3 we show how the use of "social states of beliefs" leads to RBE which have the anonymity property.

We note that in OLG models with a single, homogeneous, consumption good *old* agents do not need to optimize by allocating a budget over alternative consumption vectors. As a result, equilibrium prices do not depend upon the entire history of the economy and under our assumptions such RBE have a finite number of equilibrium prices. Since we aim to study the construction of the state space, we consider the assumption of a single consumption good as a convenient simplification. Our construction continues to hold in an economy with an infinite number of equilibrium prices but is technically more demanding. We now outline the basic model.

The agents in the economy live for two periods. At any date there are K young agents denoted by $k = 1, 2, \ldots, K$. There are also K old agents in each generation but only the young receive an endowment $\Omega_t^k \in \mathbb{R}_+$ initially assumed constant.[7] The assumption of a constant endowment stream represents, as usual in OLG models, the labor supply of each young agent.

[7]Both Kurz and Wu [28] as well as Kurz and Schneider [27] assume the endowment to be constant over time. On the other hand, Nielsen [36] and Kurz and Beltratti [24] assume, for their modeling purposes, that $\{\Omega_t^k,\ t = 1, 2, \ldots\}$ is a stochastic process for each k. In the analytical discussion of Section 3 we assume endowment to be constant but adopt the Kurz and Beltratti [24] framework in the simulations of Section 4.

Each young person is a copy of the old person who preceded him where the term "copy" refers to the *utilities*, *endowments* and *beliefs*. Hence, ours is a model of "dynasties" and we assume that there is a finite number of such dynasties. In addition to a competitive market for the consumption good, two types of financial assets are traded at each date in competitive markets in the economy. The first asset is the common stock of an infinitely lived firm and at date 1 the supply (equal to 1) of the stock is distributed among the old at that date. The infinitely lived firm is assumed to be simple: it generates *exogenously* a stochastic sequence $\{R_t \in \mathbb{R}_+,\ t = 1, 2, \dots\}$ of dividends in the form of positive quantities of the perishable homogeneous commodity. We assume that the process $\{R_t \in \mathbb{R}_+,\ t = 1, 2, \dots\}$ is a finite state Markov process which will be specified below. The second asset is a zero net supply real short term bond which is issued at t and pays at $t+1$ one unit of the consumption good.[8] The notation which we employ in this paper is as follows:

x_t^{1k} – the consumption of k when young at t;

x_{t+1}^{2k} – the consumption of k when old at $t+1$. This indicates that k was born at date t;

θ_t^k – stock purchase of young agent k at t;

B_t^k – bond purchase of young agent k at t;

Ω^k – the endowment of k when young;

p_t^c – the price of consumption goods at date t;

P_t – the price of the common stock at date t;

q_t – the price of the bond at date t;

$s_t = (R_t, p_t^c, P_t, q_t) \in S$ is the state *from the point of view of the agents*;

$\mathcal{B}(A)$ – the Borel subsets of any measurable set A in a Euclidean space.

We turn now to specify our basic assumptions.

2.1.1. Assumption. *For each k, $u^k(\cdot)$ is a strictly increasing and quasi-concave function.*

We restrict attention to a Markovian economy along the lines of [27]. Thus we assume that $\{R_t,\ t = 1, 2, \dots\}$ is an exogenous dividend process where $R_t \in D \subseteq \mathbb{R}_+$.

[8] We note that Nielsen's [36] economy has a pure fiat money used by young agents as a store of value to transfer income from t to $t+1$. Kurz and Wu [28] follow Svensson [44] and Henrotte [19] in using Price Contingent Contracts (in short, PCC) which enable an agent to contract for the delivery of a unit of the common stock at future dates contingent upon the *prices* which prevail at these future dates.

2.1.2. Assumption. *D is a finite set with $|D|$ positive quantities; the process $\{R_t,\ t = 1, 2, \dots\}$ is a stable Markov process on D with probability measure Π_D defined on $(D^\infty, \mathcal{B}(D^\infty))$ with a stationary measure m_D.*

The price process $\{(p_t^c, P_t, q_t) \in P^*,\ t = 1, 2, \dots\}$ is of interest. The measurable set $P^* \in \mathbb{R}_+^3$ of the appropriate space of feasible prices is of central importance in the analysis below. A belief of agent k is a probability on sets of sequences $\{(R_t, p_t^c, P_t, q_t) \in D \times P^*,\ t = 1, 2, \dots\}$ and, as in [27], we characterize such beliefs with the technique of *private assessment variables.* An assessment variable y_t^k is a random variable or a parameter that agent k perceives at t. The probability of assessment variables *is part of the identity of the agent in the sense that it is selected by the agent as part of his model of the market.* It is thus clearly allowed to be stochastically interdependent with other economic variables. Putting it differently, the agent has a theory about the market mechanism which is represented by the probability belief Q^k and this belief entails some assessment which will represent the state of belief of the agent. The value of an assessment variable may depend upon market observables and would thus summarize the state of belief of the agent although it will have a random component. The probability belief Q^k is then a joint probability of the observed market data and the assessment variable. We stress that the assessment is a description of the agent's *perception* and may be considered a parameter of his belief. He alone can understand its meaning, it cannot be observed or comprehended by anyone else and should not be confused with "information" or "data" with respect to which our standard rationality of belief conditions apply. As explained in [27], the method of private assessment variables is introduced to allow us a tractable description of non-stationarity in the dynamics. Given y_t^k the agent selects one from among a *finite* number of Markov matrices to apply at t and hence, for any infinite sequence of y_t^k, his effective belief is the *conditional probability of Q^k given the sequence.* The domain of y_t^k is Y^k and is assumed to be a finite subset in $\mathbb{R}$. Q^k is then a probability on the space $((D \times P^* \times Y^k)^\infty,\ \mathcal{B}((D \times P^* \times Y^k)^\infty))$. In sum we have the following.

2.1.3. Assumption. *For all k, the system*

$$((D \times P^* \times Y^k)^\infty,\ \mathcal{B}((D \times P^* \times Y^k)^\infty),\ Q^k,\ T)$$

is stationary and ergodic. Y^k is a finite subset in $\mathbb{R}$ with $|Y^k|$ elements and under Q^k agent k believes that the process $\{(R_t, p_t^c, P_t, q_t, y_t^k),\ t = 1, 2, \dots\}$ is a Markov process. The non-stationarity induced by each assessment sequence $y^k \in (Y^k)^\infty$ is a selection, at each date, of a Markov transition function *(a matrix if the set of prices is countable) which is determined by the value taken by y_t^k. Kurz and Schneider* [27, Section 4] *provides details.*

Since the *effective* belief of the agent is the conditional probability of Q^k given y_t^k, and this may be time dependent, Assumption 2.1.3 (of stationarity of the joint system) means that the description of the variables y^k *exhausts all the time dependency which the agent perceives.*

Given that Q^k is jointly stationary on $((D \times P^* \times Y^k)^\infty)$ the standard theorems of dynamic programming apply when each agent knows $(R_t, p_t^c, P_t, q_t, y_t^k)$ in the sense that he *observes* (R_t, p_t^c, P_t, q_t) while he perceives the parameter y_t^k which is *generated privately.* With this in mind we turn to the formulation of the optimization problem of the agents. The problem of agent k when young is then as follows:

$$\max_{(x^{1k}, \theta^k, B^k)} E_{Q^k}\left[u^k(x_t^{1k}, x_{t+1}^{2k}(s_{t+1})) \mid s_t, y_t^k\right] \tag{2.1}$$

subject to

$$\begin{aligned} p_t^c x_t^{1k} + P_t\theta_t^k + q_t B_t^k &= p_t^c \Omega^k \\ \theta_t^k(P_{t+1} + p_{t+1}^c R_{t+1}) + B_t^k p_{t+1}^c &= p_{t+1}^c x_{t+1}^{2k}(s_{t+1}). \end{aligned} \tag{2.2}$$

The market clearing conditions for this model are then

$$\begin{aligned} \sum_{k=1}^K \theta_t^k &= 1, \qquad t = 1, 2, \ldots \\ \sum_{k=1}^K B_t^k &= 0, \qquad t = 1, 2, \ldots . \end{aligned} \tag{2.3}$$

It follows from (2.2) and (2.3) that when markets clear,

$$\begin{aligned} p_t^c x_t^1 + P_t &= p_t^c \Omega, \qquad t = 1, 2, \ldots \\ p_t^c x_t^2 &= P_t + p_t^c R_t, \qquad t = 1, 2, \ldots, \end{aligned}$$

where x_t^1, x_t^2, and Ω are the aggregates defined by

$$x_t^i = \sum_{k=1}^K x_t^{ik}, \quad i = 1, 2; \qquad \Omega = \sum_{k=1}^K \Omega^k.$$

Under the Markov assumption, the demand functions of all generations take the form

$$\begin{aligned} x_t^{1k} &= \varphi_c^k(R_t, p_t^c, P_t, q_t, y_t^k) \\ \theta_t^k &= \varphi_\theta^k(R_t, P_t^c, p_t, q_t, y_t^k) \\ B_t^k &= \varphi_B^k(R_t, p_t^c, P_t, q_t, y_t^k). \end{aligned}$$

An equilibrium requires that conditions (2.3) be satisfied, hence

$$\begin{aligned}\sum_{k=1}^{K} &= \varphi_\theta^k(R_t, p_t^c, P_t, q_t, y_t^k) = 1\\ \sum_{k=1}^{K} &= \varphi_B^k(R_t, p_t^c, P_t, q_t, y_t^k) = 0.\end{aligned} \tag{2.4}$$

Using the notation $y_t = (y_t^1, y_t^2, \ldots, y_t^K) \in Y \equiv Y^1 \times Y^2 \times \cdots \times Y^K$ we can solve (2.4) and write the equilibrium map in the form

$$\begin{pmatrix} p_t^c \\ P_t \\ q_t \end{pmatrix} = \Phi^*(R_t, y_t) \quad \text{for } t = 1, 2, \ldots\ . \tag{2.5}$$

Solutions of the form (2.5) are also derived by Nielsen [36], Kurz and Schneider [27], and by Kurz and Wu [28]. In all these models an RBE has the property that the vector of private assessment variables influences prices and consequently the state space for equilibrium analysis is $(D \times Y)$ which is different from the state spaces $(D \times P^* \times Y^k)$ of the individual agents. Note that the number of distinct equilibrium prices cannot exceed $M = |D|\ \Pi_{k=1}^K\ |Y^k|$. Indeed, there exists a finite collection $\{(p_i^c, P_i, q_i) \in \mathbb{R}_+^3,\ i = 1, 2, \ldots, M\}$ of equilibrium price vectors such that

$$\begin{pmatrix} p_i^c \\ P_i \\ q_i \end{pmatrix} = \Phi^*(R_i, y_i) \quad \text{for } i = 1, 2, \ldots, M. \tag{2.6}$$

To complete the model we specify the true joint distribution of private assessments y_t and dividends as a probability Π_{DY} on the measurable space $((D \times Y)^\infty,\ \mathcal{B}((D \times Y)^\infty))$. This is an important part of the formulation and we need to explore the restrictions on this measure and, correspondingly, on the beliefs of the agents. First consider the vector y_t of private assessments. The probability of each y_t^k is determined by agent k and hence, each agent *knows* his own distribution. The probability of the signal as perceived by agent k is the *marginal* measure of Q^k on Y^k and we denote it by $Q_{Y^k}^k$ (where Q_Z is the marginal measure of Q on a subspace $(Z^\infty, \mathcal{B}(Z^\infty))$). Given Π_{DY}, the implied probability of y^k is $\Pi_{(DY)_{Y^k}}$. It must then be true that

$$\Pi_{(DY)_{Y^k}} = Q_{Y^k} \quad \text{for } k = 1, 2, \ldots, K. \tag{2.7}$$

The specification of Π_{DY} implies that agents may condition on prices and dividends when forecasting their own future signals. More important is the fact that *the specification permits the private assessments to be correlated with each other and such correlation may be affected by the observed prices and dividends.* Each agent does not know other agent's assessments and does not know the structure of this correlation and cannot take this structure into account in his own optimization. This leads to the emergence of an important market externality.

The fact that assessment signals are *entirely private* yet *correlated* is the result of social communication through which agents interact with each other. In addition agents observe the same data and such common observations act as correlating devices. To put it differently, y_t^j and y_t^i may be positively or negatively correlated and, in general, jointly distributed with observed data in the economy such as prices and dividends, because agents i and j communicate with each other and may influence each other's models. This correlation plays a central role in an RBE as demonstrated in [27] and in [24], in the study of the volatility of asset prices. Therefore, it would have been desirable to formulate the structure of social communication as part of the model. We have not done so and the assumption of a fixed structure of communication (implied by Π_{DY}) is a simple representation of the impact of social communication on economic fluctuations. Our assumption is then the following.

2.1.4. Assumption. *Under* Π_{DY} *the process* $\{(R_t, y_t),\ t = 1, 2, \dots\}$ *is a Markov process and the dynamical system* $((D\times Y)^\infty,\ \mathcal{B}((D\times Y)^\infty),\ \Pi_{DY},\ T)$ *is stable and ergodic with a stationary measure* m_{DY}. *We denote by* m_D *and by* m_Y *the corresponding* D *and* Y *marginal measures.*

2.1.5. Lemma. *The price process* $\{(p_t^c, P_t, q_t),\ t = 1, 2, \dots\}$ *is a stable and ergodic process on the finite state space* $D \times Y$ *with probability* Π_P *and a stationary measure* m_P. *The probability* Π_P *on* $((P^*)^\infty,\ \mathcal{B}((P^*)^\infty))$ *is defined by the probability* Π_{DY} *together with the equilibrium price map* (2.6). *The measure* m_P *is also obtained from* m_{DY} *and the map* Φ^* *in* (2.6).

To simplify we use the notation $(p^c, P, q) = \Phi^*(R, y)$ to mean $(p_t^c, P_t, q_t) = \Phi^*(R_t, y_t)$ for all t. Now, for any set $A \in \mathcal{B}(D^\infty)$ define

$$\Phi_D^*(A) = \{(p^c, P, q) \in (P^*)^\infty \ :\ (p^c, P, q) = \Phi^*(R, y) \text{ for } R \in A,\ y \in Y^\infty\} \tag{2.8}$$

and interpret (2.8) to identify the set of prices associated with any given set of infinite sequences of dividends. It then follows from the equilibrium map (2.6) that in equilibrium we must have

$$\Pi_D(A) = \Pi_P(\Phi_D^*(A)) \quad \text{for all } A \in \mathcal{B}(D^\infty) \tag{2.9}$$

and therefore

$$m_D(A) = m_P(\Phi_D^*(A)) \quad \text{for all } A \in \mathcal{B}(D^\infty). \tag{2.10}$$

(2.6), (2.7) and (2.9)–(2.10) provide the tools for stating the rationality conditions of the agents. Note that a belief Q_k is a probability on the space $((D \times P^* \times Y^k)^\infty, \mathcal{B}((D \times P^* \times Y^k)^\infty))$ since the agent is not assumed to know the map Φ^*. However, the data reveals that the empirical distribution of prices and dividends must conform to (2.10) and this condition must be satisfied by Q^k. The following is then implied by the Conditional Stability Theorem (see [27]):

2.1.6. Lemma. *Under the assumptions of Lemma* 2.1.5 Q^k *is a rational belief relative to* Π_P *if*

1. $\Pi_{(DY)_{Y^k}} = Q^k_{Y^k}$.
2. $Q^k_D(A) = Q^k_P(\Phi^*_D(A)) = m_D(A)$ *for all* $A \in \mathcal{B}(D^\infty)$.
3. $Q^k_P = m_P$.

Using Lemma 2.1.6 we can define a Rational Belief Equilibrium as follows:

2.1.7. Definition. $\{\Pi_P,$
$\{(Q^k, \theta^k_i, B^k_i)$ *for* $k = 1, 2, \ldots, K$ *and* $i = 1, 2, \ldots, M\}$, *and*
$\{(p^c_i, P_i, q_i)$ *for* $i = 1, 2, \ldots, M\}\}$ *constitute a* ***Rational Belief Equilibrium*** *(RBE) of the heterogenous agent stock market OLG economy if*

1. Q_k *is a rational belief relative to* Π_P *for* $k = 1, 2, \ldots, K$ *and* Π_P *is defined by* Π_{DY} *and by the equilibrium map induced by* $(Q^1, Q^2, \ldots, Q^K)$.
2. $(\theta^k_1, \theta^k_2, \ldots, \theta^k_M)$, $(B^k_1, B^k_2, \ldots, B^k_M)$ *are optimal agent allocations for* $k = 1, 2, \ldots, K$.
3. $\sum_{k=1}^K \theta^k_i = 1$ *for all* t *and all* i.
4. $\sum_{k=1}^K B^k_i = 0$ *for all* t *and all* i.

2.1.8. Theorem. *Under Assumptions* 2.1.1–2.1.4 *there exists an RBE.*[9]

[9] A comment on multiple and sunspot equilibria is warranted at this point. The definition of an RBE does not address directly the issue of multiple equilibria. Keep in mind that we are modeling the economy as a dynamical system in which infinite random draws are associated with *definitive* sequences of realized economic allocations. This means that if at any date the economy can have multiple market clearing outcomes, then as part of the dynamics postulated there is a procedure for selecting a *particular* one of them which, in turn, generates the data observed in the economy. This, indirectly, addresses also the issue of sunspot equilibria. Such equilibria require a device for alternating random selections from

2.2. The problem of anonymity of an RBE

We say that an RBE lacks anonymity *if equilibrium prices depend upon individual states of beliefs.* Consider the special case $D = \{R^H, R^L\}$, $K = 2$, with $Y^k = \{0, 1\}$ for $k = 1, 2$ which we use in the simulations below. Members of D identify the *state of the dividend process* while members of $Y^1 \times Y^2$ identify the *individual states of belief* of the agents. The equilibrium map (2.6) implies that the state space has 8 members and we think of $V = \{1, 2, \ldots, 8\}$ as *the price state space.* Although the equilibrium map of this RBE takes the form

$$\begin{bmatrix} p_i \\ q_i \\ q_i \end{bmatrix} = \Phi^*(R_i, y_i^1, y_i^2), \qquad i = 1, 2, \ldots, 8, \tag{2.11}$$

we can define an equivalent map Φ between the *indices* of the price states $\{1, 2, \ldots, 8\}$ and the vectors of *dividend states and states of belief* as follows:

$$\begin{bmatrix} 1 \\ 2 \\ 3 \\ 4 \\ 5 \\ 6 \\ 7 \\ 8 \end{bmatrix} = \Phi \begin{bmatrix} R_1 = R^H & y_1^1 = 1 & y_1^2 = 1 \\ R_2 = R^H & y_2^1 = 1 & y_2^2 = 0 \\ R_3 = R^H & y_3^1 = 1 & y_3^2 = 1 \\ R_4 = R^H & y_4^1 = 1 & y_4^2 = 0 \\ R_5 = R^L & y_5^1 = 1 & y_5^2 = 1 \\ R_6 = R^L & y_6^1 = 1 & y_6^2 = 0 \\ R_7 = R^L & y_7^1 = 1 & y_7^2 = 1 \\ R_8 = R^L & y_8^1 = 1 & y_8^2 = 0 \end{bmatrix}. \tag{2.12}$$

We further assume below that the marginal distributions of the assessments y^1 and y^2 are i.i.d. with $P\{y_t^k = 1\} = \alpha_k$ for $k = 1, 2$. Rationality of beliefs implies that the agents have two pairs of matrices (F_1, F_2) and (G_1, G_2) such that the beliefs Q^1 and Q^2 are characterized as follows:

$$\begin{aligned} &Q^1 \text{ for agent } 1: \quad \begin{cases} \text{adopt } F_1 \text{ if } y_t^1 = 1 \\ \text{adopt } F_2 \text{ if } y_t^1 = 0, \end{cases} \\ &Q^2 \text{ for agent } 1: \quad \begin{cases} \text{adopt } G_1 \text{ if } y_t^2 = 1 \\ \text{adopt } G_2 \text{ if } y_t^2 = 0. \end{cases} \end{aligned} \tag{2.13}$$

among multiple equilibria of some underlying economy over time. If such an equilibrium is to be realized then this selection must be part of the description of the dynamical system. Moreover, a formal coordination among agents is feasible only if one of the observable exogenous variables provides the needed signal for joint action and all agents interpret this public signal in exactly the same way. In that case we must interpret the fluctuations of the economy which are due to the publicly observed sunspot variable as *exogenously caused.*

The implied rationality conditions are $\alpha_1 F_1 + (1 - \alpha_1)F_2 = \Gamma$ and $\alpha_2 G_1 + (1-\alpha_2)G_2 = \Gamma$ where Γ is a Markov matrix which defines the stationary measure. It is then clear that RBE defined by (2.11) or (2.12) lacks anonymity since a change in the state of belief of an agent causes the equilibrium price to change. It is also clear that the equilibrium concept adopted here requires the agents to ignore their effects on equilibrium prices.

In a finite economy all competitive equilibria fail to be anonymous and hence the problem above is no different from the corresponding problems which arise in equilibria with finite number of agents. The traditional tool to explore anonymity has been the "replica economy" and this is the motivation for our adoption of this tool here. The rest of this paper is an examination of the consequences of this approach.

3. Rational Belief Equilibria with social states of belief

3.1. Anonymity in a large replica economy

We start by reconsidering the RBE concept defined in Section 2 to highlight the decreased effect of each individual's belief on the equilibrium outcome as the economy becomes large. The implication is that equilibrium prices are functions of the social *distribution* of beliefs rather than functions of the vectors of individual states of belief. This means that in large economies endogenous uncertainty impacts aggregate economic fluctuations via the *distribution* of beliefs in the economy. Such distributions define the "social states of beliefs" which are determined by *the structure of correlation among the individual beliefs.*

Consider the model of Section 2 and restrict attention to the financial structure used in [24] which consists of one stock and one bond. Suppose, however, that each one of the K agents in the model is now considered to be a "type" *with N replicas* and that instead of $2K$ agents (K young and K old) we now have $2KN$ agents. As is standard, the economy becomes large if N is large. The N replicas have the same utility, endowment and belief but not necessarily the same realized assessment. Hence, for all $n = 1, 2, \ldots, N$, $y^{k,n} \in Y^k$ where $y^{k,n}$ is the i.i.d. assessment of replica agent n of type k and the central question of interest is the joint distribution of the assessments. Since the beliefs of the agents are determined by the $y^{k,n}$ the N agents of type k may not hold the same conditional belief. In the extreme case these may be perfectly correlated so that all of them take the same value in Y^k. Indeed, one way to interpret the results of a small economy (consisting of, say, two agents) is to observe that they apply to a large economy in which the N assessments are perfectly correlated. This suggests two different types of correlations in society. The first is a correlation among $(y_t^{k,1}, y_t^{k,2}, \dots, y_t^{k,N})$ which is "within type" correlation that determines the distribution of assessments of type k agents. We define each such possible distribution as a "type–state." The second is the correlation among the type–states themselves which, in turn, determines the aggregate social states. Why are the correlations "within" a type and "across" types not the same? Without a formal model of social communication to explain this assumption we can suggest that one must visualize agents of the same type as associating with each other in a different manner than agents of different types and communicating with each other via different and more complex channels than the public channels used by agents of different types.

To be concrete suppose that for all k, $Y^k = \{1, 2, \dots, L\}$ and $y_t^{k,n}$ are i.i.d. marginally with probabilities $(\alpha_{k1}, \dots, \alpha_{kL})$ hence the model of Section 2 has $|D|L^{KN}$ individual states. It follows from the optimization (2.1)–(2.2) that the demand functions of agent (k, n) have the form

$$\theta_t^{k,n} = \varphi_\theta^k(R_t, p_t^c, P_t, q_t, y_t^{k,n}), \qquad B_t^{k,n} = \varphi_B^k(R_t, p_t^c, P_t, q_t, y_t^{k,n}). \tag{3.1}$$

(3.1) points out that all agents of type k with the same realized assessment *have the same demand for securities.* Consider any individual state i of the $|D|L^{KN}$ states. Denote by $s_i(k, \ell)$ the number of agents of type k with assessments taking the value ℓ in state i. It follows from (3.1) that the market clearing conditions in the RBE of Section 2 take the exact form

$$\begin{aligned} &\sum_{k=1}^{K} \sum_{\ell=1}^{L} \frac{s_i(k,\ell)}{N} \varphi_\theta^k(R_i, p_i^c, P_i, q_i, \ell) = 1, \qquad i = 1, 2, 3, \dots, |D|L^{NK} \\ &\sum_{k=1}^{K} \sum_{\ell=1}^{L} \frac{s_i(k,\ell)}{N} \varphi_B^k(R_i, p_i^c, P_i, q_i, \ell) = 0, \qquad i = 1, 2, 3, \dots, |D|L^{NK}. \end{aligned} \tag{3.2}$$

Now, each type has only L *different* demand functions hence variability in (3.2) is caused by the different distributions (type–states) of the assessments of the N agents of type k. To see that the number of distinct distributions $\left\{\frac{s_i(k,\ell)}{N},\ \ell = 1, 2, \dots, L\right\}$ is dramatically less than L^{KN}, consider the case $L = 2$. For each k there are 2^N permutations of the assessments but the set of distinct values that may be taken by $s_i(k, \ell)$ is $\{0, 1, \dots, N\}$ with at most $N + 1$ distributions $\{(0,1),\ (1/N,\ (N-1)/N), \dots, (n/N,\ (N-n)/N), \dots, (1,0)\}$. Jointly for all the K types there are only $(N+1)^K$ distinct distributions. For $L = 3$ the number of distinct distributions (i.e. three tuples) for each k is $\sum_{\tau=0}^{N}(\tau + 1)$ and for any L this number is $M_{NL} = \sum_{\tau_L=0}^{N} \sum_{\tau_{L-1}=0}^{\tau_L} \cdots \sum_{\tau_4=0}^{\tau_5} \sum_{\tau=0}^{\tau_4}(\tau + 1)$. The implication is that the number of *distinct* prices in (3.2) is the relatively small number of $|D|(M_{NL})^K$ rather than $|D|L^{KN}$ and hence, for large N *most of the equations in* (3.2) *are redundant.* We thus arrive at the following.

3.1.1. Observation. *Even in a finite economy the number of distinct equilibrium prices is much smaller than the number of individual states and hence the equilibrium map in terms of individual states such as* (2.11) *or* (2.12) *is generically not invertible. The number of distinct prices is determined by the number of distinct vectors of type–states of beliefs and exogenous states.*

Under the assumption that the assessments of all agents are independent, Observation 3.1.1 implies that a finite replica economy *tends to anonymity as the number of replicas increases* since the effect of the belief of each agent on the type–states becomes small. Consequently, in the case of independence a finite but large economy is approximately anonymous and we may as well assume that the agent neglects the minimal effect his belief has on the type–states and thus on prices. This conclusion is completely analogous to competitive equilibria of a replica economy.

Now suppose that N is large. In the case of independence within and across types the zero–one law implies that with probability 1 the assessments of type k have only *one* limit distribution $(\alpha_{k1}, \ldots, \alpha_{kL})$ and hence at any date the type–state of belief of type k agents is represented by the constant vector $(\alpha_{k1}, \ldots, \alpha_{kL})$. Similarly, there is only one joint distribution for all types. If the state space D of the exogenous process has a dimension $|D|$ then it follows that with probability one the system of equations (3.2) is reduced to $2|D|$ independent equations implying that there are, at most, $|D|$ distinct price vectors associated with a constant vector of distributions of beliefs. This is the case of *a single social state of belief* in the limit economy and our first task below is to explore the nature of endogenous uncertainty in such an economy. The case of correlation among the beliefs of the agents leads to very different economies and since such correlation is central to the conclusions of this paper we comment on this issue now.

Extensive work has explored in recent years the implications of alternative patterns of economic interactions[10] and the main conclusion of this literature is that relatively simple local interactions are sufficient to induce a limiting behavior which is a random variable rather than a constant. As an illustration of an explicit analysis of such interactions the reader may consult the procedure used by Brock [12] in which he utilizes the results of Kac [20] to derive the limiting behavior of the system. Given the extensive amount of interaction among participants in financial markets, one must therefore conclude that

[10]See, for example, [12, 10, 11, 9, 16, 15, 18]. For a related approach see [7, 8, 43].

the assumption of independent assessments is an extreme one and the case of correlation among beliefs of agents is the norm.

In the case of correlation we do not have general convergence results but even if the limit distributions exist, one cannot ensure anonymity. This is because the assessment of some agent may become an atom and consequently we can only make the following, self–evident, comment:

3.1.2. Observation. *Assume the existence of limit distributions of beliefs across all types in the case of correlation. The replica economy tends to anonymity as N becomes large if the limit distributions do not have an atom concentrated on the assessment of any one agent.*

When correlation among assessments is present, both the limit distributions (type–states) of each type as well as the joint distribution over all types are random variables. In the applications below we make the following assumption.

3.1.3. Assumption. *There is a finite number of type–states of beliefs in the economy.*

Assumption 3.1.3 holds in any finite economy. It would also be satisfied in an infinite economy in which the limit random variables

$$\left\{\frac{s_i(k,\ell)}{N}, \quad \ell = 1,2,\ldots,L, \ k = 1,2,\ldots,K\right\}$$

are well–defined random variables and with probability 1 take only a finite number of values. Anonymity holds if these limit random variables are not correlated with the assessment of any one agent. The thrust of Assumption 3.1.3 is that only a finite number of market clearing conditions in (3.2) are applicable to an RBE with social states. This is because almost all market clearing conditions in (3.2) apply to individual states i that occur with probabilities which tend to zero as N goes to infinity and hence are ignored. We now formulate the concept of "social states of belief."

3.2. Social states of belief as distributions

The concept of *social states of belief*, inspired by concepts of *collective risk* developed in [33, 32] and in [6], can now be defined in a natural way. Let M_L^k be the number of distinct distributions $\left\{\frac{s_i(k,\ell)}{N}, \quad \ell = 1,2,\ldots,L\right\}$ for each k. In the finite economy one computes these distributions for each of the vectors of individual states i but in the infinite economy one takes the limit as N goes to infinity and ignores individual states with zero probability. We now introduce notation to describe members of this set of distributions. Thus, for each of the M_L^k distributions we use the following notation:

$\zeta^{k,\ell}$ = the proportion of agents of type k with assessment variables taking the value ℓ.

3.2.1. Definition. *A* ***type–state*** *of agents of type k is a distribution of the form $\zeta^k = (\zeta^{k,1}, \zeta^{k,2}, \ldots, \zeta^{k,L})$ such that $\zeta^{k,\ell}$ are nonnegative numbers and $\sum_{\ell=1}^{L} \zeta^{k,\ell} = 1$. Let $S_B^k = \{\zeta^k\colon\ \zeta^k$ is a type state for type $k\}$. Then S_B^k has M_L^k members. A* ***social state of belief in the economy*** *is a vector of distributions $\zeta = (\zeta^1, \zeta^2, \ldots, \zeta^K),\quad \zeta^k \in S_B^k$.*

The set of possible social states of belief is then
$S_B = \{\zeta\colon\ \zeta = (\zeta^1, \zeta^2, \ldots, \zeta^K),\ \zeta^k \in S_B^k\}$, and this set has $M_L = \prod_{k=1}^{K} M_L^k$ members. We then define naturally:

3.2.2. Definition. *A* ***social state*** *for the economy is a pair consisting of a dividend state and a state of belief in the economy. It consists of a $|D| + KL$ tuple*

$$(d, \zeta) \equiv (d, \zeta^1, \zeta^2, \ldots, \zeta^K), \qquad d \in D,\ \zeta \in S_B.$$

Denote by M the number of possible social states and we know that $M = |D| M_L$. Now, list the M social states by the index s and this set is then defined by

$$\hat{S} = \{(d_s, \zeta_s^1, \zeta_s^2, \ldots, \zeta_s^K),\quad d_s \in D,\quad \zeta_s^k \in S_B^k,\quad s = 1, 2, \ldots, M\}. \tag{3.3}$$

Since the $y^{k,n}$ are not observable and the agents do not know the equilibrium map, one may think of social states as a *listing of the index* s of the states in (3.3) and define the price state space to be

$$S = \{1, 2, \ldots, M\}. \tag{3.4}$$

The difference between the state spaces in (3.3) and (3.4) is analogous to the distinction between the maps (2.5) and (2.6). In sum, we have the following observation:

3.2.3. Observation. *Given the market clearing conditions* (3.2) *then with probability* 1 *there are* ***at most*** *M distinct social states. They induce at most M different aggregate excess demand functions and hence there are at most M* ***distinct*** *equilibrium prices.*

It then follows that we may rewrite the system (3.2) in the form

$$\begin{aligned} \sum_{k=1}^{K}\sum_{\ell=1}^{L} \zeta_s^{k,\ell}\varphi_\theta^k(R_s, p_s^c, P_s, q_s, \ell) &= 1, \qquad s = 1,2,\ldots,M \\ \sum_{k=1}^{K}\sum_{\ell=1}^{L} \zeta_s^{k,\ell}\varphi_B^k(R_s, p_s^c, P_s, q_s, \ell) &= 0, \qquad s = 1,2,\ldots,M. \end{aligned} \tag{3.5}$$

The interpretation of (3.5) leads to the final clarification of the nature of an equilibrium with social states of belief. Conditional on their assessments agents carry out the optimization in (2.1)–(2.2) leading to demand functions which depend upon the private value of their own assessment variable. These private assessments are then aggregated into distributions $(\zeta_s^1, \zeta_s^2, \ldots, \zeta_s^K)$ which constitute social states of belief. Equilibrium prices and dividends are then maps defined on the social state space and are written in the form (R_s, p_s^c, P_s, q_s).

In the redefined economy what matters in equilibrium is the *distribution* of beliefs rather than the belief of any one agent. However, the distribution of beliefs in society may exhibit a complex structure even if the assessment of each agent is i.i.d. As a result, the aggregate implications of our approach depend decisively upon the structure of correlation among agents. It is generally difficult to study analytically the impact of different correlation structures on the long term volatility of the implied equilibria and the appropriate tool to carry out such an examination is the method of simulation. Accordingly, the simulation work of the next section aims to exhibit how the model of an RBE with social states helps the understanding of the factors which determine asset price volatility in general and the equity premium in particular. More specifically, we focus on the effect of the correlation structure within the model on the volatility characteristics of the equilibrium. The five measures of uncertainty and volatility that we focus on are: (i) the equity premium, (ii) the riskless rate, (iii) the standard deviation of the risky returns, (iv) the standard deviation of the riskless rate and (v) the variance of the price/dividend ratio.

4. Market volatility and correlation among social states of belief: simulation analysis

In order to proceed to the simulation results we need first to reformulate the above model to conform to the growth assumptions of [24]. In order to accomplish this we briefly review the assumptions made in [24].

4.1. A brief review of the [24] assumptions

The OLG model used in [24] aims to approximate the model of Mehra and Prescott [35]. Accordingly, [24] assumes an economy with two agents and no replica: $K = 2$, $N = 1$, $L = 2$, $Y^k = \{1, 0\}$ for $k = 1, 2$ with the time additive utility function $\frac{1}{1-\gamma_k} c^{1-\gamma_k}$ and a constant discount factor β_k. Also, the model conforms to the real growth assumptions made in [35]. Under these assumptions $\{R_t,\ t = 1, 2, \dots\}$ satisfies

$$R_{t+1} = d_t R_t.$$

The growth rate process $\{d_t,\ t = 1, 2, \dots\}$ is then assumed to be a stationary and ergodic Markov process on the state space $\{d^H, d^L\}$ with a transition matrix

$$\begin{bmatrix} \phi & 1-\phi \\ 1-\phi & \phi \end{bmatrix}. \tag{4.1}$$

[35] assumes that $d^H = 1.054$, $d^L = .982$, and $\phi = .43$. Since this implies that over time agents experience a rise in the level of dividends, it requires us to redefine the budget constraints. We revise the assumption that Ω^k is constant and instead assume that $\omega^k = \frac{\Omega_t^k}{R_t}$ for $k = 1, 2$ are constant over time. This in accord with the assumption often used (see [35]) that the growth rate of the output of the economy *as a whole* is a stationary Markov process with a transition matrix (4.1). Now denote by $b_t^k = \frac{B_t^k}{R_t}$ the bond/dividend ratio of agent k and by $p_t = \frac{P_t}{R_t}$ the price/dividend ratio at date t. Normalizing by setting $p_t^c = 1$ for all t and using the notation introduced, the budget equations (2.2) are now written as

$$x_t^{1k} = [\omega^k - p_t\theta_t^k - q_t b_t^k] R_t$$
$$x_{t+1}^{2k} = [\theta_t^k (p_{t+1} + 1) d_{t+1} + b_t^k] R_t.$$

The Markov assumptions imply that given assessments (y_t^1, y_t^2), the market clearing conditions are $\theta_t^1(p_t, q_t, d_t, y_t^1) + \theta_t^2(p_t, q_t, d_t, y_t^2) = 1$ and $b_t^1(p_t, q_t, d_t, y_t^1) + b_t^2(p_t, q_t, d_t, y_t^2) = 0$. It is then clear that the implied equilibrium map has the exact form specified in (2.11), with an index map such as (2.12) and a price state space $S = \{1, 2, \dots, 8\}$. [24] also assumes that the marginal distributions of the assessment variables of the two agents are i.i.d. with the probability of 1 being α_1 and α_2, respectively.

Denote by $Q^k(j \mid s, y^k)$ agent k's conditional probability of price state j given price state s and the value of y^k (but ignoring his effect on prices). The first order conditions can then be written in terms of price states for $k = 1, 2$ and $j, s = 1, 2, \ldots, 8$ as follows

$$\begin{aligned} &-(\omega^k - \theta_s^k p_s - b_s^k q_s)^{-\gamma_k} p_s \\ &\qquad + \beta_k \sum_{j=1}^{8} (\theta_s^k (p_j + 1) d_j + b_s^k)^{-\gamma_k} (p_j + 1) d_j Q^k(j \mid s, y^k) = 0, \\ &-(\omega^k - \theta_s^k p_s - b_s^k q_s)^{-\gamma_k} q_s \\ &\qquad + \beta_k \sum_{j=1}^{8} (\theta_s^k (p_j + 1) d_j + b_s^k)^{-\gamma_k} Q^k(j \mid s, y^k) = 0. \end{aligned} \tag{4.2}$$

Once (Q^k, ω^k) are specified for $k = 1, 2$ one computes the demand functions (θ_s^k, b_s^k) as a function of the 8 prices. In the equilibrium

$$\begin{aligned} \theta_s^1 + \theta_s^2 &= 1 \quad \text{for } s = 1, 2, \ldots, 8 \\ b_s^1 + b_s^2 &= 0 \quad \text{for } s = 1, 2, \ldots, 8. \end{aligned} \tag{4.3}$$

(4.2) and (4.3) constitute a system of 48 equations in prices and quantities which are the basis of the simulation results of [24].

4.2. Reformulation of the model to a replica economy with large N

We now make use of the conclusions of Section 3.1. It follows from (3.5) that for any N the first order conditions (4.2) remain the same and consequently the implied demand functions are entirely determined by the *type* of an agent and the value of his assessment variable. Since in the simulations below we assume that $K = 2$ and $L = 2$, there are two pairs of demand functions $\theta^{k,n} = \varphi_\theta^k(R, p, q, y^{k,n})$ and $b^{k,n} = \varphi_b^k(R, p, q, y^{k,n})$ for $n = 1, 2, \ldots, N$ and for $k = 1, 2$.

We also assume that the marginal distributions of $y^{k,n}$ are i.i.d. with parameters α_1 and α_2. We have noted that independence among the assessments of each type implies that for large N the type–state is almost surely constant at $(\alpha_1,\ 1 - \alpha_1)$. Hence we can express the correlation among the assessments of a type by specifying the type–states to be a random variables with distributions to be specified. Size limitations in the simulations below lead us to make simplifying assumptions on the joint distribution of the two type–states in the model below:

1. For each type there are only three possible type–states.
2. The marginal distribution of the type–states is i.i.d.

3. In most calculations we set $\alpha_1 = \alpha_2 = .57$ as will be motivated later. We use the α_k notation for general discussion.

More specifically, in all simulations we assume that for both types the support of the distribution of the type–states is

$$\{(.85,\ .15),\ (.57,\ .43),\ (.25,\ .75)\} \quad \text{for} \quad k = 1, 2.$$

This reflects the idea that correlation within each type results in probability being placed not only on the type–state (.57, .43) which is sure to occur without correlation, but also on two other states. We shall also assume that marginal distributions of the two type–states are i.i.d. with probabilities

$$(.5(1-\chi_1),\ \chi_1,\ .5(1-\chi_1))$$
$$(.5(1-\chi_2),\ \chi_2,\ .5(1-\chi_2)).$$

These assumptions are compatible with Assumptions 2.1.4 and 3.1.3 and with the standing assumption that the marginal distribution of $y^{k,n}$ for each k and n is i.i.d. The special assumption of i.i.d. type–state marginals is justified by the technical fact that the representation of correlation among social states is simplified by i.i.d. marginals of the type–states.

In all cases considered below we have 2 dividend states, 2 agent types and 3 type–states for each agent type. This implies that there are 18 possible equilibrium prices and 9 social states of belief. The equilibrium map is defined by the following: prices 1–9 are

1. $(d^H,\ (.85,\ .15),\ (.85,\ .15))$	6. $(d^H,\ (.57,\ .43),\ (.25,\ .75))$	
2. $(d^H,\ (.85,\ .15),\ (.57,\ .43))$	7. $(d^H,\ (.25,\ .75),\ (.85,\ .15))$	
3. $(d^H,\ (.85,\ .15),\ (.25,\ .75))$	8. $(d^H,\ (.25,\ .75),\ (.57,\ .43))$	(4.4)
4. $(d^H,\ (.57,\ .43),\ (.85,\ .15))$	9. $(d^H,\ (.25,\ .75),\ (.25,\ .75))$.	
5. $(d^H,\ (.57,\ .43),\ (.57,\ .43))$		

Prices 10–18 are defined similarly but with d^L replacing d^H. Turning to the stochastic structure of the joint process of dividend growth and social states of belief we assume that it is a stable Markov process. As in [24], we specify the *stationary measure* by selecting the following 9×9 matrix to be the transition among the 9 social states of belief:

$$A = \begin{bmatrix} \mathbf{a}_1 \\ \mathbf{a}_2 \\ \cdot\ \cdot \\ \mathbf{a}_9 \end{bmatrix},$$

where

$$\mathbf{a}_i = \Big[a_i^1,\ a_i^2,\ \tfrac{1}{2}(1-\chi_1) - a_i^1 - a_i^2,\ a_i^3,\ a_i^4,\ \chi_1 - a_i^3 - a_i^4,$$
$$\tfrac{1}{2}(1-\chi_2) - a_i^1 - a_i^3,\ \chi_2 - a_i^2 - a_i^4,\ a_i^1 + a_i^2 + a_i^3 + a_i^4 - \tfrac{1}{2}(\chi_1+\chi_2)\Big].$$

The marginals of this matrix conform to the specified marginal i.i.d. of the type–states. Apart from the parameters χ_1 and χ_2 which are determined by the agents, the matrix A has 36 parameters which specify the joint distribution and hence the correlation among the social states of belief. These are not free parameters and we specify below the restrictions on them. To allow for the possibility of a dividend effect on the distribution of assessments we employ a second matrix B which has the same structure as A except that it is defined by parameters b_i^j. As in [24], the stationary measure is identified by the 18×18 Markov transition matrix of the form:

$$\Gamma = \begin{bmatrix} \phi A & (1-\phi)A \\ (1-\phi)B & \phi B \end{bmatrix},$$

where A and B are 9×9 matrices as defined above. Each is characterized by the 36 parameters $a = (a^1, a^2, a^3, a^4)$ where $a^j = (a_1^j, a_2^j, \ldots, a_9^j)$, $j = 1,2,3,4$, and $b = (b^1, b^2, b^3, b^4)$ where $b^j = (b_1^j, b_2^j, \ldots, b_9^j)$, $j = 1,2,3,4$. The first 9 rows of the matrix Γ are identified with d^H and the 9 specified states of belief while the second 9 rows of Γ are identified with d^L and the 9 states of belief. With this identification Γ satisfies the required properties: the marginal of Γ on the dividends is the matrix (4.1) and the marginals on the type–states are as specified. The simulation model would be completed once we specify the beliefs of the two types of agents. The rationality conditions are similar to [24].

An inspection of the matrices A, B and Γ reveal that there are feasibility conditions which must be satisfied by the parameters a and b. More specifically there are 90 inequality constraints which the parameters must satisfy and these are as follows: for $i = 1, 2, \ldots, 9$

$$\begin{aligned} a_i^1 + a_i^2 &\le \frac{1}{2}(1-\chi_1) & b_i^1 + b_i^2 &\le \frac{1}{2}(1-\chi_1) \\ a_i^3 + a_i^4 &\le \chi_1 & b_i^3 + b_i^4 &\le \chi_1 \\ a_i^1 + a_i^3 &\le \frac{1}{2}(1-\chi_2) & b_i^1 + b_i^3 &\le \frac{1}{2}(1-\chi_2) \\ a_i^2 + a_i^4 &\le \chi_2 & b_i^2 + b_i^4 &\le \chi_2 \\ a_i^1 + a_i^2 + a_i^3 + a_i^4 &\ge \frac{1}{2}(\chi_1+\chi_2) & b_i^1 + b_i^2 + b_i^3 + b_i^4 &\ge \frac{1}{2}(\chi_1+\chi_2). \end{aligned} \tag{4.5}$$

In addition we have the 8 conditions which specify that the rows of A and B sum up to 1.

The marginal distributions of $y^{1,n}$ and $y^{2,n}$ are i.i.d. with $P\left\{y_t^{k,n}=1\right\}=\alpha_k$ for $k=1,2$. This means that as in (2.13) the agents have two pairs of matrices (F_1), (G_2) such that the conditional beliefs $Q_{1,n}$ and $Q_{2,n}$ given the assessments as follows:

$$Q_t^{1,n}(j \mid s, y_t^{1,n}) = \begin{cases} F_1^{s,j} & \text{if } y_t^{1,n}=1 \\ F_2^{s,j} & \text{if } y_t^{1,n}=0, \end{cases} \qquad Q_t^{2,n}(j \mid s, y_t^{2,n}) = \begin{cases} G_1^{s,j} & \text{if } y_t^{2,n}=1 \\ G_2^{s,j} & \text{if } y_t^{2,n}=0 \end{cases}$$

for $n=1,2,\ldots,N$ where $C_1^{s,j}$ is the (s,j) element of matrix C_1. Rationality of beliefs requires

$$\alpha_1 F_1 + (1-\alpha_1)F_2 = \Gamma \qquad \text{and} \qquad \alpha_2 G_1 + (1-\alpha_2)G_2 = \Gamma. \tag{4.6}$$

The matrices (F_1, F_2, G_1, G_2) are defined by two sets of 18 parameters $\lambda = (\lambda_1, \lambda_2, \ldots, \lambda_{18})$ and $\mu = (\mu_1, \mu_2, \ldots, \mu_{18})$ which will be interpreted later. To describe how they are constructed we introduce the notation for the row vectors of A and B:

$$A^j = (a_j^1,\ a_j^2, \ldots,\ a_j^1 + a_j^2 + a_j^3 + a_j^4 - \frac{1}{2}(\chi_1 + \chi_2))$$
$$B^j = (b_j^1,\ b_j^2, \ldots,\ b_j^1 + b_j^2 + b_j^3 + b_j^4 - \frac{1}{2}(\chi_1 + \chi_2)).$$

With this notation define the 4 matrix functions of a vector $z=(z_1, z_2, \ldots, z_{18})$ of real numbers:

$$A_1(z) = \begin{bmatrix} z_1 A^1 \\ z_2 A^2 \\ \cdots\cdot \\ z_9 A^9 \end{bmatrix}, \qquad B_1(z) = \begin{bmatrix} z_{10} B^1 \\ z_{11} B^2 \\ \cdots\cdot \\ z_{18} B^9 \end{bmatrix},$$

$$A_2^{\phi}(z) = \begin{bmatrix} \dfrac{1-\phi z_1}{1-\phi} A^1 \\ \dfrac{1-\phi z_2}{1-\phi} A^2 \\ \cdots\cdots\cdots \\ \dfrac{1-\phi z_9}{1-\phi} A^9 \end{bmatrix}, \qquad B_2^{\phi}(z) = \begin{bmatrix} \dfrac{1-(1-\phi) z_{10}}{\phi} B^1 \\ \dfrac{1-(1-\phi) z_{11}}{\phi} B^2 \\ \cdots\cdots\cdots \\ \dfrac{1-(1-\phi) z_{18}}{\phi} B^9 \end{bmatrix}. \tag{4.7}$$

Using (4.7), define the matrices

$$F_1(\lambda) = \begin{bmatrix} \phi A_1(\lambda) & (1-\phi)A_2^{\phi}(\lambda) \\ (1-\phi)B_1(\lambda) & \phi B_2^{\phi}(\lambda) \end{bmatrix}$$

$$G_1(\mu) = \begin{bmatrix} \phi A_1(\mu) & (1-\phi)A_2^{\phi}(\mu) \\ (1-\phi)B_1(\mu) & \phi B_2^{\phi}(\mu) \end{bmatrix}$$

and (F_2, G_2) determined by (4.6). The selection of the vectors (λ, μ) is restricted by 108 inequality constraints which define the feasible region. These constraints are as follows:

$$\begin{array}{lll} \lambda_s \le \dfrac{1}{\phi} & \mu_s \le \dfrac{1}{\phi} & \text{for } s = 1,2,\ldots,9 \\ \lambda_s \le \dfrac{1}{1-\phi} & \mu_s \le \dfrac{1}{1-\phi} & \text{for } s = 10,11,\ldots,18 \\ \lambda_s \le \dfrac{1}{\alpha_1} & \mu_s \le \dfrac{1}{\alpha_2} & \text{for } s = 1,2,\ldots,18 \\ \lambda_s \ge \dfrac{\alpha_1+\phi-1}{\phi\alpha_1} & \mu_s \ge \dfrac{\alpha_2+\phi-1}{\phi\alpha_2} & \text{for } s = 1,2,\ldots,9 \\ \lambda_s \ge \dfrac{\alpha_1-\phi}{(1-\phi)\alpha_1} & \mu_s \ge \dfrac{\alpha_2-\phi}{(1-\phi)\alpha_2} & \text{for } s = 10,11,\ldots,18. \end{array} \tag{4.8}$$

To motivate this construction note that the intensity parameters λ_s and μ_s are multiplied by the rows of A and B and hence are proportional changes of the conditional probabilities of the two sets of nine states $(1, 2, \ldots, 9)$ and $(10, 11, \ldots, 18)$ *relative to the stationary measure* represented by Γ. Since λ_s and μ_s are the factors of proportionality by which the probability beliefs of the agent deviate from the stationary probabilities in Γ, we refer to the parameters λ_s, μ_s as "intensity" parameters. It should be clear that up until now the assessment variables of the agents had no economic meaning. They attain meaning only when the agents specify how they interpret these variables in generating conditional probability beliefs. For example, $\lambda_s > 1$ implies *increased* probabilities of states $(1, 2, \ldots, 9)$ in F_1 relative to Γ of an agent of type 1 given that he is in state s. This means that the assessment variables induce more "optimism" or "pessimism" about the prospects of prices $(1, 2, \ldots, 9)$ at $t+1$ relative to Γ. To see why, suppose that $\lambda > 1$ and that at some date t state $s = 1$ occurs so that (p_1, q_1) is realized. In that case type 1 agents with assessments $y_t^{1,n} = 1$ use matrix F_1 to forecast prices at $t+1$ and by (4.7) they are more optimistic (relative to Γ) about the probabilities of $((p_1, q_1), (p_2, q_2), \ldots, (p_9, q_9))$ at $t+1$. The equilibrium map (4.4) shows that conditionally on (p_1, q_1), 85% of type 1 agents are then optimistic about the prospects of the first 9 prices.

We observe that conditionally upon (p_1, q_1) (i.e. in state 1), 15% of type 1 agents have an assessment $y_t^{1,n} = 0$ and consequently use matrix F_2 to forecast prices at $t+1$. If $\lambda_1 > 1$ it follows from (4.7) that they are more pessimistic (relative to Γ) about the probabilities of the nine prices $((p_1, q_1), (p_2, q_2), \ldots, (p_9, q_9))$ at $t+1$. The converse applies when $\lambda_s < 1$. We also note that the possible dependence of the deviations (λ_s, μ_s) from Γ on the state s is very important since this is a way for the agents *to condition beliefs on prices.* Formally, if λ_s or μ_s vary with s then we say that *the impact of the assessment variables on the forecasts of the agents is price dependent.* This fact is central to the interpretation of our results below.

We note in summary that a simulation model requires the specification of 108 parameters: 36 for matrix A, 36 for matrix B and 36 intensity parameters (λ, μ). However, these belief parameters are restricted by the following 242 rationality conditions:

1. 98 equality and inequality restrictions (4.5) on the matrices A and B.
2. 36 direct rationality conditions (4.6) on the structure of the matrices (F_1, F_2, G_1, G_2).
3. 108 inequality restrictions (4.8) on the choices of (λ, μ).

The simulations focus on the factors which generate endogenous uncertainty in the replica RBE with types and the determinants of the equity risk premium. There are four such factors:

1. *Deviations over time of the intensity parameters* (λ, μ) *from* 1 reflecting the non-stationarity of beliefs of the agents. Hence, aggregate volatility may be caused by the fact that the conditional probability beliefs of the agents may vary over time.
2. *Correlation of assessments within types* represented by the existence of type–states other than the type–state $(\alpha_k,\ 1-\alpha_k)$.
3. *Correlation among type–states* (i.e. across types) represented by the vectors (a, b) of parameters inducing a joint distribution of the assessments which is Markov and not i.i.d.
4. *Price dependency of the intensity variables* λ_s *and* μ_s.

The objective of the parameter specification below is to study the configurations which generate equilibrium volatility and equity premia. These specifications do not represent *illustrations* of parameters which generate volatility and premia but rather, they are the only configuration which we found to generate volatility and premia in the range observed in the U.S. economy and hence their interpretation provides an explanation of the volatility and premia which arise in the real economy. Some discussion of the results is provided below.

4.3. Simulation results

As in [24], the focus of the simulation results is the equity premium and related statistics. We thus report in each table the following key variables:

ρ – the long term mean equity risk premium; historically around 6%,

σ_r – the long term standard deviation of the risky returns on equity; historically about 18%,[11]

r^F – the long term riskless rate on one period loans; historically .5%–1.0%,

σ_F – the long term standard deviation of the short term riskless rate; historically about 6%,

σ_p^2 – the long term variance of the price/dividend ratio; observations on σ_p^2 do not correspond to the economic concept due to tax and accounting distortions in reported earnings.

The historical estimates vary depending upon definitions, data sources and periods of estimation. We disregard these fine details and focus on the order of magnitudes involved.

4.3.1. Rational expectations equilibria

To enable comparison with results obtained under rational beliefs we report in Table 1 (on the next page) the results for rational expectations equilibria.

The results are in accord with the standard results which gave rise to the equity premium puzzle debate: a very high riskless rate over 5%; a very low equity premium of less than .5% and a very low standard deviation of the risky returns on equity around 4.1%. We also report here the extremely low variance of the price dividend/ratio which we consider to be an important indicator. Price volatility is the primary phenomenon associated with endogenous uncertainty and hence we are inclined to pay particular attention to it. Table 1 also shows that the results are not sensitive to parameter values in the realistic range. This conclusion does not hold for RBE where the results are sensitive to parameter values (see [24]).

Since we focus in this paper on social states and on the effects of correlation, we restrict ourselves to the fixed set of parameters $\gamma_1 = \gamma_2 = 3.25$, $\beta_1 = \beta_2 = .92$.

[11] This corrects the confusing practice in [24] and [25] of reporting the variance of the risky returns as $\frac{1}{100}\sigma_t^2$. Thus, they report the variance as 3.42% instead of 342%.

Table 1: Rational Expectations Equilibria with selected variables

		$\gamma_2 = 2.75$ $\beta_2 = .92$	$\gamma_2 = 2.75$ $\beta_2 = .96$	$\gamma_2 = 3.25$ $\beta_2 = .92$	$\gamma_2 = 2.75$ $\beta_2 = .96$
$\gamma_1 = 2.75$ $\beta_1 = .92$	ρ	.41%	.41%	.44%	.44%
	r^F	5.16%	5.15%	5.12%	5.11%
	σ_r	4.04%	4.04%	4.07%	4.06%
	σ_F	.83%	.82%	.85%	.84%
	σ_p^2	.0049	.0049	.0056	.0056
$\gamma_1 = 2.75$ $\beta_1 = .96$	ρ	.41%	.41%	.44%	.44%
	r^F	5.14%	5.13%	5.11%	5.09%
	σ_r	4.04%	4.04%	4.06%	4.06%
	σ_F	.82%	.82%	.84%	.84%
	σ_p^2	.0049	.0049	.0056	.0056
$\gamma_1 = 3.25$ $\beta_1 = .92$	ρ	.44%	.44%	.49%	.49%
	r^F	5.12%	5.11%	5.08%	5.06%
	σ_r	4.07%	4.06%	4.09%	4.09%
	σ_F	.85%	.84%	.87%	.87%
	σ_p^2	.0056	.0056	.0065	.0064
$\gamma_1 = 3.25$ $\beta_1 = .96$	ρ	.44%	.44%	.49%	.49%
	r^F	5.11%	5.09%	5.06%	5.05%
	σ_r	4.06%	4.06%	4.09%	4.08%
	σ_F	.45%	.84%	.87%	.86%
	σ_p^2	.0056	.0056	.0064	.0064

4.3.2. Rational Belief Equilibria I: a constant, single, social state of belief and no correlation with $\chi_1 = \chi_2 = 1$

We start the study of the equity risk premium by assuming a constant social state of belief hence $\chi_1 = \chi_2 = 1$. This economy should be considered to be the limit of a replica economy under the assumption of no correlation among the assessments of the agents and no price dependency in the intensities (λ, μ) of deviation from the Markov matrix Γ. Under the assumption of independence, the single social state of belief is $((\alpha_1,\ 1-\alpha_1),\ (\alpha_2,\ 1-\alpha_2))$ and the two social states are

$$\left\{ \left(d^H,\ (\alpha_1,\ 1-\alpha_1),\ (\alpha_2,\ 1-\alpha_2)\right),\ \ \left(d^L,\ (\alpha_1,\ 1-\alpha_1),\ (\alpha_2,\ 1-\alpha_2)\right) \right\}.$$

It follows from the equilibrium map (4.4) that in such RBE there are only two prices which are associated with these two social states and this is the same number of prices as in the rational expectations equilibria reported in Table 1. This means that in such RBE endogenous uncertainty does not lead to the emergence of additional prices but rather, it changes the two rational expectations equilibrium prices. Indeed, we shall shortly see that it can induce dramatic increases in the volatility of equilibrium prices. We call such an effect a *volatility amplification* effect.

Under the assumption of no price dependency we must have $\lambda_s = \lambda^0$, $\mu_s = \mu^0$ for all s and given this assumption let us adopt the convention of selecting $\lambda^0 > 1$ and $\mu^0 > 1$. We can then interpret the model to be one in which a proportion α_k of agents of type k are *always* relatively (to Γ) optimistic about the states of high prices in the next period and a proportion $(1 - \alpha_k)$ of agents of type k are *always* relatively pessimistic. The beliefs of individual agents fluctuate over time between optimism and pessimism but over the long run every agent is relatively optimistic a fraction α_k of the time and relatively pessimistic a fraction $(1-\alpha_k)$ of the time. The parameterization of the model is then reduced to the four parameters (α_1, λ^0), (α_2, μ^0) and we need to consider the effect of the feasibility restrictions (4.5), (4.6) and (4.8).

Note that as we vary the four parameters (α_1, λ^0), (α_2, μ^0) over the feasible region we reach boundary points at which some of the inequalities in (4.5) or (4.8) are satisfied with equality. It can be checked that at these boundary points some probabilities in the matrices F_1, F_2, G_1, or G_2 become zero. More specifically, we adopt in this section the following rules:

1. For each α_1 select the *largest feasible* λ^0,
2. For each α_2 select the *largest feasible* μ^0. (4.9)

To illustrate, suppose that we select $\alpha_1 = .5$ and $\alpha_2 = .4$. A single social state of belief implies that we must select $\chi_1 = \chi_2 = 1$, $a^1 = a^2 = a^3 = b^1 = b^2 = b^3 = 0$, $a^4 = b^4 = 1$. It follows from (4.8) that we must also have the following four restrictions:

$$\lambda^0 < \frac{1}{\phi} = 2.3256 \qquad \mu^0 < \frac{1}{\phi} = 2.3256$$

$$\lambda^0 < \frac{1}{1-\phi} = 1.7544 \qquad \mu^0 < \frac{1}{1-\phi} = 1.7544$$

$$\lambda^0 < \frac{1}{\alpha_1} = 2.000 \qquad \mu^0 < \frac{1}{\alpha_2} = 2.500.$$

In this case the binding constraint is 1.7544. Other constraints will be binding if we wanted to select the *smallest* feasible λ^0 or μ^0.

To see the meaning of the criteria specified in (4.9) keep in mind that under the above specifications the matrices F_1, F_2, G_1 and G_2 are in effect all 2×2

matrices. Hence, a zero probability in, say, the matrix F_1 means that given that some state of low or high prices is obtained at date t, the agent who uses the matrix F_1 is *certain* that at date $t+1$ high or low prices will be realized. This is a rather extreme belief. Note also that given the rationality condition $\alpha_1 F_1 + (1-\alpha_1)F_2 = \Gamma$, an extreme optimism about high prices when using F_1 must be associated with extreme pessimism when using F_2. Note also that some boundary restrictions apply only to the first 9 states and others only to states 10–18 (see (4.8)). Hence, under the criteria (4.9) we know that a *positive fraction* of the agent will hold conditional probabilities with zero entries *some of the time.*

Table 2 reports the volatility results for RBE simulated under several configurations of the parameters (α_1, λ^0) and (α_2, μ^0) derived under the criteria (4.9).

Table 2: RBE with a single, constant social state of belief ($\chi_1 = \chi_2 = 1$) derived under (4.9) and no correlation

		$\lambda^0 = 1.754$, $\alpha_1 = .5$	$\lambda^0 = 1.754$, $\alpha_1 = .57$	$\lambda^0 = 1.666$, $\alpha_1 = .6$	$\lambda^0 = 1.428$, $\alpha_1 = .7$
$\mu^0 = 1.754$, $\alpha_2 = .5$	ρ	.98%	4.94%	3.92%	2.88%
	r^F	6.05%	3.55%	3.17%	3.71%
	σ_r	16.34%	23.51%	16.41%	13.32%
	σ_F	14.01%	19.65%	12.37%	9.59%
	σ_p^2	4.5417	9.8228	4.4648	2.7414
$\mu^0 = 1.754$, $\alpha_2 = .57$	ρ	4.94%	10.00%	7.69%	6.14%
	r^F	3.55%	.43%	.45%	1.25%
	σ_r	23.51%	31.00%	21.70%	18.00%
	σ_F	19.65%	24.30%	15.88%	12.87%
	σ_p^2	9.8228	16.7917	8.1316	5.4623
$\mu^0 = 1.666$, $\alpha_2 = .6$	ρ	3.92%	7.69%	5.23%	3.96%
	r^F	3.17%	.45%	1.46%	2.32%
	σ_r	16.41%	21.70%	13.43%	10.43%
	σ_F	12.37%	15.88%	8.61%	6.15%
	σ_p^2	4.4648	8.1316	2.6852	1.3887
$\mu^0 = 1.428$, $\alpha_2 = .7$	ρ	2.88%	6.14%	3.96%	2.88%
	r^F	3.71%	1.25%	2.32%	3.10%
	σ_r	13.32%	18.00%	10.43%	7.75%
	σ_F	9.59%	12.87%	6.15%	3.91%
	σ_p^2	2.7414	5.4623	1.3887	.5543

There are two important conclusions that can be drawn from the table. First, it shows that although the RBE with a single social state has only two prices, which is the same number as in the REE, the two equilibria are dramatically different. The crucial difference between them is found in the fact that in the RBE, half of the agents have optimistic probability beliefs relative to Γ about the prospects of $((p_1, q_1), (p_2, q_2), \ldots, (p_9, q_9))$ while half of the agents have pessimistic beliefs (relative to Γ) about these prices. This in contrast with the REE in which all agents hold Γ as their belief at all dates. Table 2 then demonstrates a new property of the model of the replica economy with types: volatility does not necessarily emerge as a result of an increase in the number of social states of beliefs but may arise as a result of *the nature of the distribution of beliefs* in each social state. Compare this conclusion with the observations made in the papers in the volume by Kurz [21] that endogenous uncertainty is induced by the variability, over time, in the states of belief. This idea is explained in detail in [22, page 32] and is based on RBE of models with individual states of beliefs. One of the important results of the model with types and social states is that volatility may be propagated simply by the social distribution itself and not by any variations over time in the social states of belief.

The second conclusion that we draw from Table 2 is that the amplification of volatility in RBE with a constant social state of belief can be very dramatic if agents are allowed to adopt boundary beliefs. Indeed, these are the maximal volatility measures and equity premia that this specification of the model can generate. It is interesting, however, that both at low as well as high α_k the equity premium is low and the riskless rate is high. The largest equity premium is realized in the middle of the table where α_1 and α_2 are close to .57 but in those cells the standard deviations of both the riskless rates as well as those of the risky returns are much too large. As α_1 and α_2 move away from .57 the volatility of both the riskless rate as well as the risky returns falls dramatically. As a result of these facts there is no cell which fits the historical record of all four moments ($\rho = 6\%$, $r^F = .5\%$, $\sigma_r = 18\%$, $\sigma_F = 6\%$).

Under the axioms of the theory of rational beliefs agents may hold extreme beliefs but this does not mean that such beliefs must be observed in the market. Indeed, we shall shortly argue that one may choose between two alternative hypotheses by imposing restrictions on beliefs based on known facts about the distribution of beliefs in the market. The question then becomes *which of the two alternative hypotheses performs better under the stipulated restrictions.* To motivate these restrictions we note that although high degrees of optimism or pessimism are observed in the beliefs of investors in security markets, it is evident that *certainty beliefs are rarely encountered.* We then propose to restrict the beliefs of the agents so as not to permit them to hold boundary beliefs. Formally we require

$$f_{ij}^k \geq .05\ \Gamma_{ij}, \qquad g_{ij}^k \geq .05\ \Gamma_{ij}, \qquad k = 1,2 \tag{4.10}$$

where f_{ij}^k and g_{ij}^k are the (ij) elements of the matrices F_k and G_k. (4.10) specifies that any deviations from the stationary measure should not result in probabilities which are less than 5% of the corresponding probabilities in Γ. Observe that lower bound restrictions *imply* upper bound restrictions due to the rationality conditions $\alpha_1 F_1 + (1 - \alpha_1)F_2 = \Gamma$. We call the collection of all such restrictions the 5% boundary restrictions on beliefs. It is clear that under these restrictions the beliefs used in Table 2 are not allowed.

Table 3 presents the results for RBE with the same values of (α_1, α_2) as in Table 2 but under the 5% boundary restrictions on beliefs.

Table 3: RBE with a single, constant, social state of belief ($\chi_1 = \chi_2 = 1$) and with the 5% boundary restrictions on agents' beliefs

		$\lambda^0 = 1.72$, $\alpha_1 = .5$	$\lambda^0 = 1.72$, $\alpha_1 = .57$	$\lambda^0 = 1.63$, $\alpha_1 = .6$	$\lambda^0 = 1.41$, $\alpha_1 = .7$
	ρ	1.12%	2.10%	1.85%	1.52%
$\mu^0 = 1.72$	r^F	5.05%	4.32%	4.33%	4.47%
$\alpha_2 = .5$	σ_r	10.16%	11.97%	10.09%	8.45%
	σ_F	7.02%	8.51%	6.63%	5.07%
	σ_p^2	1.3487	2.0779	1.3026	.7676
	ρ	2.10%	3.23%	2.85%	2.38%
$\mu^0 = 1.72$	r^F	4.32%	3.47%	3.56%	3.79%
$\alpha_2 = .57$	σ_r	11.97%	13.87%	11.74%	9.92%
	σ_F	8.51%	10.00%	7.93%	6.25%
	σ_p^2	2.0779	2.9743	1.9493	1.2334
	ρ	1.85%	2.85%	2.45%	2.01%
$\mu^0 = 1.63$	r^F	4.33%	3.56%	3.71%	3.96%
$\alpha_2 = .6$	σ_r	10.09%	11.74%	9.70%	7.99%
	σ_F	6.63%	7.93%	5.99%	4.42%
	σ_p^2	1.3026	1.9493	1.1457	.2668
	ρ	1.52%	2.38%	2.01%	1.62%
$\mu^0 = 1.41$	r^F	4.47%	3.79%	3.96%	4.20%
$\alpha_2 = .7$	σ_r	8.45%	9.92%	7.99%	6.41%
	σ_F	5.07%	6.25%	4.42%	2.97%
	σ_p^2	.7676	1.2334	.2668	.2720

The results reported in Table 3 represent the largest possible volatility measures and equity premia that can be generated by the RBE under the restriction of no correlation and a constant social state of belief. One can see that once the 5% restriction is imposed, the model cannot generate statistics which are even close to the historical record: the equity premia are too low, the riskless rates are too high and the volatility of the riskless rate is too low.

We remark that a comparison of the results of Tables 2 and 3 is complicated by the fact that the impact of the 5% restrictions varies across the cells of the table and each of those restrictions may affect different segments of the agents and only part of the time. However, the results in Table 3 show that in order for the RBE with a constant social state of belief to generate high volatility and large equity premia it is necessary that some of the agents hold, some of the time, conditional beliefs which are rather extreme.

One of the conclusions of this paper is that an equilibrium with a single social state cannot generate data which match all four moments under examination. However, an RBE with a constant state of belief is a relatively simple model that can provide an intuitive explanation of the mechanism which generates equity premium in the model with types. This fact is compatible with one of the aims of this paper which is to give an intuitive explanation of the mechanism which generates an equity risk premium in an RBE. Thus, before we proceed to study the model with correlation among the beliefs of agents, let us pause to explain the results reported in Tables 2 and 3 and the particular role played by the value of .57 taken by α_k.

Note at the outset two facts about the equilibrium model which generate the results in Tables 2 and 3. On the one hand, a change in α_k results in a change of the proportion of type k agents who are optimistic at any moment of time about future capital gains. Since the social state of belief is constant this proportion is constant. On the other hand, the rationality conditions $\alpha_1 F_1(\lambda)+(1-\alpha_1)F_2(\lambda) = \Gamma$ imply that as α_k changes the *intensity* of optimism and pessimism must change so as to compensate for the number of agents who are optimistic or pessimistic. "Intensity" is measured in terms of the probability with which the agents forecast higher or lower prices. The volatility characteristics of the economy are then determined by the interplay between the proportion of agents who are optimistic or pessimistic and the intensity of their optimism/pessimism. The crucial variable that needs to be understood in this connection is the behavior of the riskless rate.

To explore the behavior of the riskless rate observe at the outset that the mean risky rate of return on equity remains in the 6%–8% range for almost all cells of Tables 2 and 3; the main determinant of the premium is therefore the equilibrium value of the riskless rate. Now consider the number and intensity of belief of those agents who expect at date t a recession and hence lower prices to be realized at date $t+1$. It is clear that as α_k increases, the number of such agents decreases. However the rationality conditions induce a nonlinear relationship between the *number* of such agents and the level of their *intensity.* The structure of this nonlinear relation has three parts:

1. For small α_k the rationality conditions limit the intensity of pessimists and even if their number is larger than the optimists, the intensity of the optimists is at a very high level. Since the intensity with which the optimists want to borrow is relatively high in relation to the intensity with which the pessimists want to lend, the results are high riskless rates, low premia and low volatility.

2. As α_k increases the intensity of the pessimists rises and is maximized at (.57, .57); it cannot increase beyond that point. Around .57 the intensity of the pessimists dominates the rising number of optimists and the result is a decline in the riskless rate and a rise in the premium. The rise in the volatility of prices and risky returns in this region is a result of the fact that the intensity of both sides is at the high level and this results in more drastic changes of excess demand in response to fluctuations in the realized dividend growth.

3. As α_k increases beyond .57 the intensity of the pessimists remains constant but their number declines. As the relative number of optimists rises, their intensity declines, the level of volatility falls dramatically and the riskless rate rises again.

In sum, the equity risk premium is the result of the interplay between the number and intensity of beliefs of the optimists vs. the pessimists and hence it is determined by the distribution of beliefs in the economy. For low α_k the *intensity* of the optimists has the stronger impact and for large α_k their *number* has the dominant impact. The nonlinearity induced by the rationality conditions results in the middle region in which the *intensity and number* of the pessimists just outweighs the optimists, causing increased volatility and a lowered riskless rate. This structure is made much more complicated in a world of correlation in which there are more social states with more configurations of belief and intensities.

The alternative model with which we propose to explain the data is a model where correlation among the beliefs of agents turns the social state of belief into a random variable. Although the mechanism which generates an equity premium is more complicated, the insight provided by the model with a single state of belief remains correct. We turn now to this subject.

4.3.3. Rational Belief Equilibria II: the effect of correlation among the beliefs of agents

Correlation among the beliefs of agents is a complicated phenomenon due to the fact that it may take several forms. Hence, in order to study the effect of correlation we need to clarify the terms used to characterize it. Here are the basic terms which we use:

1. **Correlation within types** is characterized by the assumption that $\chi_1 < 1$ and $\chi_2 < 1$. Under the specifications above we have 3 type–states and hence 9 social states of belief.

2. **Correlation across types** is characterized by the fact that the matrices A and B are *not* transition matrices of a joint process of i.i.d. random variables. For each value of χ_1 and χ_2 the type–states are jointly i.i.d. if the following are the values of the parameters in A and B (which are then parameters of the matrix Γ):

$\underline{\chi_1 = \chi_2 = .5:}$ For all s, $a_s^1 = b_s^1 = .0625, \quad a_s^2 = b_s^2 = .125,$
$a_s^3 = b_s^3 = .125, \quad a_s^4 = b_s^4 = .25.$

$\underline{\chi_1 = \chi_2 = .2:}$ For all s, $a_s^1 = b_s^1 = .16, \quad a_s^2 = b_s^2 = .08,$
$a_s^3 = b_s^3 = .08, \quad a_s^4 = b_s^4 = .04.$

$\underline{\chi_1 = \chi_2 = .1:}$ For all s, $a_s^1 = b_s^1 = .2025, \quad a_s^2 = b_s^2 = .045,$
$a_s^3 = b_s^3 = .045, \quad a_s^4 = b_s^4 = .01.$

3. **Price dependency** is characterized by the fact that the parameters λ_s and μ_s are *dependent on s*.

We comment on these by noting that the conditions $\chi_1 < 1$ and $\chi_2 < 1$ could be associated with two situations. First, we may have a large but finite economy which is approximately anonymous in which the existence of multiple type–states is a natural fact. The assumption of three type–states is then an assumption about the nature of correlation (in addition to being a computational simplification). Second, we may have an infinite replica economy and the individual assessments of the agents are not i.i.d. Our assumption that the type–states are marginally i.i.d. makes sense only if there is correlation among the assessments within a type.

The distinction between correlation among the type–states and price dependency is important. The correlation among the type–states is a statistical condition stipulating that the assessments are random variables which are statistically correlated. Price dependency is not a condition of statistical correlation; rather, it stipulates the *commonality in the interpretation of the assessments by the agents.*

The terms defined above show that in order to specify a model with correlation, we need to specify feasible values of (χ_1, χ_2), (a, b), and (λ, μ). It follows from (4.5) that the parameters (a, b) depend upon (χ_1, χ_2) so that as we vary (χ_1, χ_2) we must also vary (a, b) in accord with the feasibility conditions (4.5). It is therefore impossible to isolate the net effect of varying (χ_1, χ_2). In the analysis below we assume $\chi_1 = \chi_2 = \chi$, taking the three values .5, .2, and .1. Correspondingly, we vary (a, b) in a reasonably similar manner but exact comparability is impossible. We, therefore, focus only on simulations in which (χ_1, χ_2) are fixed.

Parameter specifications. The basic specification takes the case $\alpha_1 = \alpha_2 = .57$. The corresponding RBE under the 5% boundary restrictions on beliefs is the "reference RBE." This is motivated by our aim to examine what would be the contribution of models of correlation. Hence, the reader should keep in mind the results for this reference case as reported in Table 3 (i.e. the case with $\alpha_1 = \alpha_2 = .57$, $\lambda_s = \mu_s = 1.72$). We thus compare the reference RBE with RBE under the following specifications:

1. χ takes the values .5, .2 and .1.
2. The intensity variables are specified as follows:
   (a) For RBE with i.i.d. assessments and without price dependency we specify $\lambda_s = \mu_s = 1.72$.
   (b) For RBE with *price dependency* we specify

$$\lambda_1 = \lambda_2 = \lambda_3 = .46, \qquad \lambda_4 = \lambda_5 = \lambda_6 = 1,$$
$$\lambda_7 = \lambda_8 = \lambda_9 = 1.72, \qquad \lambda_{10} = \lambda_{11} = .46,$$
$$\lambda_{12} = \lambda_{13} = \lambda_{14} = \lambda_{15} = \lambda_{16} = \lambda_{17} = \lambda_{18} = 1.72,$$
$$\mu_1 = .46, \qquad \mu_2 = \mu_3 = 1.72,$$
$$\mu_4 = \mu_5 = \mu_6 = 1,$$
$$\mu_7 = .46, \qquad \mu_8 = \mu_9 = 1.72,$$
$$\mu_{10} = .46, \qquad \mu_{11} = \mu_{12} = 1.72,$$
$$\mu_{13} = \mu_{14} = \mu_{15} = 1, \qquad \mu_{16} = \mu_{17} = \mu_{18} = 1.72.$$

3. The (a, b) parameters which are dependent upon χ are specified in the Appendix.

Table 4 presents the results for $\chi = .5$. The reference RBE under $\chi = 1$ is reproduced in Column 1. A comparison of Columns 1 and 2 of the table shows that the reference RBE with a single social state of belief exhibits about the same volatility characteristics as the RBE with correlation within types but with i.i.d. type–states. In Column 3 we see, however, that price dependency increases the premium, reduces the riskless rate but also leads to a reduction in volatility. The addition of correlation *across* type–states raises the premium to 4.18% and restores some volatility.

Table 4: RBE with correlation among beliefs, with $\chi = .5$ and with the 5% boundary restrictions on beliefs

	RBE (Reference) with a constant social state of belief $\chi_1 = \chi_2 = 1$	RBE i.i.d. type states no price dependence	RBE i.i.d. type states with price dependence	RBE correlation across types with price dependence
ρ	3.23%	2.87%	3.92%	4.18%
r^F	3.47%	3.63%	2.25%	2.10%
σ_r	13.87%	12.60%	10.92%	11.97%
σ_F	10.00%	9.68%	6.51%	7.94%
σ_p^2	2.9743	2.4287	1.8498	2.0854

Altogether, the results reported in Table 4 do not match the data very well and leads to the conclusion that if correlation is to generate more volatility, we must explore parameter configurations which place less probability on the social states of belief ((.57, .43), (.57, .43)). We thus explore the two other cases $\chi = .2$ and $\chi = .1$. Since $\alpha_1 = \alpha_2 = .57$, these specifications imply that the correlation among the assessments leads the probabilities to be "spread" away from ((.57, .43), (.57, .43)) which is the constant social state of belief that would be realized under i.i.d. assessments. For $\chi = .2$ and $\chi = .1$ most of the probability is placed on the type–states (.85, .15) and (.25, .75).

Table 5 (next page) reports the results which are our main results regarding the effects of correlation:

In Column 1 we repeat the "reference RBE" with a constant social state of belief as in Table 3.

In Column 2 we report the results for RBE with three type–states which are i.i.d. (hence with correlation within types) and without price dependence. It is evident that these specifications contribute little *by themselves.*

In Column 3 we report the results for the effect of price dependence. It is clear that in conjunction with the correlation within types and the specification $\chi \leq .2$, price dependence has a strong effect. In Column 4 we report the added effect of full correlation across types. It contributes about 1% to the premium and substantially contributes to the volatility of returns.

It is instructive to note that the introduction of correlation within types (i.e. $\chi < 1$) *by itself* contributes little to explaining volatility. However, as we add price dependency and correlation across types, the results reported in the last two columns of Table 5 emerge as a result of a *combined* effect of all three forms of correlation. This indicates *a strong interaction effect among the three factors of correlation involved.*

Table 5: RBE II under correlation among beliefs, with $\chi = .2$ and $\chi = .1$ and with the 5% boundary restrictions on beliefs

RBE (Reference) with a constant social state of belief $\chi_1 = \chi_2 = 1$			RBE i.i.d. type states no price dependence	RBE i.i.d. type states with price dependence	RBE correlation across types with price dependence
3.23%		ρ	2.77%	5.02%	5.83%
3.47%		r^F	3.61%	1.23%	.66%
13.87%	$\chi = .2$	σ_r	11.85%	11.73%	13.75%
10.00%		σ_F	9.29%	7.72%	10.52%
2.9743		σ_p^2	2.1450	2.3004	2.9487
3.23%		ρ	2.76%	5.42%	6.54%
3.47%		r^F	3.58%	.87%	.25%
13.87%	$\chi = .1$	σ_r	11.59%	12.08%	15.84%
10.00%		σ_F	9.13%	8.20%	12.81%
2.9743		σ_p^2	2.0566	2.5036	3.9960

We now offer some intuitive explanation of the specification of the matrices A and B which regulate the long term correlation across type–states.

We have already noted that variations of the parameter χ induce changes in the feasibility conditions (4.5) so that it is impossible to vary this parameter while keeping constant the parameters (a, b) of correlation across type–states. The main facts behind the selection of (a, b) is that the 9 prices associated with the states of expanding dividends are higher than the 9 prices associated with the states of declining dividends. In addition, within these two categories of states the prices $((p_1, q_1), (p_5, q_5), (p_9, q_9), (p_{10}, q_{10}), (p_{14}, q_{14}), (p_{18}, q_{18}))$ are the high prices while the "crash" prices are $((p_{12}, q_{12}), (p_{16}, q_{16}))$. Other prices are "medium" prices. The parameters $a = (a^1, a^2, a^3, a^4)$ are selected subject to feasibility so that there is high probability of transition from the very high prices to crash and medium prices. In addition, these parameters aim to maximize transition probabilities from crash prices to very high prices and from all other prices to medium and high prices. The parameters $b = (b^1, b^2, b^3, b^4)$ are selected to maximize transition probabilities to the very high prices, subject to feasibility. This parameterization of the transition probabilities contributes to price volatility. However, keep in mind that the feasibility conditions leave limited room for such selections so that the nature of these transition probabilities and the implied correlations across type–states may be very different for different values of χ (see the specifications of the (a, b) vectors for the different values of χ in the Appendix).

Recall that all simulations in Table 5 have been conducted under the 5% boundary restrictions on beliefs. Comparing the results in Columns 2–4 with the results in Column 1 or in Table 3, we conclude that the model with correlation among beliefs of agents performs much better than the model with a single state of belief. We have seen in Table 3 that the model with a single state of belief could not generate a riskless rate which is smaller than 3%–4%. These simulations were conducted under the assumption that $\lambda_s = \lambda^0$, $\mu_s = \mu^0$ for all s which means that price dependency was not allowed whereas price dependency is compatible with a single state of belief. We have sampled extensively in the parameter space and can report that allowing price dependency has not changed the essential results of Table 3: the riskless rate in all our simulations was never below 3%. We conclude that under the 5% boundary restrictions on beliefs the model with a single social state of belief cannot generate data which will match the observed values of the four moments which we have been examining.

In contrast to the above conclusion, under the same 5% boundary restrictions on beliefs, the model specification with correlation among the beliefs of agents generates statistics which match all four empirical moments rather well. The standard deviation of the risky returns is somewhat smaller than the historical record and the standard deviation of the riskless rate is somewhat larger than the record. To gain more insight into these results let us examine some variants of the case $\chi = .1$, $\alpha_1 = \alpha_2 = .57$ by perturbing α_1 and α_2 over the values .54, .57, .60.

Table 6: RBE II under correlation among beliefs with $\chi = .1$ and with the 5% boundary restrictions on beliefs

		$\alpha_1 = .54$	$\alpha_1 = .57$	$\alpha_1 = .6$
$\alpha_2 = .54$	ρ	4.87%	5.80%	5.05%
	r^F	1.58%	.82%	1.32%
	σ_r	14.51%	14.62%	12.88%
	σ_F	11.31%	11.94%	10.28%
	σ_p^2	2.7746	3.3381	2.4664

		$\alpha_1 = .54$	$\alpha_1 = .57$	$\alpha_1 = .6$
$\alpha_2 = .57$	ρ	5.83%	6.54%	5.65%
	r^F	.78%	.25%	.86%
	σ_r	14.65%	15.84%	13.92%
	σ_F	11.97%	12.81%	11.10%
	σ_p^2	3.3459	3.9960	2.9654
$\alpha_2 = .6$	ρ	5.20%	5.76%	4.82%
	r^F	1.19%	.76%	1.45%
	σ_r	13.02%	14.00%	12.06%
	σ_F	10.35%	11.07%	9.29%
	σ_p^2	2.5227	2.9908	2.0821

Table 6 shows that the results are rather sensitive to parameter values but there is a significant region in the parameter space that can give rise to statistics which are compatible with the empirical moments. Key variables that would change the results in the table are the values of the probabilities (χ_1, χ_2) and the social distribution of beliefs defined in our models by the type–states (.85, .15) and (.25, .75).

4.4. Understanding how an equity risk premium is generated under rational belief

Ever since the publication of the paper by Mehra and Prescott [35] on the equity premium, numerous theories were offered to explain the empirically observed premium. For example, Mankiw [34] proposed to explain the premium by the presence of nondiversifiable risks; Reitz [41] proposed to explain it by the introduction of big crash states; Weil [45] and Epstein and Zin [17] suggest that a non-expected utility model may be used to explain the data and Constantinides [13] initiated a large literature on the use of habit forming utility functions to explain the data. This paper complements the earlier paper by Kurz and Beltratti and proposes the theory of rational belief as an explanation of the data. The model of an RBE with types offers an intuitive explanation to which we now turn.

The basic assumption of the theory of rational belief is that agents do not observe the social states and do not know the equilibrium map. The consequence of the rationality axioms is that agents form beliefs about prices, not about social states, and may have diverse beliefs about the probabilities of future prices. The important conclusion of the theory is that *if agents disagree then their state of belief must fluctuate over time.* To understand why, observe that if agents disagree then they must deviate from the stationary measure.

However, deviations from the stationary measure at one date must be compensated by other deviations at other dates so that the time average of the deviations tends to zero in order to satisfy the rationality axioms. These fluctuations over time in the states of belief of the agents is the mechanism which generates endogenous uncertainty in an RBE and is reflected in the volatility of equilibrium prices and quantities. It then follows that the first component of explaining the risk premium in an RBE is the presence of endogenous uncertainty. All risk averse agents who perceive the extra endogenous volatility of returns will require the compensation of an added risk premium in order to be willing to hold the more risky equity. This argument is, however, insufficient since agents who disagree may be more or less optimistic with respect to future events and thus require a higher or lower premium depending upon their probability assessment. The first basic argument must be then supplemented by an explanation of how the diversity of beliefs by itself can add to equilibrium equity premium.

When some agents are optimistic and some are pessimistic, trading opportunities naturally become available but this need not have anything to do with the equity risk premium. However, when such optimism or pessimism is defined with respect to the future risky rates of return on equity then it will have an effect on the premium. For example, if at price vector 1 the level of pessimism about future equity returns of an agent increases he will select a portfolio with lower weight on equity and higher weight on riskless debt and this will tend to reduce the price of equity and increase the price of riskless debt resulting in increased premium in state 1. The situation is substantially complicated by the rationality conditions which hold that an agent who is relatively optimistic at some date must be relatively pessimistic at some other date. In a large economy with a single social state the proportions of optimists and pessimists are fixed and in the simulations above we allowed these proportions to vary across models. When the proportion of optimists changes, the rationality conditions imply that the *intensity* of optimism and pessimism must change. This shows that at any time both the proportion of pessimists as well as their intensity matter to market equilibrium. We have observed in Tables 2 and 3 that a simultaneous change in the proportions and intensities of the optimists and the pessimists (via changes in α_1 and α_2) has a nonlinear effect on market excess demand and hence on the premium. The implication of this observation is that the distribution of beliefs in the market at any date is the crucial factor which determines the equity risk premium at that date. This observation extends to the model with correlation.

In the general model with correlation we cannot think of the equilibrium premium as being determined by a fixed proportion of optimists and pessimists. Since the social state of belief is a random variable these proportions vary but the observation made in the model with a constant state of belief remains valid: *at any date the risk premium is determined by the distribution of beliefs at that date.* But then, any parameter that has an impact on the distribution of beliefs and on the frequencies at which the states of belief are realized over time will have an effect on the average premium of the economy. It is appropriate to think of time dependency and correlation among the assessments of agents as belief externalities which affect the distribution of beliefs in the following two ways:

1. Price dependence has the effect of changing the number of optimists and pessimists given any price. For example consider price vector (p_1, q_1) defined in the models above by the social state $(d^H,\ (.85,\ .15),\ (.85,\ .15))$. If λ_1 is price dependent, it will have the following effect: if $\lambda_1 > 1$ then in this first state 85% of type 1 agents are *optimistic* about high prices the next period and if $\lambda_1 < 1$ then in this first state 85% of type 1 agents are *pessimistic* about high prices the next period.
2. Correlation among type–states is an externality which can increase the frequency over time of states of beliefs which generate higher premium. The externality also creates new distributions of belief which an agent cannot deduce from his own belief. For example, although the simulations in Tables 4–5 postulate RBE in which $\alpha_1 = \alpha_2 = .57$, the correlation among beliefs leads to the emergence of social states of belief which are different from ((.57, .43), (.57, .43)) but the agents do not know the structure of this externality.

Based on these comments we suggest that the exact interpretation of the parameterizations of (A, B, λ, μ) in the various models in Tables 4–6 is less important than the function of these parameterizations in *regulating the distribution of the states of belief and the frequencies of their realization.* Correspondingly, all four moments of the distribution of the risky and riskless returns are determined by the frequencies of the realized states of belief. From this perspective the reason why models of RBE can generate theoretical moments with high volatility, low riskless rate and high equity premium can be summarized as follows:

1. In the typical RBE there are relative pessimists at all dates and there is always a range of parameter values where either the number or the intensity of the pessimists dominate and have the impact of pushing the riskless rate down and hence the premium up. The volatility in prices and returns is then a consequence of the fact that due to the rationality conditions the relative impact of the pessimists and optimists vary in such

equilibria across states and market prices naturally reflect these changes. Although the simulated RBE with a single social state have the property that the pessimists are in the minority and their intensity dominates the bond market, we cannot be certain of the generality of this conclusion since there are other forms of pessimism and optimism which we have not studied. The general principle proposed by the theory of RBE is, however, clear. At all dates there are, in the economy, optimists and pessimists and either the number or the intensity of the pessimists is dominant: it pushes the riskless rate down and the equity risk premium up.

2. The correlation among the beliefs of agents has a dual impact on an RBE. First, it can change the relative number of optimists and pessimists at each state by making the intensity parameters price dependent and this allows the attainment of a low riskless rate and higher premium even when the intensity of the pessimists is not extreme. Second, it can change the stationary distribution and hence the long run frequency at which the different price states are realized. This changes the relative probabilities of states with high premium and consequently the average premium over time.

Let us close with a methodological note. The 5% boundary restrictions on beliefs were not derived from axioms of the theory of rational belief but rather from empirical observations. Using this restriction we argued that the model with correlation among the beliefs of agents is superior to a model with i.i.d. assessments in which there is a single, constant social state of belief. Since not all rational beliefs need to be observed in our economy, in future research we may generalize this approach as follows. One needs to start by obtaining more empirical information about the social distribution of beliefs. Given such data one may then ask what could be the type configurations and the sets of parameters characterizing the beliefs of the agents that would "rationalize" the data. Given that the distribution of beliefs is approximately rationalized, one can then proceed to test if the model with the specified family of beliefs can explain the observed volatility characteristics of the market.

Appendix

Specification of the parameter (a, b) in Tables 2–5

$\chi = 1$:

$$a_s^1 = a_s^2 = a_s^3 = b_s^1 = b_s^2 = b_s^3 = 0, \quad a_s^4 = b_s^4 = 1 \qquad \text{for } s = 1, 2, \ldots, 9.$$

$\underline{\chi = .5:}$

$a^1 = (.0001,\ .0001,\ .2498,\ .0001,\ .0001,\ .0001,\ .2498,\ .0001,\ .0001),$
$a^2 = a^3 = (.2498,\ .2498,\ .0001,\ .2498,\ .2498,\ .2498,\ .0001,\ .2498,\ .2498),$
$a_4 = (.0003,\ .0003,\ .4998,\ .0003,\ .0003,\ .0003,\ .4998,\ .0003,\ .0003),$
$b^1_s = .2498, \quad b^2_s = b^3_s = .0001, \quad b^4_s = .4998 \qquad \text{for } s = 1, 2, \ldots, 9.$

$\underline{\chi = .2:}$

$a^1 = (.0001,\ .0001,\ .25,\ .0001,\ .0001,\ .0001,\ .25,\ .0001,\ .0001),$
$a^2 = a^3 = (.1998,\ .1998,\ .1480,\ .1998,\ .1998,\ .1998,\ .1480,\ .1998,\ .1998),$
$a_4 = (.0001,\ .0001,\ .0001,\ .0001,\ .0001,\ .0001,\ .0001,\ .0001,\ .0001),$
$b^1_s = .3998, \quad b^2_s = b^3_s = .0001, \quad b^4_s = .1998 \qquad \text{for } s = 1, 2, \ldots, 9.$

$\underline{\chi = .01:}$

$a^1 = (.0001,\ .0001,\ .35,\ .0001,\ .0001,\ .0001,\ .35,\ .0001,\ .0001),$
$a^2 = a^3 = (.0998,\ .0998,\ .0998,\ .0998,\ .0998,\ .0998,\ .0998,\ .0998,\ .0998),$
$a^4 = (.0001,\ .0001,\ .0001,\ .0001,\ .0001,\ .0001,\ .0001,\ .0001,\ .0001),$
$b^1_s = .4498, \quad b^2_s = b^3_s = .0001, \quad b^4_s = .0998 \qquad \text{for } s = 1, 2, \ldots, 9.$

References

1. K. J. Arrow, Exposition of the theory of choice under uncertainty, in: *Decisions and Organization* (eds. C. B. McGuire and R. Radner), North-Holland, Amsterdam, 1971, 19–55.
2. K. J. Arrow and F. H. Hahn, *General Competitive Analysis*, Holden–Day, San Francisco, 1971.
3. K. J. Arrow and G. Debreu, Existence of an equilibrium for a competitive economy, *Econometrica* 22 (1954), 265–290.
4. K. J. Arrow, Le rôle des valeurs boursières pour la rèpartition la meilleure des risques, *Économetrie*, Colloque Internationaux du C.N.R.S. 11 (1953), 41–48; English translation: The role of securities in the optimal allocation of risk bearing, *Review of Economic Studies* 31 (1964), 91–96.
5. K. J. Arrow, Alternative approaches to the theory of choice in risk–taking situations, *Econometrica* 19 (1951), 404–437.
6. D. Cass, G. Chichilnisky, and H. M. Wu, Individual risk and mutual insurance, *Econometrica* 64 (1996), 333–341.
7. A. V. Benerjee, A simple model of herd behavior, *Quarterly Journal of Economics* 107 (1992), 797–817.

8. S. Bikhchandani, D. Hirshleifer, and I. Welch, A theory of fads, fashion, custom, and cultural change as information cascades, *Journal of Political Economy* 100 (1992), 992–1026.

9. W. A. Brock and S. N. Durlauf, Discrete choice with social interactions II, Department of Economics, University of Wisconsin, Madison, Wisconsin, 1997.

10. W. A. Brock, Asset price behavior in complex environments, Report No. 9606, Department of Economics, University of Wisconsin, Madison, Wisconsin, 1996.

11. W. A. Brock and S. N. Durlauf, Discrete choice with social interactions I: Theory, Report No. 9521, Department of Economics, University of Wisconsin, Madison, Wisconsin, 1995.

12. W. A. Brock, Pathways to randomness in the economy: emergent nonlinearities and chaos in economics and finance, *Estudios Economicos* 8, no. 1 (1993), 3–55.

13. G. Constantinides, Habit formation: a resolution of the equity premium puzzle, *Journal of Political Economy* 98 (1990), 519–543.

14. G. Debreu, *Theory of Value*, Wiley, New York, 1959.

15. S. N. Durlauf, Neighborhood feedbacks, endogenous stratification, and income inequality, Department of Economics, University of Wisconsin, Madison, Wisconsin, 1994.

16. S. N. Durlauf, Nonergodic economic growth, *Review of Economic Studies* 60 (1993), 349–366.

17. L. G. Epstein and S. E. Zin, "First–order" risk aversion and the equity premium puzzle, *Journal of Monetary Economics* 26 (1990), 387–407.

18. H. Föllmer, Random economies with many interacting agents, *Journal of Mathematical Economics* 1 (1974), 51–62.

19. P. Henrotte, Construction of a state space for interrelated securities with an application to temporary equilibrium theory, *Economic Theory* 8 (1996), 423–459.

20. M. Kac, Mathematical mechanisms of phase transitions, in: *Statistical Physics: Phase Transitions and Superfluidity*, Vol. 1 (eds. M. Chretien, E. Gross, and S. Deser), Brandeis University Summer Institute in Theoretical Physics (1966), 1968, 241–305.

21. M. Kurz (ed.), *Endogenous Economic Fluctuations: Studies in the Theory of Rational Belief*, Studies in Economic Theory No. 6, Springer–Verlag, Berlin and New York, 1997.

22. M. Kurz, Endogenous economic fluctuations and rational beliefs: a general perspective, in: *Endogenous Economic Fluctuations: Studies in the Theory of Rational Belief*, Chapter 1 (ed. M. Kurz), Studies in Economic Theory No. 6, Springer–Verlag, Berlin and New York, 1997.

23. M. Kurz, Asset prices with rational beliefs, in: *Endogenous Economic Fluctuations: Studies in the Theory of Rational Belief*, Chapter 9 (ed. M. Kurz), Studies in Economic Theory No. 6, Springer–Verlag, Berlin and New York, 1997.

24. M. Kurz and A. Beltratti, The equity premium is no puzzle, in: *Endogenous Economic Fluctuations: Studies in the Theory of Rational Belief*, Chapter 11 (ed. M. Kurz), Studies in Economic Theory No. 6, Springer–Verlag, Berlin and New York, 1997.

25. M. Kurz, On the volatility of foreign exchange rates, in: *Endogenous Economic Fluctuations: Studies in the Theory of Rational Belief*, Chapter 12 (ed. M. Kurz), Studies in Economic Theory No. 6, Springer–Verlag, Berlin and New York, 1997.

26. M. Kurz, Social states of belief, rational belief equilibrium and the structure of the equity premium, Draft dated July 9, 1997, Department of Economics, Stanford University, Stanford, CA, 1997.

27. M. Kurz and M. Schneider, Coordination and correlation in Markov rational belief equilibria, *Economic Theory* **8** (1996), 489–520.

28. M. Kurz and H. M. Wu, Endogenous uncertainty in a general equilibrium model with price contingent contracts, *Economic Theory* **8** (1996), 461–488.

29. M. Kurz, On the structure and diversity of rational beliefs, *Economic Theory* **4** (1994), 877–900.

30. M. Kurz, On rational belief equilibria, *Economic Theory* **4** (1994), 859–876.

31. M. Kurz, The Kesten–Stigum model and the treatment of uncertainty in equilibrium theory, in: *Essays on Economic Behavior Under Uncertainty* (eds. M. S. Balch, D. L. McFadden, and S. Y. Wu), North–Holland, Amsterdam, 1974, 389–399.

32. E. Malinvaud, Markets for an exchange economy with individual risks, *Econometrica* **41** (1973), 383–409.

33. E. Malinvaud, The allocation of individual risks in large markets, *Journal of Economic Theory* **4** (1972), 312–328.

34. G. N. Mankiw, The equity premium and the concentration of aggregate shocks, *Journal of Monetary Economics* 15 (1986), 145–161.

35. R. Mehra and E. C. Prescott, The equity premium: a puzzle, *Journal of Monetary Economics* 15 (1985), 145–162.

36. C. K. Nielsen, Rational belief structures and rational belief equilibria, *Economic Theory* 8 (1996), 399–422.

37. R. Radner, Equilibrium under uncertainty, in: *Handbook of Mathematical Economics*, Vol. II, Chapter 20 (eds. K. J. Arrow and M. D. Intriligator), North–Holland, Amsterdam, 1982.

38. R. Radner, Rational expectations equilibrium: generic existence and the information revealed by prices, *Econometrica* 47 (1979), 655–678.

39. R. Radner, Existence of equilibrium of plans, prices, and price expectations in a sequence of markets, *Econometrica* 40 (1972), 289–303.

40. R. Radner, Competitive equilibrium under uncertainty, *Econometrica* 36 (1968), 31–58.

41. T. A. Reitz, The equity premium: a solution, *Journal of Monetary Economics* 22 (1988).

42. L. J. Savage, *The Foundations of Statistics*, Wiley, New York, 1954.

43. D. S. Scharfstein and J. C. Stein, Herd behavior and investment, *American Economic Review* 80 (1990), 465–479.

44. L. E. O. Svensson, Efficiency and speculation in a model with price–contingent contracts, *Econometrica* 49 (1981), 131–151.

45. P. Weil, The equity premium puzzle and the riskfree rate puzzle, *Journal of Monetary Economics* 24 (1989), 401–422.

Mordecai Kurz
Department of Economics
Serra Street at Galvez
Stanford University
Stanford, CA 94305-6072
USA
email: mordecai@leland.stanford.edu

QUASI-EQUILIBRIUM AND EQUILIBRIUM IN A LARGE PRODUCTION ECONOMY WITH DIFFERENTIATED COMMODITIES

KONRAD PODCZECK*

Abstract. A general equilibrium model of economies with differentiated commodities and infinitely many producers and consumers is developed. In particular, results on the existence of quasi-equilibria and equilibria are proved. The key assumption for the quasi-equilibrium existence result is that preferences and production sets are uniformly proper.

1. Introduction

This paper establishes results on the existence of competitive (quasi-)equilibria for economies with infinitely many differentiated commodities and infinitely many consumers and producers. Our model extends the approaches by Mas-Colell [15] and Jones [12] from large exchange economies with differentiated commodities to large production economies.

Economies with infinitely many producers and infinitely many differentiated commodities appear in some models of monopolistic competition. See e.g. Hart [6, 7, 8, 9], Jones [14], and Pascoa [20, 21]. The model in our paper is related to that in Hart [6, 7]. In [6, 7] economies are considered where firms, in a Cournot fashion, simultaneously choose quantities and characteristics of their products. It is shown that, under certain conditions, Cournot equilibria in sequences of such economies become approximately Walrasian if one approaches a limit economy where individual producers are insignificant relative to the size of the whole economy. Thus, the notion of a competitive equilibrium for an economy with infinitely many firms and infinitely many commodities plays an important role. However, no existence results are given in [6, 7]. The existence result of our paper captures, in particular, the limit economies in [6, 7]. More precisely, we prove the existence of competitive equilibria for economies which arise as limits of sequences of economies with finitely many consumers and producers where the numbers of both types of agents go to infinity but where the quotient of these numbers tends to a strictly positive real number. Such sequences are explicitly treated in Hart [7].

As in the models of Mas-Colell [15], Hart [6, 7], and Jones [12], we specify the set of commodities as a compact metrizable space G, and model commodity bundles as measures on this space of commodities. As in [6, 7] and [12], but different to [15], it is assumed that all commodities are divisible. The space

*I wish to thank E. Dierker, M. Nermuth, J. Ostroy, and W. Zame for valuable comments.

of all commodity bundles is then $M(G)$, the space of all finite (signed) Borel measures on G. Following [15] and [12], we describe economies by distributions on appropriate spaces of agents' characteristics: the production sector is specified by a distribution on a space of production sets and the consumption sector by a joint distribution on spaces of preferences, endowments, and profit shares. Of course, without an explicit set of firms in the model we cannot proceed in the usual way in describing shares. We will argue that it is appropriate in our context to model the shares of a consumer as a measure on the space of production sets. Furthermore, we will assume that the mass of producers in an economy relative to the mass of consumers is finite and greater than zero. Thus, as intended, we cover economies which are limits of increasing sequences of economies with a finite number of firms and consumers but where the quotient of both numbers converges to some strictly positive real number. Finally, as in the models of [15] and [12], we shall use the concept of a Walrasian equilibrium distribution as equilibrium notion and demonstrate the existence of equilibria where price systems are continuous functions on the space of all (pure) commodities. Thus, similar commodities will have similar prices in equilibrium.

We will not make convexity assumptions with respect to individual production sets and consumers' preferences. Thus we allow for production sets implying set up costs or implying that only one good out of a large number of potential products can be produced at the same time, specifications that are typical for the monopolistic competition literature. In particular, the production sets treated in Hart [6] are covered.

The existence proof in Mas-Colell [15] and Jones [12] consists of a limit argument based on equilibria in economies with finitely many commodities, belonging to a sequence which approximates a given economy with differentiated commodities. In the framework of our paper it would be too complicated to construct a suitable approximating sequence of economies with finitely many commodities. Instead, following a development in the literature on infinite–dimensional commodity spaces initiated by Aliprantis and Brown [1], we focus on the lattice structure of the commodity space and its dual. Using this structure together with the notions of uniform properness due to Mas-Colell [16] (for consumers' preferences) and Richard [25] (for production sets), we are able to construct a suitable compact set of price systems which serves as a basis for a fixed point argument. That it is possible to use fixed point arguments in infinite–dimensional spaces directly to solve equilibrium existence problems, rather than to proceed by finite–dimensional approximations, has been demonstrated by Yannelis [29].

Concerning uniform properness, we shall argue that an appropriate topology on $M(G)$ to make properness assumptions in a context of commodity differentiation is the bounded weak*–topology. In fact, we shall show that the smoothness assumptions imposed on consumers' preferences in Jones [12] imply them to be uniformly proper for this topology.

Equilibrium existence results explicitly based on models with commodity differentiation can also be found in Jones [13], Yannelis and Zame [30], Fradera [5], Ostroy and Zame [19], and Podczeck [22, 23]. In [30] exchange economies with finitely many consumers are considered. A special feature of the model in [30] is that preferences need not be ordered or monotone. In [13] production economies with finitely many agents are considered. Accordingly, it is assumed that preferences and production sets are convex. Thus, types of production sets mentioned above as typical for the monopolistic competition literature are excluded. Apart from this, the production sets in the model of [13] are general, in particular differentiated commodities being inputs in production are allowed for. To handle this aspect, a smoothness condition is imposed on production sets. As we will show, a slight strengthening of this condition implies that production sets are uniformly proper. The model in [23] describes economies with finitely many agents where production is characterized by constant returns and no joint production and where only finitely many primary factors exist. However, differentiated commodities are used as intermediate products. In this framework instead of properness or smoothness conditions the weaker requirement that similar commodities can be produced by similar input combinations is sufficient for equilibrium existence. Also, properness conditions with respect to consumers' preferences are not needed. In [5] and [22] large production economies are considered However, in both models there is only one commodity which can be used as an input in production. Moreover, in [5] joint production is excluded. On the other hand, the model in [5] allows for free entry of firms, i.e., there is no a priori bound on the mass of firms that could possibly operate in an economy. In [19] large exchange economies are studied. In contrast to other models of large economies with differentiated commodities, which describe economies merely by distributions on spaces of agents' characteristics, the economy of [19] is specified as a mapping from an atomless measure space of consumers into a space of consumers' characteristics. Moreover, allowing for prices systems being just measurable functions on the set of commodities, equilibrium existence is demonstrated under weaker conditions concerning marginal rates of substitution. On the other hand, convexity of preferences is assumed.

Finally, we mention the papers of Ostroy [18] and Zame [31] on large exchange economies with abstract commodity spaces. These papers, however, do not cover $M(G)$ as a model of commodity differentiation which is our concern.

2. Notation and definitions

Let K be any compact metric space. By $C(K)$ we denote the set of all continuous functions on K and by $M(K)$ the set of all finite (signed) Borel measures on K; $C_+(K)$ and $M_+(K)$ denote the positive cones of $C(K)$ and $M(K)$, respectively. Note that $C(K)$ and $M(K)$ are vector lattices. Given elements x, y of $C(K)$ or of $M(K)$, x^+, x^-, $|x|$, $x \wedge y$, and $x \vee y$ have the usual lattice theoretical meaning. Let $p \in C(K)$ and $z \in M(K)$. Then $\|p\|_\infty$ denotes the sup–norm of p and $\|z\|_V$ the variation norm of z. Note that $\|z\|_V = |z|(K) = z^+(K) + z^-(K)$. If $f : K \to \mathbb{R} \cup \{+\infty\}$ is a z–integrable or a nonnegative and z–measurable function we will mostly write $f \cdot z$ instead of $\int f\, dz$. Viewing $C(K)$ as endowed with the sup–norm, $M(K)$ is the dual space of $C(K)$ and the value of z at p is equal to $p \cdot z$ with the just introduced meaning (Riesz representation theorem). By the *support* of z, denoted by $\operatorname{supp} z$, we mean the smallest closed subset F of K such that $|z|(K \setminus F) = 0$. Given any $k \in K$ we write δ_k for the Dirac measure at k. Finally, 1_K denotes the indicator function of K, i.e. $1_K(k) = 1$ for each $k \in K$.

Recall that the *w^*–topology* (weak*–topology) of $M(K)$ is the topology of pointwise convergence on the elements of $C(K)$, and that the bw^*–topology (bounded weak*–topology) of $M(K)$ is the strongest topology on $M(K)$ agreeing with the w^*–topology on every w^*–compact set. Note that the bw^*– and the w^*–topology generate the same Borel σ–algebra.

If A is any non-empty subset of $M(K)$ we denote by i_A the inclusion of A into $M(K)$. When $A = M(K)$, we simply write i_M instead of $i_{M(K)}$. Thus, i_M is the identity on $M(K)$. Finally, given some non-empty $A \subset M(K)$ endowed with the (relativized) bw^*– or w^*–topology, and given some finite positive Borel measure v on A, $\int i_A\, dv$ means the Gelfand integral of i_A relative to v; by the phrase "$\int i_A\, dv$ exists" we indicate or require (depending on the context) that this integral exists; similarly, if we write $\int i_A\, dv = z$ this indicates or requires that the Gelfand integral of i_A relative to v exists and is equal to z. Recall from the definition of the Gelfand integral that $\int i_A\, dv = z$ for some $z \in M(K)$ means that for each $p \in C(K)$, $\int |p \cdot z'|\, dv(z') < \infty$, and $p \cdot z = \int p \cdot z'\, dv(z')$.

3. The model and the results

3.1. The commodity space

The commodity space (i.e., the set of possible commodity bundles) for the economies we are going to consider is $M(G)$ where G is a given compact metrizable space. Each point in G has the interpretation of representing a complete description of all relevant characteristics of one unit of a certain commodity, and two commodities are regarded as similar if they can be identified with points in G which are topologically close. If a commodity bundle $x \in M(G)$ describes a consumption activity then x specifies for each Borel subset B of G the total consumption of all those commodities having characteristics so that their units can be identified with points in B. Similarly, if $y \in M(G)$ stands for a production activity then, for each Borel subset B of G, $y^+(B)$ measures the total output and $y^-(B)$ the total input of all those commodities whose units belong to B. Note that by letting every element of $M(G)$ represent a possible commodity bundle it is assumed, as in the models of Jones [12, 13] but different to that in Mas-Colell [15], that all commodities are perfectly divisible.

3.2. Consumption set and preferences

We assume that the consumption set of an individual consumer is equal to $M_+(G)$ and that his preferences can be described by a binary relation $\succ$ on $M_+(G)$ satisfying the following standard assumption:

(C1) $\succ$ is irreflexive and transitive.

Notice that under this assumption preferences may not be complete. Also, we will not assume that preferences are convex or monotone.

Concerning continuity properties of preferences, we assume that a preference relation $\succ$ satisfies the following condition:

(C2) $\succ$ is bw^*–continuous, i.e., $\succ$ is relatively open in $M_+(G) \times M_+(G)$ with respect to the bw^*–topology of $M(G)$.

Condition (C2) ensures, in particular, that not only consumption bundles that contain the same commodities in similar quantities are treated as close by preferences but also bundles that contain similar commodities in similar quantities. E.g., if $x + \alpha\delta_g \succ x$ then also $x + \alpha'\delta_{g'} \succ x$ if α' is sufficiently close to α and g' is sufficiently close to g in G. That is, commodities which are nearby as points in G must be treated as good substitutes. As argued in Mas-Colell [15] and Jones [12, 13], it is appropriate in a context of commodity differentiation to assume that preferences display this property. (But see also Ostroy and Zame [19] where it is shown that in order to construct a model of a nonatomic thin markets economy one has to weaken this assumptions.) Note that the bw^*– and the w^*–topology agree on $M_+(G)$ (see Fact 4.1.11 in Section 4.1). Thus with (C2) we assume, in fact, that preferences are w^*–continuous (as it is assumed in the models of Mas-Colell [15] and Jones [12, 13]). However, for reasons of notation which will become clear in the following paragraph, we prefer to use the bw^*–topology in the statement of (C2).

As it is clear from the literature on infinite–dimensional commodity spaces, we have to make, finally, an assumption concerning *marginal rates of substitution*. We will, however, formulate such an assumption not in the way it is, following Jones [12, 13], usually done in commodity differentiation models. We find it more convenient to use the notion of *uniform properness* due to Mas-Colell [16]. To be specific, we assume in our model that a preference relation $\succ$ is uniformly proper with respect to the bw^*–topology of $M(G)$:

(C3) $\succ$ is uniformly bw^*–proper. That is to say, there is a bw^*–open convex cone $\Gamma \subset M(G)$, with $\Gamma \cap M_+(G) \neq \emptyset$, such that if $x \geq 0$, $\gamma \in \Gamma$, and $x + \gamma \geq 0$ then $x + \gamma \succ x$.

The reason for using the bw^*–topology in the formulation of our properness assumption on preferences is that for this topology (but not for the w^*–topology) a uniform properness condition is closely related to the usual assumptions about marginal rates of substitution in models of commodity differentiation. In fact, the assumptions in Jones [12] imply uniform properness of preferences for the bw^*–topology (but not for the w^*–topology):

Let $\succeq$ be a preference/indifference relation on $M_+(G)$ and let d_G be a metric for the topology of G. The following uniform substitutability assumption (which we label as (USC) for later reference) is made in [12]:

(USC) For every number $\alpha > 1$ there is a number $\beta(\alpha) > 0$ such that for all numbers $\lambda \geq 0$ and all $g, g' \in G$ with $d_G(g, g') < \beta(\alpha)$, if $x \geq 0$ and $x + \lambda\alpha\delta_g - \lambda\delta_{g'} \geq 0$ then $x + \lambda\alpha\delta_g - \lambda\delta_{g'} \succeq x$.

Informally, (USC) postulates that, uniformly over the consumption set, commodities which are nearby as points in G have a marginal rate of substitution close to one. The following proposition shows that, for a reflexive, transitive, complete, and bw^*–continuous preference/indifference relation, (USC) together with the existence of some always desirable commodity bundle implies uniform properness with respect to the bw^*–topology of $M(G)$.

3.2.1. Proposition. *Let d_G be a metric for the topology of G and let $\succeq$ be a preference/indifference relation on $M_+(G)$. Suppose the following:*

(a) *$\succeq$ is reflexive, transitive, and complete;*

(b) *$\succeq$ is bw^*–continuous, i.e., $\succeq$ is closed in $M_+(G) \times M_+(G)$ with respect to the bw^*–topology;*

(c) *there exists $v \geq 0$ such that $x + \lambda v \succ x$ for all $x \geq 0$ and all real numbers $\lambda > 0$;*

(d) (USC) *is satisfied with respect to d_G.*

Then $\succ$, the asymmetric part of $\succeq$, is uniformly bw^*–proper in the sense of (C3) *(and thus $\succeq$ is uniformly bw^*–proper in the sense of the original formulation of uniform properness in* [16]*).*

(See Section 4.3 for the proof.) Note that the continuity and monotonicity conditions imposed on preferences in Jones [12] imply (b) respectively (c) of Proposition 3.2.1. Thus, the preferences considered in [12] are uniformly proper for the bw^*–topology. On the other hand, note that condition (USC) implies weak monotonicity for a transitive and bw^*–continuous preference/indifference relation (because for the bw^*–topology the set of all elements of $M_+(G)$ having a finite support is dense in $M_+(G)$), which is note the case for the condition of uniform bw^*–properness. Thus the latter condition is more general than (USC) and gives more flexibility. We close this section by presenting an example which shows that the conditions of Proposition 3.2.1 do not imply uniform properness with respect to the w^*–topology. In particular, this example shows that uniform bw^*–properness does not imply uniform w^*–properness. This completes our argumentation that the bw^*–topology is an appropriate choice for making a properness assumption on preferences in a context of commodity differentiation.

3.2.2. Example. Let $C_{++}(G) = \{p \in C(G): \quad p(g) > 0 \text{ for all } g \in G\}$. Choose some $v \in M_+(G)$ with $\operatorname{supp} v = G$ (since G is separable, such v exists) and then choose a subset B of $C(G)$ such that:

(i) $B \subset C_{++}(G)$;

(ii) B is norm–compact;

(iii) the linear hull of B is infinite–dimensional (we assume, of course, that G is infinite);

(iv) $p \cdot v = 1$ for each $p \in B$.

Let $u : M_+(G) \to \mathbb{R}$ be the utility function given by

$$u(x) = \min \{p \cdot x : \; p \in B\}, \quad x \in M_+(G).$$

From the properties of B one easily sees that the preference/indifference relation given by u satisfies (USC) (with respect to any metric for the topology of G) as well as conditions (b) and (c) of Proposition 3.2.1. Let

$$\Gamma = \{\gamma \in M(G): \quad p \cdot \gamma > 0 \text{ for each } p \in B\}.$$

Then Γ is a convex cone with $\Gamma \cap M_+(G) \neq \emptyset$ (note that v belongs to this intersection) and from Fact 4.1.2 in Section 4.1 it follows that Γ is bw^*–open.

Clearly, if $x \in M_+(G)$, $\gamma \in \Gamma$, and $x + \gamma \geq 0$ then we have $u(x+\gamma) > u(x)$ and thus the preferences given by u are uniformly bw^*–proper.

Pick any $\gamma \in \Gamma$ and consider any finite subset E of $C(G)$. According to (iii) above we can choose $\bar{p} \in B$ which does not belong to the linear hull of E, and by the separation theorem we can then find $z \in M(G)$ such that $q \cdot z = 0$ for each $q \in E$ and $\bar{p} \cdot (\gamma + z) < 0$. It follows that no element of Γ can have a w^*–neighborhood which is contained in Γ, i.e., the w^*–interior of Γ is empty.

Now observe that

$$\{x \in M_+(G): \quad u(x) > u(v)\} = (\{v\} + \Gamma) \cap M_+(G), \tag{3.1}$$

since $u(v) = 1 = p \cdot v$ for each $p \in B$ by construction. Let $A(v)$ be the order ideal generated by v, i.e.

$$A(v) = \{z \in M(G): \quad \lambda(-v) \leq z \leq \lambda v \quad \text{for some } \lambda > 0\}.$$

By construction, $\operatorname{supp} v = G$ whence $A(v)$ is w^*–dense in $M(G)$. Consider any non-empty and w^*–open cone $\Gamma' \subset M(G)$. Since the w^*–interior of Γ is empty there is a $\gamma' \in \Gamma'$ such that $\gamma' \notin \Gamma$. That is, for some $p \in B$ we have $p \cdot \gamma' \leq 0$. Thus, since Γ' is w^*–open and $A(v)$ is w^*–dense in $M(G)$, we can assume that $\gamma' \in A(v)$. But then for some real number $\lambda > 0$ we have $v + \lambda\gamma' \geq 0$ and it follows from (3.1) that $u(v + \lambda\gamma') \leq u(v)$. Thus, the preferences represented by u are not uniformly w^*–proper. ■

3.3. Production sets

A producer is characterized by his *production set* Y, which is a subset of $M(G)$ satisfying the following standard assumptions:

(P1) $0 \in Y$;

(P2) If $y \in Y$, $z \in M_+(G)$, and $z \leq y^+$ then $z - y^- \in Y$ (free disposal of outputs).

A full free disposal assumption will be made later on with respect to the aggregate production set of an economy. Notice also that we will not make a convexity assumption with respect to individual production sets.

Concerning closedness properties of the production set Y of a single producer we assume that

(P3) Y is closed in the bw^*–topology of $M(G)$.

Note that this is weaker than to assume Y to be w^*–closed because Y need not be convex. On the other hand, (P3) still fits to the context of differentiated commodities: e.g., if $\{g^n\}$ is a sequence in G, converging to some $g \in G$, and if

$y + \delta_{g^n} \in Y$ for some $y \in M(G)$ and each n, it should be that also $y + \delta_g \in Y$, and under (P3) this is the case, indeed.

Finally, we have to make an assumption about the marginal rates of substitution of a production set. As for a preference relation, we use the notion of uniform properness, formulated with respect to the bw^*–topology:

(P4) Y is uniformly bw^*–proper. That is to say, there is a convex bw^*–open cone $\Gamma \subset M(G)$, with $\Gamma \cap M_+(G) \neq \emptyset$, such that if $y \in Y$, $\gamma \in \Gamma$, and $y - \gamma \leq y^+$, then $y - \gamma \in Y$.

(The notion of uniform properness for a production set has been developed by Mas-Colell [17] and Richard [25]. The version we use is that of [25]. Note, however, that for the case of a commodity space which is not a topological vector lattice, Richard [26] has used the version of [17] which involves a so called pretechnology condition. Note that $M(G)$ is not a topological vector lattice in the bw^*–topology. Nevertheless, we do not need a pretechnology condition. For an approach do deal with marginal rates of substitution of production sets in a way alternative to uniform properness see Zame [32].)

The conditions used by Jones [13] to handle the problem of marginal rates of substitution of production sets are closely related to properness in the bw^*–topology:

Let d_G be a metric for the topology of G and let $Y \subset M(G)$ be a production set. Consider the following condition:

(USP) For every number $\alpha > 1$ there is a number $\beta(\alpha) > 0$ such that for all numbers $\lambda \geq 0$ and all $g, g' \in G$ with $d_G(g, g') < \beta(\alpha)$, if $y \in Y$ and $y - \lambda\alpha\delta_g + \lambda\delta_{g'} \leq y^+$ then $y - \lambda\alpha\delta_g + \lambda\delta_{g'} \in Y$.

Informally, (USP) says that input commodities with close characteristics have marginal rates of substitution close to one, uniformly over the production set. Note that the requirements concerning marginal rates of substitution between similar commodities in the related assumption FS1 in Jones [13] are made uniformly only over w^*–compact subsets of a production set. However, a strengthening of FS1 in [13] to a condition which makes a statement about marginal rates of substitution uniformly over a production set would, together with the free disposal condition F2(c) in [13], exactly amount to (USP). The following proposition relates (USP) to the notion of uniform properness.

3.3.1. Proposition. *Let d_G be a metric for the topology of G and let $Y \subset M(G)$ be a production set fulfilling* (P3), *i.e., Y is bw^*–closed in $M(G)$. Suppose the following*:

(a) *There exists a countable dense set $D \subset G$ with the following property: For every $y \in Y$ there is a sequence $\{y_n\}$ in Y such that the support of y_n is a finite subset of D for each n and such that $y_n \to y$.*

(b) (USP) *holds with respect to d_G.*

(c) *The set of isolated points of G is finite.*

Then Y is uniformly bw^*-proper.

(See Section 4.4 for the proof.) Condition (a) of this proposition is contained in the model of Jones [13] also. Condition (c) can be dropped; however, this would complicate the proof. In essence, the assumptions on production sets made in [13] tend to imply uniform properness for the bw^*-topology.

Finally, let us consider production sets with only finitely many potential input commodities. Production sets of this type can be found in many models of monopolistic competition with differentiated commodities (e.g. Hart [6]). Suppose $G = F \cup K$ where F is a finite set all of whose points are isolated in G. Let $\Gamma = \{\mu \in M(G): \ \mu(g) > 0 \text{ for all } g \in F\}$. Then Γ is a bw^*-open (in fact, w^*-open) convex cone in $M(G)$ with $\Gamma \cap M_+(G) \neq \emptyset$. Moreover, $\gamma \in \Gamma$, $y \in M(G)$, and $(y-\gamma) \leq y^+$ together imply $(y-\gamma) \leq y^+ - y^-|_F$, where $y^-|_F$ denotes the restriction of y^- to F.[1] Thus we have the following.

3.3.2. Proposition. *Suppose $G = F \cup K$ where F is a finite set, each point of which is isolated in G. Let $Y \subset M(G)$ be a production set satisfying the following conditions:*

(a) *$y \in Y$ implies $y^+ - y^-|_F \in Y$,*

(b) *if $y \in Y$ and $z \in M_+(G)$ then $y - z \in Y$.*

Then Y is uniformly bw^*-proper.

Thus, production sets where the potential inputs are commodities out of a finite set of isolated points of G are automatically uniformly bw^*-proper (in fact, uniformly w^*-proper) if free disposal is fulfilled. In particular, if G is a finite set then free disposal implies uniform properness.

3.4. Economies

For the rest of this paper, the following convention holds.

3.4.1. Convention. *$M(G)$ is endowed with the bw^*-topology; subsets of $M(G)$ are always regarded as endowed with the relativization of this topology (unless, temporarily, something else is declared).*

[1] That is, $(y^-|_F)(B) = y^-(B \cap F)$ for all Borel subsets B of G.

We are going to describe economies by distributions of agents' characteristics. For this, we first have to construct appropriate spaces of those characteristics. Let us start with consumers' preferences and production sets. In addition to the assumptions discussed in the previous two subsections we shall require uniform properness to hold uniformly over consumers' preferences and the production sets of the producers (see the remarks at the end of this subsection). We choose and fix, for the remainder of Section 3, an open convex cone $\Gamma \subset M(G)$ with $\Gamma \cap M_+(G) \neq \emptyset$.

3.4.2. Definition. *The set of allowed preference relations, denoted by* $\mathcal{P}_\Gamma$, *is the set of all preference relations* $\succ$ *on* $M_+(G)$ *satisfying* (C1), (C2), *and, with respect to the given* Γ, *the requirements in* (C3), *i.e., if* $x \in M_+(G)$, $\gamma \in \Gamma$, *and* $x + \gamma \geq 0$ *then* $x + \gamma \succ x$.

3.4.3. Definition. *The set of allowed production sets, denoted by* $\mathcal{Y}_\Gamma$, *is the set of all* $Y \subset M(G)$ *satisfying* (P1), (P2), (P3), *and, with respect to the given* Γ, *the requirements in* (P4), *i.e., if* $y \in Y$, $\gamma \in \Gamma$, *and* $y - \gamma \leq y^+$ *then* $y - \gamma \in Y$.

We have to endow $\mathcal{P}_\Gamma$ and $\mathcal{Y}_\Gamma$ with a suitable topology. To this end, let $\mathcal{Z}$ be the space of all closed subsets of $M_+(G) \times M_+(G)$, endowed with the topology of closed convergence. According to Facts 4.1.12 and 4.1.13 in Section 4.1, $M_+(G)$ is locally compact, separable, and metrizable, and hence so is $M_+(G) \times M_+(G)$. Thus, $\mathcal{Z}$ is a compact metrizable space (see Hildenbrand [10, p. 19, Theorem 2]). Now let $\mathcal{P}$ be the set of all preference relations on $M_+(G)$ fulfilling (C1) and (C2). Because of (C2) we can identify $\mathcal{P}$ with a subset of $\mathcal{Z}$, and in this way we can endow $\mathcal{P}$ with the relativized topology of closed convergence. As in [10, p. 97] it follows that thereby $\mathcal{P}$ becomes a compact metrizable space. Moreover, we have:

3.4.4. Lemma. $\mathcal{P}_\Gamma$ *is a compact subset of* $\mathcal{P}$.

(See Section 4.5 for the proof.) For the rest of this paper the following convention is made:

3.4.5. Convention. $\mathcal{P}_\Gamma$ *is endowed with the (relativized) topology of closed convergence and is thus a compact metrizable space (by Lemma* 3.4.4).

Next, we will construct a topology on $\mathcal{Y}_\Gamma$. Let $\mathcal{Y}$ be the set of all nonempty closed subsets of $M(G)$. Because of (P3) we have $\mathcal{Y}_\Gamma \subset \mathcal{Y}$. However, since $M(G)$ is not locally compact (unless G is finite), the topology of closed convergence on $\mathcal{Y}$ is not separated. (See [10, p. 18]). Nevertheless, a suitable metric on $\mathcal{Y}$ can be defined as follows. As noted above, $\mathcal{Z}$ is a compact metrizable space. Let $d_{\mathcal{Z}}$ be a corresponding metric. Now let $\kappa : \mathcal{Y} \to \mathcal{Z}$ be the mapping given by $\kappa(Y) = \{(a,b) \in M_+(G) \times M_+(G) : \ (a-b) \in Y\}$. Evidently, κ is injective, so setting $d_{\mathcal{Y}}(Y,Y') = d_{\mathcal{Z}}(\kappa(Y), \kappa(Y'))$ gives a metric $d_{\mathcal{Y}}$, on $\mathcal{Y}$. The following lemma characterizes convergence in $\mathcal{Y}$ with respect to this metric.

3.4.6. Lemma. *A sequence $\{Y_n\}$ in $\mathcal{Y}$ converges to some $Y \in \mathcal{Y}$ with respect to $d_{\mathcal{Y}}$ if and only if the following two conditions are satisfied:*

(a) *If $\{Y_{n_k}\}$ is a subsequence of $\{Y_n\}$ and $y_k \in Y_{n_k}$ for each k so that the sequence $\{y_k\}$ converges to some $y \in M(G)$, then $y \in Y$.*

(b) *If $y \in Y$ then there exists a sequence $\{y_n\}$ in $M(G)$ with $y_n \in Y_n$ for each n, such that $y_n^+ \to y^+$ and $y_n^- \to y^-$.*

(See Section 4.6 for the proof.) In view of this lemma, $d_{\mathcal{Y}}$ is, indeed, a suitable metric in order to define a topology on $\mathcal{Y}_\Gamma$. Moreover, we have:

3.4.7. Lemma. *$\mathcal{Y}_\Gamma$ is a compact subset of $\mathcal{Y}$ with respect to the metric $d_{\mathcal{Y}}$.*

(See Section 4.7 for the proof.) For the rest of the paper the following convention holds.

3.4.8. Convention. *$\mathcal{Y}_\Gamma$ is endowed with the metric $d_{\mathcal{Y}}$ constructed above, and is thus a compact metric space (by Lemma 3.4.7).*

Next, we have to model *profit distribution* among consumers. Since we will not specify an explicit set of firms in our model of an economy, we cannot proceed in the usual way in describing profit shares. However we are interested in competitive (quasi-)equilibria and consequently assume all producers being price takers. Hence we can base the modeling of shares on the fact that all producers with the same production set earn equal profits, given a price system. Suppose a consumer holds shares in a finite set of firms $J = \{1, \dots, n\}$, given by $\theta = (\theta_1, \dots, \theta_n)$, $0 < \theta_j \leq 1$ for each $j \in J$. We can regard θ as a measure on J. Let $g : J \to \mathcal{Y}_\Gamma$ be the mapping which describes the production sets of the firms in J. Suppose (given some price system) that $\bar{\bar{\pi}} : J \to \mathbb{R}_+$ describes the profits earned by the firms in J and let $\pi : \mathcal{Y}_\Gamma \to \mathbb{R}_+ \cup \{+\infty\}$ be the mapping which assigns to every production set the supremum of attainable profits. Under the price taking hypothesis we then have $\bar{\bar{\pi}}(j) = \pi(g(j))$ for all $j \in J$. Let $s = \theta \circ g^{-1}$ be the image measure of θ under g. Then $\pi \cdot s = \sum_{j=1}^n \pi(g(j))\theta_j = \sum_{j=1}^n \bar{\bar{\pi}}(j)\theta_j$. Thus, for our purposes, the measure s on $\mathcal{Y}_\Gamma$ describes the shares of the consumer in question as well as $\theta = (\theta_1, \dots, \theta_n)$. Therefore, we will describe the shares of a consumer simply by a nonnegative finite Borel measure s on $\mathcal{Y}_\Gamma$.

3.4.9. Definition. *The set of all possible shares is* $S_\Gamma \equiv M_+(\mathcal{Y}_\Gamma)$.

3.4.10. Convention. S_Γ *is endowed with the relativized* bw^**-topology of* $M(\mathcal{Y}_\Gamma)$ *and is thus a separable completely metrizable space. (See Fact* 4.1.12 *in Section* 4.1.)

Concerning initial endowments of consumers we postulate the following.

3.4.11. Definition. *The space of allowed initial endowments is* $\Omega \equiv M_+(G)$. *(By Fact* 4.1.12 *in Section* 4.1 *and Convention* 3.4.1, Ω *is a separable completely metrizable space.)*

An individual consumer is characterized by his preferences, his initial endowment, and his profit shares. That is:

3.4.12. Definition. *The space of consumers' characteristics is* $C_\Gamma \equiv \mathcal{P}_\Gamma \times \Omega \times S_\Gamma$. C_Γ *is endowed with the product topology (and is thus a separable completely metrizable space).*

Finally, we state the following.

3.4.13. Definition. *An* ***economy*** *is a pair* (ϱ, ν) *where*

(a) ϱ *is a positive finite Borel measure on* $\mathcal{Y}_\Gamma$,

(b) ν *is a positive finite Borel measure* C_Γ,

such that

(i) $\int i_\Omega \, d\nu_\Omega$ *exists, and*

(ii) $\int i_S \, d\nu_S = \varrho$.

(ν_S *and* ν_Ω *denote the marginal distributions of* ν *on* S_Γ *and* Ω, *respectively. For the rest of the notation in* (i) *and* (ii) *see Section* 2.)

Thus, up to normalization, ϱ is the distribution of producers' characteristics and ν the distribution of consumers' characteristics. The numbers $\varrho(\mathcal{Y}_\Gamma)$ and $\nu(C_\Gamma)$ represent the total mass of producers and consumers, respectively. Since both of these numbers are assumed to be greater than zero and finite, an economy according to our definition can be interpreted as limit of an increasing sequence of economies with a finite number of firms and consumers but where the quotient of both numbers converges to some strictly positive real number. Condition (i) says that the aggregate endowment in the consumption sector of the economy is finite, i.e., can be represented by an element of $M_+(G)$. Condition (ii) is an adaptation of the classical definition of a private ownership production economy to our framework. It ensures that the total profit earned in the production sector is equal to the total profit income in the consumption sector (see Lemma 4.10.6 in Section 4.10).

Finally, note that Definition 3.4.13 includes the hypothesis that the preferences of all consumers and the production sets of all producers of an economy are uniformly proper with respect to the same properness cone. This is, of course, a very strong hypothesis. On the other hand, as shown by Jones [12], without an assumption that restricts the marginal rates of substitution of all agents of an economy in a uniform way, equilibrium existence can fail. (See also Zame [31] concerning this point.) It could be argued, however, that it is acceptable to assume that all consumers' preferences are uniformly proper with respect to a common properness cone, say Γ^1, and that all production sets are uniformly proper with respect to a common properness cone, say Γ^2, but that it would not be acceptable to assume $\Gamma^1 = \Gamma^2$. However, given that on the one side uniform properness holds uniformly over consumers, and on the other side uniformly over producers, it follows that uniform properness holds uniformly over both types of agents, provided one strengthens the free disposal of outputs condition (P2) to a condition of full free disposal:

(P2′) If $y \in Y$ and $z \in M_+(G)$ then $y - z \in Y$.

Indeed, let Γ^1 and Γ^2 be convex open cones according to the definitions of uniform properness for preferences and production sets, respectively, and assume that the space of allowed preferences is $\mathcal{P}_{\Gamma^1}$ and that of allowed production sets $\mathcal{Y}_{\Gamma^2}$. Since Γ^1 is open and $\Gamma^1 \cap M_+(G) \neq \emptyset$, we can find $\bar{v} \in \Gamma^1 \cap M_+(G)$ with $\operatorname{supp} \bar{v}$ finite. By the lemma below, there is an open convex cone $\Gamma^3 \subset \Gamma^2$, containing some $v' \geq 0$, such that every $Y \in \mathcal{Y}_{\Gamma^2}$ is proper also with respect to the cone $\Gamma^4 \equiv \Gamma^3 + \{\lambda\bar{v} \in M_+(G) : \ \lambda \text{ is a real number with } \lambda \geq 0\}$. Then, since Γ^1 is open, $\lambda v' + \bar{v} \in \Gamma^5 \equiv \Gamma^1 \cap \Gamma^4$ for a small enough $\lambda > 0$, and it follows that every $Y \in \mathcal{Y}_{\Gamma^2}$ and every $\succ \in \mathcal{P}_{\Gamma^1}$ is uniformly proper also with respect to the open convex cone Γ^5.

3.4.14. Lemma. *Let the production set $Y \subset M(G)$ satisfy* (P2′), (P3), *and* (P4). *Let Γ be a properness cone according to* (P4) *and let $\bar{v}$ be an element of $M_+(G)$ with $\operatorname{supp} \bar{v}$ finite. Then there exists an open convex cone $\Gamma' \subset \Gamma$, containing some $v' \geq 0$, such that Y is proper also with respect to $\Gamma'' = \Gamma' + \{\lambda\bar{v} : \lambda \in \mathbb{R}_+\}$.*

(See Section 4.8 for the proof.)

3.5. Quasi-equilibrium and equilibrium

Let (ϱ, ν) be an economy, constructed according to Definition 3.4.12. The economy (ϱ, ν) is kept fixed throughout this subsection. To simplify some notation, we make, without loss of generality, the following convention.

3.5.1. Convention. *The measures ϱ and ν are probability measures.*

We wish to prove the existence of equilibria where, as in the models Mas-Colell [15] or Jones [12], commodity prices are continuous with respect to the characteristics of commodities. That is, similar commodities should have similar prices in an equilibrium (or quasi-equilibrium). Thus, we are looking for equilibrium price systems in $C(G)$, the set of continuous functions on G. To describe equilibrium allocations, we also follow [15] and [12] and use the Hart–Hildenbrand–Kohlberg concept of an equilibrium distribution, however, extended by us to cover production economies. For this purpose, for every $p \in C(G)$ and every $Y \in \mathcal{Y}_\Gamma$ let

$$\pi(p, Y) = \sup\{p \cdot y : \ y \in Y\}.$$

Note that for each $p \in C(G)$, given any $Y \in \mathcal{Y}_\Gamma$, we have $\pi(p, Y) \in \mathbb{R}_+ \cup \{+\infty\}$ because $0 \in Y$. Moreover, note that for each $p \in C(G)$ the mapping $\pi(p, \cdot)$ is lower semi-continuous, as it can easily be verified by using Lemma 3.4.6. Hence the expression $\pi(p, \cdot) \cdot s$ in the definition below is well-defined. It can take the value $+\infty$.

3.5.2. Definition. *A* ***quasi-equilibrium*** *of the economy (ϱ, ν) is a triple (p, η, τ) where $p \in C(G)$, with $p \neq 0$, and where η and τ are (Borel) probability measures on $\mathcal{Y}_\Gamma \times M(G)$ and $C_\Gamma \times M(G)$, respectively, such that:*

(a) *$\eta_Y = \varrho$ and $\tau_C = \nu$;*

(b) *for η–almost all $(Y, y) \in \mathcal{Y}_\Gamma \times M(G)$ we have $y \in Y$ and $p \cdot y = \pi(p, Y)$;*

(c) *for τ–almost all $(\succ, \omega, s, x) \in C_\Gamma \times M(G)$ we have*
$x \in M_+(G)$, $p \cdot x \leq p \cdot \omega + \pi(p, \cdot) \cdot s$, and
if $x' \succ x$ then $p \cdot x \geq p \cdot \omega + \pi(p, \cdot) \cdot s$;

(d) *$\int i_M \, d\eta_M$ and $\int i_M \, d\tau_M$ exist and $\int i_M \, d\tau_M = \int i_\Omega \, d\nu_\Omega + \int i_M \, d\eta_M$.*

(τ_C and τ_M denote the marginal distributions of τ on C_Γ and $M(G)$, respectively, and η_Y and η_M denote the marginal distributions of η on $\mathcal{Y}_\Gamma$ and $M(G)$, respectively.)

3.5.3. Definition. *An* ***equilibrium*** *of the economy (ϱ, ν) is a triple (p, η, τ) where $p \in C(G)$, with $p \neq 0$, and where η and τ are (Borel) probability measures on $\mathcal{Y}_\Gamma \times M(G)$ and $C_\Gamma \times M(G)$, respectively, such that* (a), (b), *and* (d) *of Definition* 3.5.2 *are satisfied and such that*

(c′) *for* τ*–almost all* $(\succ, \omega, s, x) \in C_\Gamma \times M(G)$ *we have*
$x \in M_+(G)$, $p \cdot x \leq p \cdot \omega + \pi(p, \cdot) \cdot s$, *and*
if $x' \succ x$ *then* $p \cdot x > p \cdot \omega + \pi(p, \cdot) \cdot s$.

To prove the existence of a quasi-equilibrium we need assumptions concerning the aggregate level of an economy. In particular, we need the notion of an aggregate production set. To this end, let

$$\Upsilon(\varrho) = \Big\{\eta : \quad \eta \text{ is a probability measure on } \mathcal{Y}_\Gamma \times M(G) \text{ such that: } \text{(i) } \eta_Y = \varrho, \text{ (ii) } \textstyle\int i_M \, d\eta_M \text{ exists, (iii) } y \in Y \text{ for } \eta\text{–almost all } (Y, y) \in \mathcal{Y}_\Gamma \times M(G)\Big\}$$

3.5.4. Definition. *The* ***aggregate production set*** *of the economy* (ϱ, ν) *is the set*

$$Y^a(\varrho) \equiv \Big\{y \in M(G) : \quad y = \int i_M \, d\eta_M \text{ for some } \eta \in \Upsilon(\varrho)\Big\}.$$

We make the following two standard assumptions about $Y^a(\varrho)$:

(A1) $Y^a(\varrho) \cap M_+(G) = \{0\}$ (no free production).

(A2) $-M_+(G) \subset Y^a(\varrho)$ (free disposal).

(Recall that on the level of individual producers we have only assumed free disposal of outputs.) Moreover, we assume:

(A3) $Y^a(\varrho)$ is closed.

(A3) can be viewed as a condition which requires that ϱ–almost everywhere non-convexities of individual production sets are bounded in some uniform way. This is illustrated by the following proposition which provides conditions sufficient for $Y^a(\varrho)$ to be closed. (Recall: $\|\cdot\|_V$ denotes the variation norm.)

3.5.5. Proposition. *For each* $y \in Y^a(\varrho)$ *let* $\zeta(y) = \{\eta \in \Upsilon(\varrho) : \int i_M \, d\eta_M = y\}$. *Moreover, for every* $m \in \mathbb{N}$ *let* $M^m(G) = \{z \in M(G) : \ z^+(G) \leq m\}$. *Suppose that the following two conditions are satisfied:*

(*) *For every* $k \in \mathbb{N}$ *there exists an* $m_k \in \mathbb{N}$ *such that if* $y \in Y^a(\varrho)$ *with* $\|y\|_V \leq k$ *then there exists* $\eta \in \zeta(y)$ *with* $\eta_M(M^{m_k}(G)) = 1$.

(**) *If* $y \in Y^a(\varrho)$ *and* $z \in M_+(G)$ *then* $y - z \in Y^a(\varrho)$.

Then $Y^a(\varrho)$ is closed.

(See Section 4.9 for the proof.) We can now state the main result of our paper.

3.5.6. Theorem. *If the economy (ϱ, ν) satisfies* (A1) *to* (A3) *then it has a quasi-equilibrium.*

(The proof is contained in Section 4.10.)

To show that a quasi-equilibrium turns out to be an equilibrium we need another set of assumptions. It would be enough, of course, to require that for ν–almost all $(\succ, \omega, s)$ the preference relation $\succ$ is strictly monotone, or, given that a quasi-equilibrium price–system is nonnegative (which is true if (A2) holds), that for ν–almost all $(\succ, \omega, s)$ the support of ω is equal to G. However, it is questionable if this would fit to a production economy, and for this reason we adapt some irreducibility–type conditions to our setting.

Of course, we have to ensure that in a quasi-equilibrium there is a non-negligible group of consumers being able to dispose some income. The following assumption does the job:

(A4) There exists a $\bar{y} \in Y^a(\varrho)$ such that $\bar{y} + \int i_\Omega \, d\nu_\Omega \geq 0$ and $\operatorname{supp}(\bar{y} + \int i_\Omega \, d\nu_\Omega) = G$.

That is, there is a production plan $\bar{y} \in Y^a(\varrho)$ such that together with the mean endowment in the consumption sector all commodities are available in the aggregate.

The next condition postulates that there is a set G_D of always desired commodities and also a set G_P of always productive commodities such that the endowment of almost all consumers contains some commodities in $G_D \cup G_P$. It is an adaptation to our framework of an assumption developed by Arrow and Debreu [3].

(A5) There exist two Borel subsets G_D and G_P of G, and a Borel set $D \subset C_\Gamma$, with $\nu(D) = 1$, such that the following conditions hold:

(a) For each $(\succ, \omega, s) \in D$ and every $x \in M_+(G)$ we have: if $v \in M_+(G)$ is such that $v(G_D) > 0$ and $v(G \setminus G_D) = 0$ then $x + v \succ x$.

(b) For every $y \in Y^a(\varrho)$ with $y + \int i_\Omega \, d\nu_\Omega \geq 0$ (i.e., y is a attainable production) and each $g \in G_P$, there exist elements z and u of $M_+(G)$ such that (i) $z(G_D) > 0$ and $z(G \setminus G_D) = 0$, (ii) $u(G \setminus \{g\}) = 0$, and (iii) $y + z - u \in Y^a(\varrho)$.

(c) $\omega(G_D \cup G_P) > 0$ for each $(\succ, \omega, s) \in D$.

(Note that (b) and (c) together imply that G_D is non-empty.)

Alternatively to (A5) we can proceed with the following assumption which is a convenient adaptation of standard *irreducibility* conditions to our context:

(A6) If $y \in Y^a(\varrho)$ and if τ is a probability measure on $C_\Gamma \times M(G)$ such that (i) $\tau_C = \nu$, (ii) $x \in M_+(G)$ for τ–almost all $(c, x) \in C_\Gamma \times M(G)$, and (iii) $\int i_M \, d\tau_M$ exists and $\int i_M \, d\tau_M = \int i_\Omega \, d\nu_\Omega + y$, then given any Borel set A in C_Γ with $\nu(A) > 0$ there exist elements z and u of $M_+(G)$ such that

(a) $\operatorname{supp} u \subset \bigcup_{(\succ,\omega,s)\in A} \operatorname{supp} \omega$.

(b) $y + z - u \in Y^a(\varrho)$.

(c) For τ–almost all $(\succ, \omega, s, x) \in C_\Gamma \times M(G)$ there exists $v \in M_+(G)$ (which may depend on $(\succ, \omega, s, x)$) with $v \leq z$ such that $x + v \succ x$.

Note that (A5) as well as (A6) hold if, for ν–almost all $(\succ, \omega, s) \in C_\Gamma$, $\succ$ is strictly monotone and $\omega \neq 0$. Clearly, (A6) holds because, given any $u \in M_+(G)$ with $u \neq 0$, (b) and (c) of that assumption are satisfied since we can set $z = u$ and $v = z$; concerning (A5), we can set $G_D = G_P = G$ and then, obviously, (a) and (c) of that assumption are fulfilled, but also (b) is satisfied because, given $g \in G$, we can set $z = u = \delta_g$.

We can now state the following theorem.

3.5.7. Theorem. *If the economy* (ϱ, ν) *satisfies* (A4) *and, besides,* (A5) *or* (A6) *then every quasi-equilibrium is an equilibrium.*

(The proof is contained in Section 4.11.)

As a corollary of Theorem 3.5.6 and Theorem 3.5.7 we then have the following.

3.5.8. Theorem. *If the economy* (ϱ, ν) *satisfies* (A1) *to* (A4) *and, besides,* (A5) *or* (A6) *then it has an equilibrium.*

Instead of (A4) we can also assume, as it is frequently done the literature on infinite–dimensional commodity spaces (see e.g. Richard [26]), that the aggregate endowment is an element of the properness cone of the economy; or, more generally:

(A4′) There exists $\bar{y} \in Y^a(\varrho)$ such that $\int i_\Omega \, d\nu_\Omega + \bar{y} \in \Gamma \cap M_+(G)$.

This yields the following.

3.5.9. Theorem. *If the economy* (ϱ, ν) *satisfies* (A1) *to* (A3), (A4'), *and, besides,* (A5) *or* (A6), *then it has an equilibrium.*

(See Section 4.12 for the proof.)

4. Proofs

4.1. Preliminaries

Let K be a compact metric space. (For the notation used in the sequel, see Section 2.) Recall the following facts about the bw^*–topology of $M(K)$ (see Holmes [11, Corollary 1, p. 70, Corollary 2, p. 72, Corollary b, p. 135, Lemma, p. 150, p. 151, Corollary, p. 152, Corollary 1, p. 154], and note that $C(K)$, the predual of $M(K)$, is a separable Banach space).

Fact 4.1.1. *A subset F of $M(K)$ is bw^*–closed if and only if $F \cap B$ is w^*–compact for every w^*–compact subset B of $M(K)$.*

Fact 4.1.2. *The collection of all subsets U of $M(K)$ of the form*

$$U = \{z \in M(K) : \quad |f \cdot z| \leq 1 \text{ for all } f \in B\},$$

where B, a $\|\cdot\|_\infty$–compact subset of $C(K)$, is a local base for the bw^*–topology on $M(K)$.

Fact 4.1.3. *$M(K)$ endowed with the bw^*–topology is a locally convex space with dual space $C(K)$.*

Fact 4.1.4. *Every ball $\{z \in M(K) : \ \|z\|_V \leq b\}$, b a positive real number, is bw^*–compact and bw^*–metrizable; in particular:*

Fact 4.1.5. *Every $\|\cdot\|_V$–bounded sequence in $M(K)$ contains a bw^*–convergent subsequence.*

Fact 4.1.6. *A sequence $\{z^\ell\}$ in $M(K)$ is bw^*–convergent to some $z \in M(K)$ if and only if $f^\ell \cdot z^\ell \to f \cdot z$ for every sequence $\{f^\ell\}$ in $C(K)$, $\|\cdot\|_\infty$–converging to some $f \in C(K)$; if and only if $\{z^\ell\}$ is w^*–convergent to z, i.e. $f \cdot z^\ell \to f \cdot z$ for each $f \in C(K)$.*

Fact 4.1.7. *A subset of $M(K)$ is bw^*–compact if and only if it is bw^*–closed and $\|\cdot\|_V$–bounded; in particular:*

Fact 4.1.8. *A bw^*–convergent sequence is $\|\cdot\|_V$–bounded; and*

Fact 4.1.9. *A bw^*–compact subset of $M(K)$ is bw^*–metrizable.*

From the definition of the bw^*–topology, Fact 4.1.1, and Fact 4.1.9, it follows that:

Fact 4.1.10. *A subset of $M(K)$ is bw^*–closed if and only if it is bw^*–sequentially closed.*

Note that a w^*–convergent net in $M_+(K)$ must be eventually $\|\cdot\|_V$–bounded. Thus, from the definition of the bw^*–topology and Fact 4.1.4 it follows that:

Fact 4.1.11. *The bw^*–topology and the w^*–topology agree on $M_+(K)$.*

Since K is a compact metric space, $M_+(K)$ in the (relativized) w^*–topology is separable and completely metrizable. (See, e.g., Varadarajan [28, Theorem 3.4]). Hence, from Fact 4.1.11 it follows that:

Fact 4.1.12. *$M_+(K)$ in the (relativized) bw^*–topology is separable and completely metrizable.*

Finally, since $\{z \in M_+(K): \ 1_K \cdot z \leq b\} = \{z \in M_+(K): \ \|z\|_V \leq b\}$ for every real number $b \geq 0$, we get from Fact 4.1.4 the following:

Fact 4.1.13. *$M_+(K)$ in the (relativized) bw^*–topology is locally compact.*

For the rest of this subsection $M(K)$ is considered as endowed with the bw^*–topology.

Since $M(K)$ is the countable union of compact and metrizable subsets, we have the following:

Fact 4.1.14. *Every Borel probability measure on $M(K)$ or on the product of a separable complete metric space with $M(K)$ has a support.*

Let λ be a finite positive Borel measure on $M(K)$. Then (see Section 2 for notation):

Fact 4.1.15. *If $\int |f \cdot z| \, d\lambda(z) < \infty$ for all $f \in C(K)$ then $\int i_M \, d\lambda$ exists; in particular:*

Fact 4.1.16. *If $\int \|z\|_V \, d\lambda(z) < \infty$ then $\int i_M \, d\lambda$ exists.*

(See Diestel and Uhl [4, p. 53].)

4.2. Additional notation and conventions for Section 4

$C(G)$ is endowed with the $\|\cdot\|_\infty$–topology; subsets of $C(G)$ are always regarded as endowed with the relative topology. E.g., if we write "$p^n \to p$ in $C_+(G)$" this means the sequence $\{p^n\}$ in $C_+(G)$ converges to some $p \in C_+(G)$ with respect to $\|\cdot\|_\infty$; if we write "B is a compact subset of $C(G)$" then this means compact with respect to $\|\cdot\|_\infty$.

In the sequel we shall use, in particular, Facts 4.1.3, 4.1.6, and 4.1.14 so frequently that at most places we shall not state their references explicitly. Recall also from Convention 3.4.1 that $M(G)$ is regarded as endowed with the bw^*–topology.

We need some further notation:

$$M^f(G) \equiv \{z \in M(G) : \ \operatorname{supp} z \text{ is finite}\}$$
$$M^f_+(G) \equiv \{z \in M^f(G) : \ z \geq 0\}.$$

4.3. Proof of Proposition 3.2.1

For each $\alpha > 1$ choose $\beta(\alpha) > 0$ according to condition (USC) and let

$$Z = \Big\{z \in M(G) : \quad z = \lambda(\alpha\delta_g - \delta_{g'}) \text{ for some } \lambda \geq 0,\ \alpha > 1, \\ \text{and } g, g' \in G \text{ with } d_G(g, g') < \beta(\alpha)\Big\}.$$

Thus

$$\text{if} \quad x \in M_+(G), \quad z \in Z, \quad \text{and} \quad x + z \geq 0 \quad \text{then} \quad x + z \succeq x. \tag{4.1}$$

Note that $\lambda\delta_g \in Z$ for each $g \in G$ and each $\lambda \geq 0$. Hence, from (4.1) and the transitivity of $\succeq$ (condition (a)) it follows that

$$\text{if} \quad x \in M_+(G) \quad \text{and} \quad \gamma \in M^f_+(G) \quad \text{then} \quad x + \gamma \succeq x. \tag{4.2}$$

Let $\widehat{\Gamma}$ be the convex hull of Z. We claim:

$$\text{if} \quad x \in M_+(G), \quad \gamma \in \widehat{\Gamma}, \quad \text{and} \quad x + \gamma \geq 0 \quad \text{then} \quad x + \gamma \succeq x. \tag{4.3}$$

To see this, we introduce the following notation: if $x' \in M_+(G)$ and $z \in Z$ then we let $\rho(z, x') = \max\{\lambda \in [0,1] : \ x' + \lambda z \geq 0\}$. Now consider any $x \in M_+(G)$ and some $\gamma \in \widehat{\Gamma}$ and suppose $x + \gamma \geq 0$. By the definition of $\widehat{\Gamma}$, we can choose a finite family $(z^1, \ldots, z^k)$ of elements of Z such that $\gamma = \sum_{i=1}^k z^i$.

Pick a member z^j of $(z^1, \ldots, z^k)$ so that $\rho(z^j, x)(z^j)^-(G) \geq \rho(z^i, x)(z^i)^-(G)$ for every $i = 1, \ldots, k$. Set $x_1 = x + \rho(z^j, x)z^j$. Then $x_1 \in M_+(G)$, and from (4.1) we have $x_1 \succeq x$. Moreover, set $z_1^j = (1 - \rho(z^j, x))z^j$ and $z_1^i = z^i$ if $i \neq j$, $i = 1, \ldots, k$. Then $x_1 + \sum_{i=1}^k z_1^i = x + \sum_{i=1}^k z^i$, and by the definition of Z we have $z_1^i \in Z$ for every $i = 1, \ldots, k$. Repeat this construction with $(z_1^1, \ldots, z_1^k)$ in place of $(z^1, \ldots, z^k)$ and x_1 in place of x to produce x_2 and $(z_2^1, \ldots, z_2^k)$. Continuing in this fashion, we obtain a sequence $\{x_n\}$ in $M_+(G)$ and, for each $i = 1, \ldots, k$, a non-increasing sequence $\{\lambda_n^i\}$ in $[0,1]$ such that at every stage n:

$$x_n \succeq x \quad \text{(by the transitivity of } \succeq\text{)},$$

$$x_n + \sum_{i=1}^k \lambda_n^i z^i = x + \sum_{i=1}^k z^i,$$

$$\sum_{i=1}^k (\lambda_n^i z^i)^-(G) - \sum_{i=1}^k (\lambda_{n+1}^i z^i)^-(G) = \\ \max\big\{\rho(\lambda_n^i z^i, x_n)(\lambda_n^i z^i)^-(G) : \quad i = 1, \ldots, k\big\}.$$

Since $\{\lambda_n^i\}$ is non-increasing in n, we have $\lambda_n^i z^i \to \lambda^i z^i$ for some $\lambda^i \in [0,1]$ as $n \to \infty$, $i = 1, \ldots, k$. Consequently, $x_n \to \bar{x}$ for some $\bar{x} \in M_+(G)$, and from the continuity of $\succeq$ we have $\bar{x} \succeq x$. Besides, $\bar{x} + \sum_{i=1}^k \lambda^i z^i = x + \sum_{i=1}^k z^i = x + \gamma$. For each n let $\lambda_n = \max\{\rho(\lambda_n^i z^i, x_n)(\lambda_n^i z^i)^-(G) : \ i = 1, \ldots, k\}$. An easy calculation (using the facts that $\lambda_n^i z^i \in Z$ for each $i = 1, \ldots, k$ and that $x_n + \sum_{i=1}^k \lambda_n^i z^i \geq 0$) shows that $\lambda_n \geq (1/k^2)(\sum_{i=1}^k \lambda_n^i z^i)^-(G)$ for every n. On the other hand, since $\{\lambda_n^i\}$ is non-increasing in n for each $i = 1, \ldots, k$, the sequence $n \mapsto \sum_{i=1}^k (\lambda_n^i z^i)^-(G)$ converges and thus we have $\lambda_n \to 0$. It follows that $(\sum_{i=1}^k \lambda_n^i z^i)^-(G) \to 0$ whence $\sum_{i=1}^k \lambda^i z^i \geq 0$. Hence $\bar{x} + \sum_{i=1}^k \lambda^i z^i \succeq \bar{x}$ according to (4.2), and thus, by the transitivity of $\succeq$, we have $x + \gamma \succeq x$. Thus (4.3) has been shown.

Now let

$$B = \left\{p \in C(G) : \quad \|p\|_\infty = 1 \text{ and } p \cdot z \geq 0 \text{ for all } z \in Z\right\}.$$

From the definition of Z it follows that $1_G \in B$ and thus B is non-empty. As noted above, we have $\delta_g \in Z$ for each $g \in G$, whence $p \geq 0$ for each $p \in B$. Furthermore, also from the definition of Z, we conclude that B is equicontinuous. It follows that B is a (norm) compact. Let

$$\Gamma = \{z \in M(G) : \quad p \cdot z > 0 \text{ for each } p \in B\}.$$

Then Γ is a convex cone in $M(G)$, and since B is (norm) compact it follows that Γ is open with respect to the bw^*-topology. (See Fact 4.1.2.) Because G is separable there is $\bar{v} \in M_+(G)$ with $\operatorname{supp} \bar{v} = G$, and since B is contained in $C_+(G) \setminus \{0\}$ we have $p \cdot \bar{v} > 0$ for each $p \in B$. Thus Γ is non-empty. In particular, $\Gamma \cap M_+(G) \neq \emptyset$. We claim:

$$\text{if} \quad \gamma \in \Gamma \cap M^f(G) \quad \text{then} \quad \gamma \in \widehat{\Gamma}. \tag{4.4}$$

To see this, let $\gamma \in M^f(G)$ and suppose $\gamma \notin \widehat{\Gamma}$. Let $\mathfrak{D}$ be the collection of all finite subsets D of G with $\operatorname{supp} \gamma \subset D$. Furthermore, for each $D \in \mathfrak{D}$ let $M(D) = \{z \in M(G) : \ \operatorname{supp} z \subset D\}$. Pick any $D' \in \mathfrak{D}$. By Minkowski's theorem, we can choose $q^{D'} : D' \to \mathbb{R}$ with $\left\|q^{D'}\right\|_\infty = 1$ such that $q^{D'} \cdot \gamma \leq q^{D'} \cdot z$ for all $z \in \widehat{\Gamma} \cap M(D')$. Note that since $\widehat{\Gamma} \cap M(D')$ is a cone, we have $q^{D'} \cdot z \geq 0$ for all $z \in \widehat{\Gamma} \cap M(D')$ and $q^{D'} \cdot \gamma \leq 0$. In particular, since $\delta_g \in \widehat{\Gamma} \cap M(D')$ for each $g \in D'$, we have $q^{D'}(g) \geq 0$ for all $g \in D'$, and hence $q^{D'}(g) = 1$ for some $g \in D'$. Extend $q^{D'}$ to a function $f^{D'}$ on G so that $\left\|f^{D'}\right\|_\infty = 1$. In this way, construct q^D and f^D for every $D \in \mathfrak{D}$. Since $\mathfrak{D}$ is directed by inclusion, the family $(f^D)_{D\in\mathfrak{D}}$ is a net. Since $\|f^D\|_\infty = 1$ for each $D \in \mathfrak{D}$, by Tychonoff's theorem we can assume (passing to a subnet if necessary) that there is function $f : G \to \mathbb{R}$ such that $f^D(g) \to f(g)$ for each $g \in G$. Since $\operatorname{supp}\gamma$ is finite, $f \cdot \gamma$ is well–defined and we have $f^D \cdot \gamma \to f \cdot \gamma$ whence $f \cdot \gamma \leq 0$. Consider any $z' \in \widehat{\Gamma}$. According to the definition of $\widehat{\Gamma}$, $\operatorname{supp} z'$ is finite. Thus $f \cdot z'$ is well–defined. Moreover, there is $D^{z'} \in \mathfrak{D}$ such that $f^D \cdot z' = q^D \cdot z' \geq 0$ for all $D \in \mathfrak{D}$ with $D \supset D^{z'}$. It follows that $f \cdot z \geq 0$ for each $z \in \widehat{\Gamma}$. In particular, we have $f \cdot z \geq 0$ for each $z \in Z$ and hence f is continuous by virtue of the definition of Z. By construction, for each $D \in \mathfrak{D}$ there is a $g^D \in D$ such that $f^D(g^D) = q^D(g^D) = 1$. Passing to a subnet of $(f^D, g^D)_{D\in\mathfrak{D}}$ if necessary, we may assume $g^D \to \bar{g}$ for some $\bar{g} \in G$. Fix a $D^0 \in \mathfrak{D}$ with $\bar{g} \in D^0$. Then, since $q^D(g^D) = 1$ for each $D \in \mathfrak{D}$ and $g^D \to \bar{g}$, for every $\epsilon > 0$ we can find a $D^\epsilon \supset D^0$ such that $q^D(\bar{g}) \geq 1 - \epsilon$ for all $D \in \mathfrak{D}$ with $D \supset D^\epsilon$, again by using the definition of Z. It follows that $f(\bar{g}) = 1$ and hence that $\|f\|_\infty = 1$. Thus, $f \in B$ since f is continuous and since $f \cdot z \geq 0$ for each $z \in \widehat{\Gamma}$. Summing up, if $\gamma \in M^f(G)$ but $\gamma \notin \widehat{\Gamma}$ then there is an $f \in B$ such that $f \cdot \gamma \leq 0$, and this implies (4.4).

We will now show that Γ is a properness cone for the preference relation $\succeq$ under consideration. Recall that $M_+(G)$ is metrizable in the relativized bw^*–topology (Fact 4.1.12) and note that $M_+^f(G)$ is dense in $M_+(G)$ for the bw^*–topology. Thus, given any $z \in M_+(G)$, we can find a sequence $\{z^n\}$ in $M_+^f(G)$ with $z^n \to z$. (Recall: all topological notions concerning $M(G)$ are with respect to the bw^*–topology if nothing else is declared.) Consider first some $x \in M_+^f(G)$. Pick any $\gamma \in \Gamma$ and suppose $x + \gamma \geq 0$ which is equivalent to the statement $\gamma^- \leq x$. Thus, in particular, $\gamma^- \in M_+^f(G)$. Choose a sequence $\{a^n\}$ in $M_+^f(G)$ with $a^n \to \gamma^+$. For each n set $\gamma^n = a^n - \gamma^-$. Then $\gamma^n \to \gamma$ and, for each n, $x + \gamma^n \geq 0$ and $\gamma^n \in M^f(G)$. Since Γ is open, we have $\gamma^n \in \Gamma$ for n large enough. Hence, from (4.4), $\gamma^n \in \widehat{\Gamma}$ and so, by (4.3), $x + \gamma^n \succeq x$ for n large enough, whence $x + \gamma \succeq x$ since $\succeq$ is continuous according to condition (b) of the proposition. Now consider any $x \in M_+(G)$, some $\gamma \in \Gamma$, and suppose $x + \gamma \geq 0$. Choose a sequence $\{x^n\}$ in $M_+^f(G)$ with $x^n \to x$. Set $\gamma^n = x - x^n + \gamma$ for each n. Then, for every n, $x^n + \gamma^n = x + \gamma \geq 0$. Moreover, $\gamma^n \to \gamma$, hence $\gamma^n \in \Gamma$ for n large enough. By the argument before, $x^n + \gamma^n \succeq x^n$ for n large enough since $x^n \in M_+^f(G)$ for all n, whence $x + \gamma \succeq x$ since $\succeq$ is continuous. Finally, let v be the element from condition (c). Thus $v \geq 0$, and for every $x \in M_+(G)$ and every $\lambda > 0$ we have $x + \lambda v \succ x$. Again pick $x \in M_+(G)$ and $\gamma \in \Gamma$ and assume that $x + \gamma \geq 0$. Since Γ is open there is $\bar{\lambda} > 0$ such that $\gamma - \bar{\lambda} v \in \Gamma$, too. Since $(x + \bar{\lambda} v) + (\gamma - \bar{\lambda} v) = x + \gamma \geq 0$, it follows that $x + \gamma \succeq x + \bar{\lambda} v$. Since, on the other hand, $x + \bar{\lambda} v \succ x$ and

$\succeq$ is transitive, we conclude that $x + \gamma \succ x$. The proof of the proposition is complete. ■

4.4. Proof of Proposition 3.3.1

For each $\alpha > 1$ choose $\beta(\alpha) > 0$ according to condition (USP), and then construct Z in the same way as in the proof of Proposition 3.2.1. Thus, if $y \in Y$, $z \in Z$, and $y - z \leq y^+$ (which is equivalent to $y^- + z \geq 0$) then $y - z \in Y$. Note that $\lambda\delta_g \in Z$ for all $g \in G$ and all $\lambda \geq 0$. Hence

$$\text{if} \quad y \in Y \quad \text{and} \quad z \in M_+^f(G) \quad \text{then} \quad y - z \in Y. \tag{4.5}$$

Let I denote the set of all isolated points of G and let

$$Z' = \{z \in Z : \ \operatorname{supp} z \subset G \setminus I\} \cup \{\lambda\delta_g : \ \lambda > 0 \text{ and } g \in I\}.$$

Then $Z' \subset Z$ and thus

$$\text{if} \quad y \in Y, \quad z \in Z', \quad \text{and} \quad y^- + z \geq 0 \quad \text{then} \quad y - z \in Y. \tag{4.6}$$

Now let $\widehat{\Gamma}$ be the convex hull of Z'. We claim:

$$\text{if} \quad y \in Y, \quad \gamma \in \widehat{\Gamma}, \quad \text{and} \quad y^- + \gamma \geq 0 \quad \text{then} \quad y - \gamma \in Y. \tag{4.7}$$

To see this, pick any $y \in Y$ and any $\gamma \in \widehat{\Gamma}$ and suppose $y^- + \gamma \geq 0$. According to the definition of $\widehat{\Gamma}$, we can choose a finite family $(z^1, \ldots, z^k)$ of elements of Z' so that $\gamma = \sum_{i=1}^k z^i$.

We shall first consider the special case where $(z^i)^- \wedge y^+ = 0$ for each $i = 1, \ldots, k$. As in the proof of Proposition 3.2.1, for every $x \in M_+(G)$ and every $z \in Z'$ let $\rho(z, x) = \max\{\lambda \in [0,1] : \ x + \lambda z \geq 0\}$.

Take z^j out of $(z^1, \ldots, z^k)$ so that $\rho(z^j, y^-)(z^j)^-(G) \geq \rho(z^i, y^-)(z^i)^-(G)$ for every $i = 1, \ldots, k$. Set $y_1 = y - \rho(z^j, y^-)z^j$. According to (4.6), $y_1 \in Y$. Set $z_1^j = (1 - \rho(z^j, y^-))z^j$ and $z_1^i = z^i$ if $i \neq j$, $i = 1, \ldots, k$. Then $y_1 - \sum_{i=1}^k z_1^i = y - \sum_{i=1}^k z^i$. Observe that $y_1^- = y^- + \rho(z^j, y^-)z^j - (\rho(z^j, y^-)z^j)^+ \wedge y^+$ because $y^- + \rho(z^j, y^-)z^j \geq 0$. Consequently, $y_1^- + \sum_{i=1}^k z_1^i \geq 0$ since $y^- + \sum_{i=1}^k z^i \geq 0$ and since we have assumed $(z^i)^- \wedge y^+ = 0$ for each $i = 1, \ldots, k$. Repeat this argument with $(z_1^1, \ldots, z_1^k)$ in place of $(z^1, \ldots, z^k)$ and y_1 in place of y to produce y_2 and $(z_2^1, \ldots, z_2^k)$. Note that we must have $(z_1^i)^- \wedge y_1^+ = 0$ for each $i = 1, \ldots, k$. Thus, in particular, $y_2^- + \sum_{i=1}^k z_2^i \geq 0$. In this manner we obtain a sequence $\{y_n\}$ in Y and, for each $i = 1, \ldots, k$, a non-increasing sequence

$\{\lambda_n^i\}$ in $[0,1]$ such that for every n:

$$y_n - \sum_{i=1}^k \lambda_n^i z^i = y - \sum_{i=1}^k z^i,$$

$$y_n^- + \sum_{i=1}^k \lambda_n^i z^i \geq 0,$$

$$\sum_{i=1}^k (\lambda_n^i z^i)^-(G) - \sum_{i=1}^k (\lambda_{n+1}^i z^i)^-(G) = \\ = \max\left\{\rho(\lambda_n^i z^i, y_n^-)(\lambda_n^i z^i)^-(G) : \quad i = 1, \ldots, k\right\}.$$

Since $y_n^- + \sum_{i=1}^k \lambda_n^i z^i \geq 0$ for each n, it follows as in the argument which has led to statement (4.3) in the proof of Proposition 3.2.1 that $\sum_{i=1}^k \lambda_n^i z^i \to \sum_{i=1}^k \lambda^i z^i$ for some $\lambda^i \in [0,1]$, $i = 1, \ldots, k$, and that $\sum_{i=1}^k \lambda^i z^i \geq 0$. In particular, $y_n \to \bar{y}$ for some $\bar{y} \in Y$ since Y is closed. From (4.5), $\bar{y} - \sum_{i=1}^k \lambda^i z^i \in Y$. Consequently, because $\bar{y} - \sum_{i=1}^k \lambda^i z^i \equiv y - \sum_{i=1}^k z_i$, we have $y - \sum_{i=1}^k z_i \in Y$.

We will now consider to the general case. For this purpose, let $C = G \setminus I$ and $E = \{g \in C : \ y^+(\{g\}) > 0\}$ (recall: I is the set of isolated points of G). Moreover, let $F = E \cap \bigcup_{i=1}^k \operatorname{supp} z^i$. If $F = \emptyset$ then the definition of Z' implies $(z^i)^- \wedge y^+ = 0$ for each member z^i of $(z^1, \ldots, z^k)$ and from above, we are done. Thus suppose $F \neq \emptyset$. We can write $F = \{g^1, \ldots, g^m\}$ for some $m \in \mathbb{N}$. Note that (in the relativized topology) C is compact and metrizable (recall that I is finite by hypothesis). Since E is (at most) countable and C has no isolated points, it follows that $C \setminus E$ is dense in C. Hence, for each $j = 1, \ldots, m$, we can choose a sequence $\{g_\ell^j\}$ in C such that $g_\ell^j \notin E$ for all ℓ and $g_\ell^j \to g^j$ as $\ell \to \infty$. Then, for every ℓ, let $\phi_\ell : G \to G$ be the mapping given by $\phi_\ell(g^j) = g_\ell^j$, $j = 1, \ldots, m$, and $\phi_\ell(g) = g$ if $g \notin F$. We have $z^i = a^i \delta_{\bar{g}^i} - b^i \delta_{\underline{g}^i}$ for some $a^i, b^i \in \mathbb{R}_+$ and some $\bar{g}^i, \underline{g}^i \in G$, $i = 1, \ldots, k$. Set $z_\ell^i = a^i \delta_{\phi_\ell(\bar{g}^i)} - b^i \delta_{\phi_\ell(\underline{g}^i)}$ for each ℓ and each $i = 1, \ldots, k$. Then $(z_\ell^i)^- \wedge y^+ = 0$ for every ℓ and every $i = 1, \ldots, k$ (note that if $g \in I$ then $z^i(g) \geq 0$ from the definition of Z', whence $z_\ell^i(\{g\}) \geq 0$). Moreover, since $y^- + \sum_{i=1}^k z^i \geq 0$ we have $\sum_{i=1}^k z^i(\{g\}) \geq 0$ for every $g \in F$, and this implies $y^- + \sum_{i=1}^k z_\ell^i \geq 0$. Finally, for each $i = 1, \ldots, k$, we have $z_\ell^i \to z^i$ by construction, hence $z_\ell^i \in Z'$ for ℓ large enough by virtue of the definitions of Z' and Z. By what has been established above, it follows that $y - \sum_{i=1}^k z_\ell^i \in Y$ if ℓ is large enough, whence $y - \sum_{i=1}^k z^i \in Y$ since Y is closed by hypothesis. Thus (4.7) has been shown.

Now let

$$B = \left\{p \in C(G) : \quad \|p\|_\infty = 1 \text{ and } p \cdot z \geq 0 \text{ for all } z \in Z'\right\}$$

and then let

$$\Gamma = \{z \in M(G): \quad p \cdot z > 0 \text{ for each } p \in B\}.$$

Since I, the set of isolated points of G, is finite the definitions of Z' and Z imply that B is equicontinuous. Using the same arguments as in the proof of Proposition 3.2.1, we conclude that Γ is open, that $\Gamma \cap M_+(G) \neq \emptyset$, and that $\Gamma \cap M^f(G) \subset \widehat{\Gamma}$.

Finally, in order to show that Γ is a properness cone for Y, pick any $y \in Y$ and any $\gamma \in \Gamma$, and suppose $y - \gamma \leq y^+$ which is equivalent to $y^- + \gamma \geq 0$ and also equivalent to $\gamma^- \leq y^-$. Thus, if $\operatorname{supp} y^-$ is finite then so is $\operatorname{supp} \gamma^-$ and it follows with arguments analogous to those in the proof of Proposition 3.2.1 that $y - \gamma \in Y$ (using (4.7) and the facts that $\Gamma \cap M^f(G) \subset \widehat{\Gamma}$ and that Y is closed). When $\operatorname{supp} y^-$ is not finite, condition (a) of this proposition ensures that we can choose a sequence $\{y_n\}$ in Y such that $\operatorname{supp} y_n^-$ is finite for each n and $y_n \to y$. By Facts 4.1.8 and 4.1.5 we can assume $y_n^+ \to y^+ + c$ and $y_n^- \to y^- + c$ for some $c \in M_+(G)$. Now, for each n, set $\gamma_n = y^- - y_n^- + \gamma + c$. Then $y_n^- + \gamma_n = y^- + \gamma + c \geq 0$ for each n. Moreover, $\gamma_n \to \gamma$, hence $\gamma_n \in \Gamma$ if n is large enough. Thus, by virtue of what has been stated above, since $\operatorname{supp} y_n^-$ is finite for each n, we have $y_n - \gamma_n \in Y$ when n is large, whence $y - \gamma \in Y$ since Y is closed. We conclude that Γ is a properness cone for Y, as was to be shown. ∎

4.5. Proof of Lemma 3.4.4

We have to show that $\mathcal{P}_\Gamma$ is closed in $\mathcal{P}$. Suppose $\succ \in \mathcal{P} \setminus \mathcal{P}_\Gamma$ and let $\{\succ^n\}$ be a sequence in $\mathcal{P}$ converging to $\succ$. Since $\succ \in \mathcal{P} \setminus \mathcal{P}_\Gamma$ there is an $x \in M_+(G)$ and a $\gamma \in \Gamma$ such that $x + \gamma \geq 0$ and $x + \gamma \not\succ x$. By the properties of the topology of closed convergence, we can find sequences $\{x^n\}$ and $\{y^n\}$ in $M_+(G)$ with $x^n \to x$, $y^n \to x + \gamma$, and such that $y^n \not\succ^n x^n$ for each n. Set $\gamma^n = y^n - x^n$. Then $\gamma^n \to \gamma$. Thus $\gamma^n \in \Gamma$ for n large enough since Γ is open. It follows that $\mathcal{P} \setminus \mathcal{P}_\Gamma$ is open, i.e. that $\mathcal{P}_\Gamma$ is closed as was to be shown. ∎

4.6. Proof of Lemma 3.4.6

For the following recall that, by convention, all topological notions concerning $M(G)$ or $M_+(G) \times M_+(G)$ are with respect to the bw^*–topology. We need the following prelemma:

Prelemma. *Let $\{\mu^\ell\}$ be a sequence in $M_+(G)$ with $\mu^\ell \to \mu \in M_+(G)$ and let z be an element of $M_+(G)$ with $z \leq \mu$. Then there is a sequence $\{z^\ell\}$ in $M_+(G)$ with $z^\ell \to z$ such that $z^\ell \leq \mu^\ell$ for each ℓ.*

Proof. Since $0 \le z \le \mu$, z is absolutely continuous with respect to μ. Therefore, by the Radon–Nikodym Theorem, it has a density with respect to μ. Let h be a version of this density with $0 \le h(g) \le 1$ for all $g \in G$. Suppose h is continuous. For each ℓ let z^ℓ be the element of $M_+(G)$ given by $z^\ell(B) = \int_B h d\mu^\ell$ for each Borel subset B of G. Then $0 \le z^\ell \le \mu^\ell$. Pick any $f \in C(G)$. Then the function fh is continuous on G,[2] and since $\mu^\ell \to \mu$ we have

$$\int f\, dz^\ell = \int (fh)\, d\mu^\ell \to \int (fh)\, d\mu = \int f\, dz\,,$$

whence $z^\ell \to z$. Thus:

If z has a density with respect to μ with a continuous version, the claim of the lemma is true. (4.8)

In the following, given any $x \in M_+(G)$ we denote by $[0, x]$ the order interval $\{x' \in M(G) : \ 0 \le x' \le x\}$. We claim:

The set of all $z \in [0, \mu]$ having a density with respect to μ with a continuous version is dense in $[0, \mu]$. (4.9)

To see this, pick any $z \in [0, \mu]$ and let h be a version of a density of z with respect to μ such that $0 \le h(g) \le 1$ for all $g \in G$. By Lusin's theorem, for every natural number m we can find a compact subset C^m of G such that the restriction of h to C^m is continuous and $\mu(G \setminus C^m) \le 1/m$. By the Tietze extension theorem we can then choose, for each m, a continuous function h^m on G which coincides with h on C^m and satisfies $0 \le h^m(g) \le 1$ for all $g \in G$. We have thus constructed a sequence $\{h^m\}$ in $C(G)$ which converges to h in μ–measure. In particular, if f is any element of $C(G)$ then $\{fh^m\}$ converges to fh in μ–measure and for each m we have $\|fh^m\|_\infty \le \|f\|_\infty$. Now, for each m, let γ^m be the element of $M_+(G)$ given by $\gamma^m(B) = \int_B h^m d\mu$ for each Borel subset B of G. Then $0 \le \gamma^m \le \mu$ for each m, and by the Lebesgue Convergence Theorem

$$\int f\, d\gamma^m = \int (fh^m)\, d\mu \to \int (fh)\, d\mu = \int f\, dz$$

for every $f \in C(G)$. Thus (4.9) holds.

By Fact 4.1.7 there is a real number a such that $[0, \mu] \subset U^a$ and, for each ℓ, $[0, \mu^\ell] \subset U^a$, where $U^a = \{\mu \in M(G) : \ \|\mu\|_V \le a\}$. According to Fact 4.1.4, U^a is compact and metrizable by some metric d. Thus, given any $z \in [0, \mu]$

[2] By fh we mean the function on G given by $(fh)(g) = f(g)h(g)$ for each $g \in G$.

we can choose, for every ℓ, $z^\ell \in [0, \mu^\ell]$ such that $d(z^\ell, z) = \min\{d(z', z) : 0 \leq z' \leq \mu^\ell\}$. Using (4.8) and (4.9), it is easy to see that $z^\ell \to z$. ■

Now note that $Z_n \to Z$ in $\mathcal{Z}$ if and only if the following two conditions are satisfied:

If $\{Z_{n_k}\}$ is a subsequence of $\{Z_n\}$ and $z_k \in Z_{n_k}$ for each k so that $z_k \to z$ for some $z \in M_+(G) \times M_+(G)$ then $z \in Z$. (4.10)

If $z \in Z$ then there is a sequence $\{z_n\}$ in $M_+(G) \times M_+(G)$ and a number $\bar{n}$ such that $z_n \in Z_n$ for every $n \geq \bar{n}$ and $z_n \to z$. (4.11)

(See Hildenbrand [10, p. 19 and p. 15].) Next, recalling that a convergent sequence in $M(G)$ is $\|\cdot\|_V$–bounded and that a $\|\cdot\|_V$–bounded sequence in $M(G)$ has a convergent subsequence (Facts 4.1.8 and 4.1.5), one easily verifies the following two sentences:

If $y_n \to y$ in $M(G)$ then there is a subsequence of $\{(y_n^+, y_n^-)\}$, say $\{(y_k^+, y_k^-)\}$, such that $(y_k^+, y_k^-) \to (y^+ + c, y^- + c)$ for some $c \in M_+(G)$. (4.12)

If $(a_n, b_n) \to (a, b)$ in $M_+(G) \times M_+(G)$ and $a \wedge b = 0$ then $(a_n - b_n)^+ \to a$ and $(a_n - b_n)^- \to b$. (4.13)

For convenient reference, we also note the following:

Let $Y \in \mathcal{Y}$. If $(a, b) \in \kappa(Y)$ then also $(a + c, b + c) \in \kappa(Y)$ for every $c \in M_+(G)$ (which follows directly from the definition of κ). (4.14)

We will now prove Lemma 3.4.6. To this end, fix any $Y \in \mathcal{Y}$ and consider any sequence $\{Y_n\}$ in $\mathcal{Y}$ (and recall that $\mathcal{Y}$ has been defined as the set of all *non-empty* closed subsets of $M(G)$). Suppose that $\{Y_n\}$ converges to Y with respect to $d_{\mathcal{Y}}$. Then $\kappa(Y_n) \to \kappa(Y)$ in $\mathcal{Z}$ by the definition of $d_{\mathcal{Y}}$. Let $\{Y_{n_k}\}$ be a subsequence of $\{Y_n\}$, let $y_k \in Y_{n_k}$ for each k, and assume $y_k \to y$ for some $y \in M(G)$. By virtue of (4.12), we can assume $(y_k^+, y_k^-) \to (a, b)$ for some $(a, b) \in M_+(G) \times M_+(G)$. From the definition of κ, $(y_k^+, y_k^-) \in \kappa(Y_{n_k})$ for each k, hence $(a, b) \in \kappa(Y)$ from (4.10), whence $y \in Y$. Thus (a) of Lemma 3.4.6 holds. Suppose $y \in Y$. Then $(y^+, y^-) \in \kappa(Y)$ and by (4.11) we can find $(a_n, b_n) \in \kappa(Y_n)$ for each n such that $(a_n, b_n) \to (y^+, y^-)$. For each n set $y_n = a_n - b_n$. Then $y_n \in Y_n$ for each n, and according to (4.13) we have $y_n^+ \to y^+$ and $y_n^- \to y^-$. Thus we have (b) of Lemma 3.4.6. Conversely, assume that (a) and (b) of Lemma 3.4.6 are satisfied. Then (4.10) and (4.11) above must hold with $\kappa(Y)$ in place of Z and $\kappa(Y_n)$ in place of Z_n. Indeed, from (4.14) and (b) it is immediate that (4.11) is satisfied. To see that (4.10) holds, let $\{\kappa(Y_{n_k})\}$ be a subsequence of $\{\kappa(Y_n)\}$. Let $(a_k, b_k) \in \kappa(Y_{n_k})$ for each k and assume $(a_k, b_k) \to (a, b)$ for some $(a, b) \in M_+(G) \times M_+(G)$. Then $(a_k - b_k) \in Y_{n_k}$ for each k and $(a_k - b_k) \to (a - b)$, hence $(a - b) \in Y$ according to (a), whence $(a, b) \in \kappa(Y)$. Thus also (4.10) is satisfied. It follows that $\kappa(Y_n) \to \kappa(Y)$ in $\mathcal{Z}$ and hence that $Y_n \to Y$ with respect to $d_{\mathcal{Y}}$ (by the definition of $d_{\mathcal{Y}}$). This completes the proof of Lemma 3.4.6. ■

4.7. Proof of Lemma 3.4.7

Let $\{Y_n\}$ be a sequence in $\mathcal{Y}_\Gamma$. We have to show that some subsequence converges to an element Y of $\mathcal{Y}_\Gamma$ with respect to $d_{\mathcal{Y}}$. Let $Z_n = \kappa(Y_n)$ for each n. Since $\mathcal{Z}$ is compact and metrizable, $\{Z_n\}$ has a subsequence, say $\{Z_k\}$, converging to some $Z \in \mathcal{Z}$. Let $\{Y_k\}$ be the corresponding subsequence of $\{Y_n\}$, i.e. $Z_k = \kappa(Y_k)$ for each k. Note that $0 \in Y_k$ hence $(0,0) \in Z_k$ for each k. Consequently $(0,0) \in Z$ by (4.10) and thus Z is non-empty. Moreover, from (4.11), (4.14), and (4.10) it follows that

$$\text{if} \quad (a,b) \in Z \quad \text{and} \quad c \in M_+(G) \quad \text{then} \quad (a+c,\, b+c) \in Z. \tag{4.15}$$

Next, we claim:

$$\text{if} \quad (a,b) \in Z \quad \text{then also} \quad \big((a-b)^+,\, (a-b)^-\big) \in Z. \tag{4.16}$$

To see this, pick some $(a, b) \in Z$. Choose $(a_k, b_k) \in Z_k$ for each k so that $(a_k, b_k) \to (a, b)$. Since $Z_k \to Z$, this is possible by virtue of (4.11). Set $y_k = a_k - b_k$ for each k and $y = a - b$. Then $y_k \in Y_k$ for all k from the definition of κ. Also, $y_k \to y$ and by (4.12) we may assume that $(y_k^+, y_k^-) \to (y^+ + c,\ y^- + c)$ for some $c \in M_+(G)$ Applying the prelemma, we can then choose a sequence $\{\tilde{a}_k\}$ in $M_+(G)$ with $\tilde{a}_k \le y_k^+$ for each k so that $\tilde{a}_k \to y^+$. Pick some $v \in \Gamma \cap M_+(G)$ (which is possible according to the hypotheses concerning Γ) and some real number $\lambda > 0$. For each k let $\gamma_k = \lambda v + y^- - y_k^- + y_k^+ - \tilde{a}_k$. Then $\gamma_k \to \lambda v$ as $k \to \infty$, hence $\gamma_k \in \Gamma$ for k large enough since Γ is open. Moreover, for each k, we have $y_k^- + \gamma_k = \lambda v + y^- + y_k^+ - \tilde{a}_k \ge 0$ whence $y_k - \gamma_k \le y_k^+$. Thus, $y_k - \gamma_k \in Y_k$ for k large enough because $Y_k \in \mathcal{Y}_\Gamma$ for all k. By construction, $y_k - \gamma_k = \tilde{a}_k - \lambda v - y^-$, and so, for k large enough, $(\tilde{a}_k,\ \lambda v + y^-) \in Z_k$ from the definition of κ since $Z_k = \kappa(Y_k)$ for each k. Using (4.10), it follows that $(y^+,\ \lambda v + y^-) \in Z$, and since λ can be arbitrarily small we thus obtain $(y^+, y^-) \in Z$ because Z being an element of $\mathcal{Z}$ is closed in $M_+(G) \times M_+(G)$. Thus (4.16) has been established (recall that we have set $y = a - b$).

Now let $Y = \{y \in M(G): \ y = a - b \text{ for some } (a,b) \in Z\}$. Then, in particular, $0 \in Y$ because $(0,0) \in Z$, and thus Y is non-empty. Since Z is closed in $M_+(G) \times M_+(G)$, it follows from (4.16) and (4.12) that Y is sequentially closed in $M(G)$. Hence, according to Fact 4.1.10, Y is closed in $M(G)$. Thus $Y \in \mathcal{Y}$. From the definitions of κ and Y we clearly have $Z \subset \kappa(Y)$. Conversely, let $(a', b') \in \kappa(Y)$. Then, according to the definitions of Y and κ again, $a' - b' = a - b$ for some $(a,b) \in Z$, whence, from (4.16) and (4.15), $(a', b') \in Z$. Thus $\kappa(Y) \subset Z$. We conclude $\kappa(Y) = Z$. In particular, $\kappa(Y_k) \to \kappa(Y)$ in $\mathcal{Z}$ and it follows from the definition of $d_{\mathcal{Y}}$ that $\{Y_k\}$ converges to Y with respect to $d_{\mathcal{Y}}$.

Thus it remains to show that $Y \in \mathcal{Y}_\Gamma$. As already noted, $0 \in Y$. To see that Y satisfies also (P2) and (P4), pick any $y \in Y$. By Lemma 3.4.6 (b) we can choose $y_k \in Y_k$ for each k so that $y_k^+ \to y^+$ and $y_k^- \to y^-$. Consider some $z \in M_+(G)$ with $z \leq y^+$. Using the prelemma, we can find a sequence $\{z_k\}$ in $M_+(G)$ with $z_k \leq y_k^+$ for each k and $z_k \to z$. In particular, then, $z_k - y_k^- \in Y_k$ for each k, whence, from (a) of Lemma 3.4.6, $z - y^- \in Y$. Thus Y satisfies (P2). To see that (P4) holds, i.e. that Y is uniformly proper with respect to Γ, pick some $\gamma \in \Gamma$ and suppose $y - \gamma \leq y^+$ which is equivalent to $y^- + \gamma \geq 0$. Set $\gamma_k = y^- - y_k^- + \gamma$ for each k. Then $\gamma_k \in \Gamma$ for k large enough since Γ is open. Moreover, $y_k^- + \gamma_k = y^- + \gamma$, whence $y_k - \gamma_k \leq y_k^+$. Consequently, $y_k - \gamma_k \in Y_k$ for k large enough, hence $y - \gamma \in Y$ by Lemma 3.4.6 (a). Thus Y is uniformly proper with respect to Γ, and we have shown that $Y \in \mathcal{Y}_\Gamma$. The proof of Lemma 3.4.7 is thus complete. ∎

4.8. Proof of Lemma 3.4.14

Set $B = \operatorname{supp} \bar{v}$, let $B' = \{g \in B: \ g \text{ is an isolated point in } G\}$, and let $\Gamma' = \Gamma \cap \{\mu \in M(G): \ \mu(\{g\}) > 0, \ \forall g \in B'\}$. Evidently Γ' is an open convex cone with $\Gamma' \cap M_+(G) \neq \emptyset$ (because Γ has all these properties) and Y is proper also with respect to Γ' (because $\Gamma' \subset \Gamma$).

Pick any number $\lambda > 0$ and let $b = \lambda \bar{v}$. Consider some $y \in Y$ and some $\gamma \in \Gamma'$ and suppose $y - b - \gamma \leq y^+$. We wish to show that $y - b - \gamma \in Y$. For this purpose, let

$$B'' = \{g \in B: \quad y(\{g\}) > 0 \ \text{ and } \ \gamma(\{g\}) < 0\}.$$

Assume first that $B'' = \emptyset$. Thus, since $b \geq 0$ by hypothesis, $y^+ \wedge b \wedge \gamma^- = 0$. Hence $(y^+ \wedge b) + \gamma^- = (y^+ \wedge b) \vee \gamma^-$. On the other hand, $y - b - \gamma \leq y^+$ implies $y^- + b + \gamma \geq 0$ whence $\gamma^- \leq b + y^-$. It follows that $\gamma^- \leq b + y^- - (y^+ \wedge b)$. From $b \geq 0$ again, $(y - b)^- = y^- + b - (y^+ \wedge b)$ and thus $\gamma^- \leq (y - b)^-$ which is equivalent to $(y - b) - \gamma \leq (y - b)^+$. From the free disposal condition (P2′), $(y - b) \in Y$, whence $y - b - \gamma \in Y$ since Y is proper with respect to Γ'.

Now suppose $B'' \neq \emptyset$. Write $B'' = \{g_1, \ldots, g_m\}$ and let d_G be a metric for the topology of G. By construction, $B'' \subset B \setminus B'$. That is, no element of B'' can be an isolated point of G. Therefore, for every integer $\ell > 0$ we can find points $g_1^\ell, \ldots, g_m^\ell \in G$ such that $g_i^\ell \notin B$, $d(g_i^\ell, g_i) \leq 1/\ell$, and $y^+(\{g_i^\ell\}) \leq 1/\ell$, $i = 1, \ldots, m$. For each ℓ set

$$y^\ell = y - \sum_{i=1}^{m} y^+(\{g_i^\ell\})\delta_{g_i^\ell}$$

$$\gamma^\ell = \gamma - \sum_{i=1}^{m} \gamma(\{g_i\})\delta_{g_i} + \sum_{i=1}^{m} \gamma(\{g_i\})\delta_{g_i^\ell}$$

$$b^\ell = b + \sum_{i=1}^{m} \gamma(\{g_i\})\delta_{g_i} - \sum_{i=1}^{m} \gamma(\{g_i\})\delta_{g_i^\ell}.$$

Then $y^\ell \in Y$ for each ℓ because of (P2′), and as $\ell \to \infty$ we have $y^\ell \to y$, $b^\ell \to b$, and $\gamma^\ell \to \gamma$. In particular, $\gamma^\ell \in \Gamma'$ for ℓ large enough since Γ' is open. Moreover, for each ℓ, $(y^\ell)^- = y^-$ and $\gamma^\ell + b^\ell = \gamma + b$, hence $(y^\ell)^- + b^\ell + \gamma^\ell \geq 0$. In particular, $b^\ell \geq 0$ because $\gamma^\ell(\{g\}) = (y^\ell)^-(\{g\}) = 0$ for each $g \in B''$. Furthermore, $y^\ell(\{g\}) > 0$ together with $b^\ell(\{g\}) > 0$ implies $\gamma^\ell(\{g\}) \geq 0$, hence $(y^\ell)^+ \wedge b^\ell \wedge (\gamma^\ell)^- = 0$. Thus, for ℓ large enough so that $\gamma^\ell \in \Gamma'$ we get, analogous to above, $y^\ell - b^\ell - \gamma^\ell \in Y$, whence $y - b - \gamma \in Y$ since Y is closed according to (P3). It follows that Y is proper with respect to $\Gamma' - \{\lambda\bar{v} : \lambda \geq 0\}$, as was to be shown. ■

4.9. Proof of Proposition 3.5.5

Recall that the *narrow topology* on the set of Radon probability measures on a completely regular topological space D is the topology of pointwise convergence on the set of real–valued continuous bounded functions defined on D. We need the following lemma:

4.9.1. Lemma. *Let D be a completely regular topological space, let $f : D \to \mathbb{R}$ be a lower semi-continuous function on D, and let $\{\mu^\alpha\}$ be a net of Radon probability measures on D, narrowly converging to some Radon probability measure μ on D. Assume there is a real number b and a closed subset K of D, with $\mu^\alpha(K) = 1$ for all α, such that $f(x) \geq b$ for all $x \in K$. Then $\int f\, d\mu \leq \underline{\lim}_\alpha \int f\, d\mu^\alpha$.*

Proof. For every $m \in \mathbb{N}$ let $f^m : D \to \mathbb{R}$ be the function given by

$$f^m(x) = \begin{cases} \min\{m, f(x)\} & \text{if } x \in K \\ m & \text{if } x \in D \setminus K\,. \end{cases}$$

Then, for each m, f^m is bounded and lower semi-continuous and hence we have $\int f^m \, d\mu \leq \underline{\lim}_\alpha \int f^m \, d\mu^\alpha$. (See Schwartz [27, p. 369, Proposition 1].) Since $\mu^\alpha(D \setminus K) = 0$ for all α, it follows that $\int f^m \, d\mu \leq \underline{\lim}_\alpha \int f \, d\mu^\alpha$ for each m. Moreover, we must have $\mu(D \setminus K) = 0$ and thus, using the monotone convergence theorem, we conclude that $\int f \, d\mu \leq \underline{\lim}_\alpha \int f \, d\mu^\alpha$. ■

Now let

$$T = \{(Y, y) \in \mathcal{Y}_\Gamma \times M(G) : \quad y \in Y\}.$$

Furthermore, for each $r \in \mathbb{N}$ let

$$U^r = \{z \in M(G) : \ \|z\|_V \leq r\} \quad \text{and} \quad T^r = T \cap (\mathcal{Y}_\Gamma \times U^r).$$

Recall that $\mathcal{Y}_\Gamma$ is a compact metric space by construction (see Convention 3.4.8), and note that each U^r is compact and metrizable in the relativized topology (see Fact 4.1.4). Using Lemma 3.4.6 (a) it follows, for each r, that T^r is closed in $\mathcal{Y}_\Gamma \times U^r$ and therefore that it is compact. In particular, it follows that T is a Borel subset of $\mathcal{Y}_\Gamma \times M(G)$. Finally, it follows that every finite Borel measure on $\mathcal{Y}_\Gamma \times M(G)$ is a Radon measure.

In order to show that $Y^a(\varrho)$ is closed it is enough to show that it is sequentially closed (see Fact 4.1.10). Thus let $\{y^n\}$ be a sequence in $Y^a(\varrho)$, converging to some $y \in M(G)$. By the definition of $Y^a(\varrho)$, we can choose a sequence $\{\eta^n\}$ of probability measures on $\mathcal{Y}_\Gamma \times M(G)$ such that

(i) $\eta^n_Y = \varrho$ for each n,

(ii) $\eta^n(T) = 1$ for each n,

(iii) $\int i_M \, d\eta^n_M = y^n$ for each n.

Since $\{y^n\}$ is convergent, it is $\|\cdot\|_V$–bounded. Thus, by the hypotheses of this proposition, we can assume for some $m \in \mathbb{N}$:

(iv) $\eta^n_M(M^m(G)) = 1$ for all n

(recall: $M^m(G) = \{z \in M(G) : \quad z^+(G) \leq m\}$).
An easy calculation shows that $\|z\|_V \leq 2m - z(G)$ for each $z \in M^m(G)$. Hence, by (iv), we have $\int \|z\|_V \, d\eta^n_M(z) \leq 2m + \int (-1_G) \cdot z \, d\eta^n_M(z)$ for all n. From (iii), $\int (-1_G) \cdot z \, d\eta^n_M(z) = (-1_G) \cdot y^n$, and since $y^n \to y$ it follows that there is a real number $b > 0$ such that $\int \|z\|_V \, d\eta^n_M(z) \leq b$ for all n. But therefore, for every $\epsilon > 0$ we can find an integer r such that, for each n, $\eta^n_M(U^r) \geq 1 - \epsilon$ and thus, by (ii), $\eta^n(T^r) \geq 1 - \epsilon$. That is, the sequence $\{\eta^n\}$ is (equally) tight. Consequently, there is a subnet of $\{\eta^n\}$, say $\{\eta^{n_\alpha}\}$, which narrowly converges to some probability measure on $\mathcal{Y}_\Gamma \times M(G)$, say η. (See Schwartz [27, p. 379, Theorem 3].) In particular, for each r, we have $\eta(T^r) \geq \overline{\lim}_\alpha \eta^{n_\alpha}(T^r)$ since T^r

is closed. It follows that for every $\epsilon > 0$ there is an r such that $\eta(T^r) \geq 1 - \epsilon$. Consequently, $\eta(T) = 1$ since $T^r \subset T$ for all r. Clearly, the nets $\{\eta_Y^{n_\alpha}\}$ and $\{\eta_M^{n_\alpha}\}$ of marginal distributions narrowly converge to η_Y and η_M, respectively. Thus $\eta_Y = \varrho$ from (i). Moreover, since $\|\cdot\|_V$ is lower semi-continuous on $M(G)$, using Lemma 4.9.1 we get $\int \|z\|_V \, d\eta_M(z) \leq \underline{\lim}_\alpha \int \|z\|_V \, d\eta_M^{n_\alpha}(z) \leq b$ where b is the number from above. Hence, by Fact 4.1.16, $\int i_M \, d\eta_M$ exists, and since $\eta_Y = \varrho$ and $\eta(T) = 1$ we have $\int i_M \, d\eta_M \in Y^a(\varrho)$. By virtue of the free disposal condition hypothesized in this proposition, it now suffices to show that $y \leq \int i_M \, d\eta_M$. To this end, pick any $p \in C_+(G)$. Then $z \mapsto (-p) \cdot z$, $z \in M(G)$, is bounded from below by $-m\|p\|_\infty$ on $M^m(G)$. Hence, by Lemma 4.9.1 again, $\int (-p) \cdot z \, d\eta_M(z) \leq \underline{\lim}_\alpha \int (-p) \cdot z \, d\eta_M^{n_\alpha}(z)$ because of (iv) since, obviously, $M^m(G)$ is closed in $M(G)$. On the other hand, from (iii) we have $\int (-p) \cdot z \, d\eta_M^n(z) = (-p) \cdot y^n$ for each n, and since $y^n \to y$ we conclude that $\int p \cdot z \, d\eta_M(z) \geq p \cdot y$ whence, since $p \in C_+(G)$ was arbitrary, $\int i_M \, d\eta_M \geq y$, as was to be shown. ■

4.10. Proof of Theorem 3.5.6

The idea of the proof is as follows: After a suitable truncation of individual production sets we are able to define (well behaved) aggregate demand and supply correspondences on the set of strictly positive price systems. Applying the hypothesis of uniform properness, we then construct a compact set Δ of positive price systems with the following property: if z is a vector of excess demand for some $p \in \Delta$ satisfying $q \cdot z \leq 0$ for all $q \in \Delta$ then it must be an element of $-M_+(G)$. (Actually z will be a limit of a sequence of vectors of excess demand.) The existence of such z will be derived from a fixed point argument. In a final step we remove the truncation of individual production sets and by a limit argument construct a quasi-equilibrium.

Recall from Section 4.2 that we shall use Facts 4.1.3, 4.1.6, and 4.1.14 without specific invocation. Some technical preliminaries are necessary. In the following, if μ is a measure on a product space $E \times Z$ then μ_E denotes the marginal distribution on E and μ_Z that on Z.

4.10.1. Lemma. *Let E and Z be separable complete metric spaces, v a probability measure on E, and F a closed subset of $E \times Z$ with* $\operatorname{proj}_E F = E$. *Then there exists a probability measure μ on $E \times Z$ with $\mu_E = v$ and* $\operatorname{supp} \mu \subset F$.

Proof. Let $\mathcal{B}(E)$, $\mathcal{B}(Z)$, and $\mathcal{B}(E \times Z)$ denote the Borel σ–algebras of E, Z, and $E \times Z$, respectively. Note that $\mathcal{B}(E \times Z) = \mathcal{B}(E) \otimes \mathcal{B}(Z)$. By the Aumann measurable selection theorem there is a $\mathcal{B}(E) - \mathcal{B}(Z)$–measurable mapping $f : E \to Z$ such that $(e, f(e)) \in F$ for v almost all $e \in E$. Let i_E be the identity on E and $h = i_E \times f$. Then h is $\mathcal{B}(E) - \mathcal{B}(E \times Z)$–measurable.

Let $\mu = v \circ h^{-1}$, i.e., μ is the image measure of v under h. Then $\mu_E = v$. Moreover, $\mu(F) = 1$ and since F is closed it follows that $\operatorname{supp}\mu \subset F$. ∎

Recall that a sequence $\{\mu^n\}$ of probability measures on a topological space D is called (equally) *tight* if for every real number $\epsilon > 0$ there is a compact subset K of D such that $\mu^n(K) \geq 1 - \epsilon$ for every n. The following fact can easily be deduced from the material in Hildenbrand [10, pp. 48–52].

Fact 4.10.1. *Let E and Z be separable complete metric spaces and $\{\mu^n\}$ a sequence of probability measures on $E \times Z$. Let v be a probability measure on E and suppose $\mu^n_E = v$ for each n. Suppose further that the sequence $\{\mu^n_Z\}$ is tight (which holds in particular if Z is compact). Then there exists a probability measure μ on $E \times Z$ together with a subsequence $\{\mu^{n_k}\}$ of $\{\mu^n\}$ such that the following hold:*

(a) $\mu_E = v$.

(b) *For every $w \in \operatorname{supp}\mu$ there is a sequence $\{w^k\}$ in $E \times Z$ with $w^k \to w$ such that $w^k \in \operatorname{supp}\mu^{n_k}$ for each k.*

(c) *Let $f : Z \to \mathbb{R}$ be a continuous function. If Z is compact* **or** *(more generally) if there is a v–integrable function $\rho : E \to \mathbb{R}_+$ such that, for each k, $(e, z) \in \operatorname{supp}\mu^{n_k}$ implies $|f(z)| \leq \rho(e)$, then f is μ_Z–integrable and we have $\int f\, d\mu_Z = \lim_k \int f\, d\mu^{n_k}_Z$.*

(d) *Let $f : Z \to \mathbb{R}_+$ be continuous. Then $\int f\, d\mu_Z \leq \underline{\lim}_k \int f\, d\mu^{n_k}_Z$.*

4.10.2. Lemma. *Let E be a separable complete metric space and v a probability measure on E. Moreover, let $\{\mu^n\}$ be a sequence of probability measures on $E \times M(G)$ with $\mu^n_E = v$ for each n. Assume* **one** *of the following conditions:*

(i) *There is a compact $A \subset M(G)$ such that $\operatorname{supp}\mu^n_M \subset A$ for each n.*

(ii) *For all n, $\operatorname{supp}\mu^n_M \subset M_+(G)$. Moreover, there is a v–integrable function $\rho : E \to \mathbb{R}_+$ such that, for each n, $(e, x) \in \operatorname{supp}\mu^n$ implies $x(G) \leq \rho(e)$.*

Then

(α) *$\int i_M\, d\mu^n_M$ exists for each n.*

Moreover, there exists a probability measure μ on $E \times M(G)$ together with a subsequence $\{\mu^{n_k}\}$ of $\{\mu^n\}$ such that the following hold:

(β) $\mu_E = v$.

(γ) *For every $w \in \operatorname{supp}\mu$ there is a sequence $\{w^k\}$ in $E \times M(G)$ with $w^k \to w$ such that $w^k \in \operatorname{supp}\mu^{n_k}$ for each k.*

(δ) *$\int i_M\, d\mu_M$ exists and, moreover, $\int i_M\, d\mu_M = \lim_k \int i_M\, d\mu^{n_k}_M$.*

Proof. Suppose first that (i) is satisfied and let $A \subset M(G)$ be as hypothesized. In particular, A is compact. Hence $\int |p \cdot x| \, d\mu_M^n(x) = \int [d\mu_M^n(x) \, A] |p \cdot x| < \infty$ for each $p \in C(G)$ and all n. Thus (α) holds in view of Fact 4.1.15. Note that A is metrizable according to Fact 4.1.9. Hence, temporarily considering every μ^n as a probability measure on $E \times A$, Fact 4.10.1 applies to the sequence $\{\mu^n\}$. It follows that there is a subsequence $\{\mu^{n_k}\}$ and a probability measure μ on $E \times M(G)$ such that (β) and (γ) are satisfied and such that (focusing on (c) of Fact 4.10.1)

$$\begin{aligned} &\int |p \cdot x| \, d\mu_M(x) < \infty \qquad \text{and} \\ &\int p \cdot x \, d\mu_M^{n_k}(x) \to \int p \cdot x \, d\mu_M(x) \quad \forall p \in C(G). \end{aligned} \tag{4.17}$$

Thus, $\int i_M \, d\mu_M$ exists according to Fact 4.1.15, and for each $p \in C(G)$ we have $p \cdot \int i_M \, d\mu_M = \lim_k p \cdot \int i_M \, d\mu_M^{n_k}$, i.e. $\int i_M \, d\mu_M^{n_k} \to \int i_M \, d\mu_M$. Thus also ($\delta$) is satisfied, and the lemma is proved under condition (i).

Now suppose that (ii) holds. Let $\rho : E \to \mathbb{R}_+$ be as hypothesized and then for each $p \in C(G)$ let $\rho_p = \|p\|_\infty \rho$. Thus, since $x(G) = \|x\|_V$ if $x \in M_+(G)$,

$$\int \|x\|_V \, d\mu_M^n(x) \le \int \rho \, dv < \infty \quad \text{for all } n \tag{4.18}$$

and hence also

$$\int |p \cdot x| \, d\mu_M^n(x) \le \int \rho_p \, dv < \infty \quad \text{for each } p \in C(G) \text{ and all } n. \tag{4.19}$$

In particular, (α) holds according to Fact 4.1.15. Moreover, since for every real number b the set $\{x \in M(G) : \ \|x\|_V \le b\}$ is compact, (4.18) implies that the sequence $\{\mu_M^n\}$ is tight. Therefore, since $M_+(G)$ in the induced topology is completely metrizable and separable, temporarily considering every μ^n as a probability measure on $E \times M_+(G)$, Fact 4.10.1 applies to the sequence $\{\mu^n\}$ again. Since $M_+(G)$ is a closed in $M(G)$, it follows that there is a subsequence $\{\mu^{n_k}\}$ and a probability measure μ on $E \times M(G)$ such that (β) and (γ) are satisfied, and, taking into account (4.19) (again focusing on (c) of Fact 4.10.1), such that (4.17) holds. As above, the latter yields (δ). ■

4.10.3. Lemma. *Let E and v be as in Lemma 4.10.2. Moreover, let $\{\mu^n\}$ be a sequence of probability measures on $E \times M(G)$ and suppose for each n that $\mu_E^n = v$, that $\operatorname{supp} \mu_M^n \subset M_+(G)$, and that $\int i_M \, d\mu_M^n$ exists. Suppose further that $\int i_M \, d\mu_M^n \to \bar{x}$ for some $\bar{x} \in M_+(G)$. Then there exists a probability measure μ on $E \times M(G)$ together with a subsequence $\{\mu^{n_k}\}$ of $\{\mu^n\}$ such that:*

(a) (β) *and* (γ) *of Lemma* 4.10.2 *hold;*

(b) $\int i_M \, d\mu_M$ *exists and, moreover,* $\int i_M \, d\mu_M \leq \bar{x}$.

Proof. By hypothesis there is an $\bar{x} \in M_+(G)$ such that

$$\int p \cdot x \, d\mu_M^n(x) \to p \cdot \bar{x} \quad \forall p \in C(G). \tag{4.20}$$

In particular, since $\|x\|_V = 1_G$ for $x \in M_+(G)$ and since $\operatorname{supp} \mu_M^n \subset M_+(G)$ for each n, there is a number b such that $\int \|x\|_V \, d\mu_M^n(x) \leq b$ for all n. As in the proof of Lemma 4.10.2, it follows that the sequence $\{\mu_M^n\}$ is tight. Therefore, again as in the proof of Lemma 4.10.2, temporarily considering every μ^n as a probability measure on $E \times M_+(G)$, Fact 4.10.1 applies to the sequence $\{\mu^n\}$. It follows that there is a probability measure μ on $E \times M(G)$, with $\operatorname{supp} \mu_M \subset M_+(G)$, and a subsequence $\{\mu^{n_k}\}$ of $\{\mu^n\}$ such that (a) of this lemma is satisfied and such that (focusing on (d) of Fact 4.10.1) $\int p \cdot x \, d\mu_M(x) \leq \underline{\lim}_k \int p \cdot x \, d\mu_M^{n_k}(x)$ for each $p \in C_+(G)$. In view of (4.20), the latter yields $\int p \cdot x \, d\mu_M(x) \leq p \cdot \bar{x}$ for all $p \in C_+(G)$. In particular, $\int \|x\|_V \, d\mu_M(x) < \infty$ since $\operatorname{supp} \mu_M \subset M_+(G)$. Thus, $\int i_M \, d\mu_M$ exists according to Fact 4.1.16, and we must have $\int i_M \, d\mu_M \leq \bar{x}$. ■

We turn now to the given economy (ϱ, ν). We have to introduce some additional notation and to make a further convention:

Notation. $C_{++}(G) \equiv \{p \in C(G) : \; p(g) > 0 \text{ for all } g \in G\}$.

Thus, $C_{++}(G)$ represents the space of strictly positive price systems.

4.10.4. Convention. *The space* $C(\mathcal{Y}_\Gamma)$ *is endowed with the sup–norm topology.*

We first consider the production sector of the economy. For every $m \in \mathbb{N}$, every $p \in C_{++}(G)$, and every $Y \in \mathcal{Y}_\Gamma$ let

$$M^m(G) = \{z \in M(G) : \; z^+(G) \leq m\}$$

$$\pi^m(p, Y) = \sup \{p \cdot y : \quad y \in Y \cap M^m(G)\}$$

$$T^m(p) = \{(Y, y) \in \mathcal{Y}_\Gamma \times M(G) : \quad y \in Y \cap M^m(G) \text{ and } p \cdot y = \pi^m(p, Y)\}$$

$$\phi^m(p) = \left\{ z \in M(G) : \; \begin{array}{l} z = \int i_M \, d\eta_M \text{ for some probability measure } \eta \\ \text{on } \mathcal{Y}_\Gamma \times M(G) \text{ with } \eta_Y = \varrho \text{ and } \operatorname{supp} \eta \subset T^m(p) \end{array} \right\}.$$

Thus, for a certain truncation of individual production sets and a given price system $p \in C_{++}(G)$, $\pi^m(p,\cdot)$ describes the profits in the production sector and $\phi^m(p)$ is the aggregate supply set.

4.10.5. Lemma. *The following statements hold for all $m \in \mathbb{N}$:*

(a) *$\phi^m(p)$ is non-empty and convex for each $p \in C_{++}(G)$.*

(b) *$\phi^m(p) \subset M^m(G)$ for each $p \in C_{++}(G)$.*

(c) *If $p_i \to p$ in $C_{++}(G)$ and $z_i \in \phi^m(p_i)$ for each i then a subsequence of $\{z_i\}$ converges to some $z \in \phi^m(p)$.*

(d) *For each $p \in C_{++}(G)$, $\pi^m(p,\cdot) \in C_+(\mathcal{Y}_\Gamma)$, i.e., $\pi^m(p,\cdot)$ is nonnegative and continuous.*

(e) *If $p_n \to p$ in $C_{++}(G)$ then $\pi^m(p_n,\cdot) \to \pi^m(p,\cdot)$ in $C(\mathcal{Y}_\Gamma)$, i.e., uniformly.*

(f) *For all $p \in C_{++}(G)$, if $z \in \phi^m(p)$ then $p \cdot z = \pi^m(p,\cdot) \cdot \varrho$.*

Proof. Fix any m. Let $Y_n \to Y$ in $\mathcal{Y}_\Gamma$ and $y_n \to y$ in $M(G)$, and suppose $y_n \in Y_n \cap M^m(G)$ for each n. Then $y \in M^m(G)$ because $M^m(G)$ is closed in $M(G)$, and from Lemma 3.4.6 (a) we have $y \in Y$. Thus $y \in Y \cap M^m(G)$. Consider any $y' \in Y \cap M^m G$. By Lemma 3.4.6 (b) and the free disposal condition (P2) it follows that there is a sequence $\{y'_n\}$ in $M^m(G)$ with $y'_n \in Y_n$ for each n and $y'_n \to y'$. Since $p_n \cdot z_n \to p \cdot z$ if $p_n \to p$ in $C(G)$ and $z_n \to z$ in $M(G)$, we obtain:

$$\text{If } p_n \to p \text{ in } C_{++}(G) \text{ and } (Y_n, y_n) \to (Y,y) \text{ in } \mathcal{Y}_\Gamma \times M(G) \text{ such that } (Y_n,y_n) \in T^m(p_n) \text{ for each } n, \text{ then } (Y,y) \in T^m(p). \tag{4.21}$$

Now let B be any non-empty compact subset of $C_{++}(G)$. There is a real number $a > 0$ such that $y \in M^m(G)$, $p \in B$, and $p \cdot y \geq 0$ together imply $\|y\|_V \leq a$. Fix such an a and let $A = \{z \in M(G) : \|z\|_V \leq a\}$. According to Fact 4.1.4, A is compact. On the other hand, by hypothesis, $0 \in Y$ for each $Y \in \mathcal{Y}_\Gamma$. Thus:

$$\text{If } p_n \to p \text{ in } C_{++}(G) \text{ then there is a compact subset } A \text{ of } M(G) \text{ such that } T^m(p_n) \subset \mathcal{Y}_\Gamma \times A \text{ for all } n. \tag{4.22}$$

Moreover, since each $Y \in \mathcal{Y}_\Gamma$ is closed in $M(G)$ (whence $Y \cap M^m(G) \cap A$ is compact):

$$\text{For each } p \in C_{++}(G) \text{ and every } Y \in \mathcal{Y}_\Gamma \text{ there is } y \in M(G) \text{ such that } (Y,y) \in T^m(p); \text{ moreover, } p \cdot y \geq 0 \text{ for such } y. \tag{4.23}$$

Now recall that $\mathcal{Y}_\Gamma$ is a compact metric space by construction (Convention 3.4.8) and note that compact subsets of $M(G)$ are metrizable (Fact 4.1.9). Hence (4.21), (4.22), and (4.23) together imply (d) and (e) of this lemma. Moreover, given any $p \in C_{++}(G)$, it follows from (4.21) and (4.22) that $T^m(p)$ is a closed subset of $\mathcal{Y}_\Gamma \times A_p$ for some compact subset A_p of $M(G)$ (hence $T^m(p)$ is closed also in $\mathcal{Y}_\Gamma \times M(G)$), and by Lemma 4.10.1 it follows from (4.23), then, that a probability measure η on $\mathcal{Y}_\Gamma \times M(G)$ with $\eta_Y = \varrho$ and $\operatorname{supp} \eta \subset T^m(p)$ exists. Using Lemma 4.10.2, we can then conclude from (4.21) and (4.22) that (a) and (c) of this lemma are true. (The convexity part in (a) follows from the definition of the Gelfand integral.) Finally, to see that (b) and (f) hold, fix any $p \in C_{++}(G)$, consider some $z \in \phi^m(p)$, and let η be a probability measure on $\mathcal{Y}_\Gamma \times M(G)$ for which $\int i_M \, d\eta_M = z$ and which satisfies the requirements in the definition of $\phi^m(p)$. In particular, then, $\operatorname{supp} \eta_M \subset M^m(G)$ and since $M^m(G)$ is convex and closed in $M(G)$ it follows from the separation theorem (and the definition of the Gelfand integral again) that $z \in M^m(G)$, which is (b). Also, for η–almost all $(Y, y) \in \mathcal{Y}_\Gamma \times M(G)$ we must have $p \cdot y = \pi^m(p, Y)$ whence $\int p \cdot y \, d\eta_M(y) = \int \pi^m(p, \cdot) \, d\varrho \equiv \pi^m(p, \cdot) \cdot \varrho$. That is, $p \cdot z = \pi^m(p, \cdot) \cdot \varrho$ and thus also (f) has been verified. ∎

Recall from Section 3.5 that

$$\pi(p, Y) = \sup \{p \cdot y : \ y \in Y\}, \quad p \in C(G), \ Y \in \mathcal{Y}_\Gamma .$$

Concerning the mappings $\pi(p, \cdot)$ we need the following lemma.

4.10.6. Lemma. *Let $p \in C(G)$. Then:*

(a) *$\pi(p, \cdot)$ is lower semi-continuous and takes values in $\mathbb{R}_+ \cup \{+\infty\}$. In particular, $\pi(p, \cdot) \cdot s \equiv \int \pi(p, \cdot) \, ds$ is well-defined for each $s \in S_\Gamma \equiv M_+(\mathcal{Y}_\Gamma)$.*

(b) *The mapping $s \mapsto \pi(p, \cdot) \cdot s$ is lower semi-continuous on S_Γ and takes values in $\mathbb{R}_+ \cup \{+\infty\}$. In particular, $\int \pi(p, \cdot) \cdot s \, d\nu_S(s)$ is well-defined.*

(c) *$\pi(p, \cdot) \cdot \varrho \equiv \int \pi(p, \cdot) \, d\varrho = \int \pi(p, \cdot) \cdot s \, d\nu_S(s)$.*

(d) *If $\{p_k\}$ is a sequence in $C_{++}(G)$ with $p_k \to p$ (which implies $p \in C_+(G)$), $\{s_k\}$ a sequence in S_Γ converging to some s, and $\{m_k\}$ an increasing sequence in $\mathbb{N}$, then $\underline{\lim}_k \pi^{m_k}(p_k, \cdot) \cdot s_k \geq \pi(p, \cdot) \cdot s$.*

Proof. Since each $Y \in \mathcal{Y}_\Gamma$ contains 0 it is clear that $\pi(p, \cdot)$ is nonnegative, and by Lemma 3.4.6 (b) it is clear that $\pi(p, \cdot)$ is lower semi-continuous. Thus (a) holds. In particular, since $\mathcal{Y}_\Gamma$ is a compact metric space by construction (Convention 3.4.8), we can choose an increasing sequence $\{h^i\}$ in $C_+(\mathcal{Y}_\Gamma)$ converging pointwise to $\pi(p, \cdot)$. Then $s \mapsto h^i \cdot s$ is continuous on S_Γ for each i, and (b) follows by the monotone convergence theorem. By the definition of an economy, $\varrho = \int i_S \, d\nu_S$. Thus $h^i \cdot \varrho = \int h^i \cdot s \, d\nu_S(s)$ for each i and it follows that (c) is true, again by using the monotone convergence theorem.

Finally, to see that (d) holds, let $\{p_k\}$ be a sequence in $C_{++}(G)$ with $p_k \to p$ and let $\{m_k\}$ be an increasing sequence in $\mathbb{N}$. Consider any subsequence $\{p_{k_n}\}$ of $\{p_k\}$. Note that if a sequence $\{y_n\}$ in $M(G)$ converges, say to y, then $\{y_n\}$ is $\|\cdot\|_V$–bounded and we must have $p_{k_n} \cdot y_n \to p \cdot y$ (Facts 4.1.8 and 4.1.6). Hence, by Lemma 3.4.6 (b), if $Y_n \to Y$ in $\mathcal{Y}_\Gamma$ then $\underline{\lim}_n \pi^{m_{k_n}}(p_{k_n}, Y_n) \geq \pi(Y, p)$. Since $\mathcal{Y}_\Gamma$ is a compact metric space, it follows that

$$\begin{aligned}&\textit{for each } h^i \textit{ and every number } \epsilon > 0 \textit{ there is } k_0 \in \mathbb{N} \textit{ such that}\\ &\textit{if } k \geq k_0 \textit{ then } \pi^{m_k}(p_k, Y) > h^i(Y) - \epsilon \textit{ for all } Y \in \mathcal{Y}_\Gamma,\end{aligned} \tag{4.24}$$

because each h^i is continuous and $\leq \pi(p, \cdot)$.

Now let $\{s_k\}$ be a sequence in S converging to some s. Then, for each i, $\lim_k h^i \cdot s_k = h^i \cdot s$ since $h^i \in C(\mathcal{Y}_\Gamma)$. Hence, by (4.24), $\underline{\lim}_k \pi^{m_k}(p_k, \cdot) \cdot s_k \geq h^i \cdot s$ for each i, whence $\underline{\lim}_k \pi^{m_k}(p_k, \cdot) \cdot s_k \geq \pi(p, \cdot) \cdot s$ by the monotone convergence theorem and the choice of the sequence $\{h^i\}$. Thus also (d) has been verified.

■

Next, we will consider the consumption sector of the economy. For each $m \in \mathbb{N}$ and each $p \in C_{++}(G)$ let

$$\begin{aligned} H^m(p) = \Big\{ (\succ, s, \omega, x) \in C_\Gamma \times M(G) : \\ x \in M_+(G), \quad p \cdot x \leq p \cdot \omega + \pi^m(p, \cdot) \cdot s, \\ \text{if} \quad x' \succ x \quad \text{then} \quad p \cdot x' > p \cdot \omega + \pi^m(p, \cdot) \cdot s \Big\} \end{aligned}$$

and

$$\varphi^m(p) = \left\{ z \in M(G) : \begin{array}{l} z = \int i_M \, d\tau_M \text{ for some probability measure } \tau \\ \text{on } C_\Gamma \times M(G) \text{ with } \tau_C = \nu \text{ and } \operatorname{supp} \tau \subset H^m(p) \end{array} \right\}.$$

The next lemma collects some facts about the aggregate demand sets $\varphi^m(p)$ which correspond to a given truncation in the production sector.

4.10.7. Lemma. *The following statements are true for all $m \in \mathbb{N}$:*

(a) *$\varphi^m(p)$ is non-empty and convex for each $p \in C_{++}(G)$.*

(b) *$\varphi^m(p) \subset M_+(G)$ for each $p \in C_{++}(G)$.*

(c) *If $p_i \to p$ in $C_{++}(G)$ and $z_i \in \varphi^m(p_i)$ for each i then a subsequence of $\{z_i\}$ converges to some $z \in \varphi^m(p)$.*

(d) *For all $p \in C_{++}(G)$, if $z \in \varphi^m(p)$ then $p \cdot z = p \cdot \int i_\Omega \, d\nu_\Omega + \pi^m(p, \cdot) \cdot \varrho$.*

Proof. Fix any m. Let $p_n \to p$ in $C_{++}(G)$ and let $\underline{a} = \inf\{p_n(g) : g \in G, n \in \mathbb{N}\}$ and $\bar{a} = \sup\{p_n(g) : g \in G, n \in \mathbb{N}\}$. Then $0 < \underline{a}$ and $\bar{a} < \infty$. Note that the definition of $\pi^m(p_n, Y)$ implies $\pi^m(p_n, Y) \leq \bar{a}m$ for all $Y \in \mathcal{Y}_\Gamma$ and all n. Hence, setting $\rho(c) = (1/\underline{a})(\bar{a}\omega(G) + \bar{a}ms(\mathcal{Y}_\Gamma))$ for each $c = (\succ, \omega, s) \in C_\Gamma$, we have $x(G) \leq \rho(c)$ for all $(c, x) \in H^m(p_n)$ and all n. Moreover, by the definition of an economy, $\rho(\cdot)$ is ν–integrable. Thus:

$$\text{If } \{p_n\} \text{ is a sequence in } C_{++}(G), \text{ converging to some } p \in C_{++}(G), \text{ then there is a } \nu\text{–integrable function } \rho : C_\Gamma \to \mathbb{R}_+ \text{ such that, for each } n, (c, x) \in H^m(p_n) \text{ implies } x(G) \leq \rho(c). \tag{4.25}$$

Now let, in addition, $(\succ_n, \omega_n, s_n, x_n) \to (\succ, \omega, s, x)$ in $C_\Gamma \times M(G)$ and assume $(\succ_n, \omega_n, s_n, x_n) \in H^m(p_n)$ for each n. Then $x \in M_+(G)$, $p_n \cdot x_n \to p \cdot x$, $p_n \cdot \omega_n \to p \cdot \omega$, and, by Lemma 4.10.5 (e), $\pi^m(p_n, \cdot) \cdot s_n \to \pi^m(p, \cdot) \cdot s$. Moreover, if $x' \succ x$ then $x' \succ_n x_n$ for n large enough (since the space of preferences is endowed with the topology of closed convergence). By a standard argument (using the continuity of $\succ$ or, in case that $p \cdot \omega + \pi^m(p, \cdot) \cdot s = 0$, the irreflexivity of $\succ$) we conclude that $(\succ, \omega, s, x) \in H^m(p)$. Thus:

$$\text{If } p_n \to p \text{ in } C_{++}(G) \text{ and } (c_n, x_n) \to (c, x) \text{ in } C_\Gamma \times M(G) \text{ such that } (c_n, x_n) \in H^m(p_n) \text{ for each } n, \text{ then } (c, x) \in H^m(p). \tag{4.26}$$

Furthermore, given any $c = (\succ \omega, s) \in C_\Gamma$ and any $p \in C_{++}(G)$, we have $p \cdot \omega + \pi^m(p, \cdot) \cdot s \geq 0$ (since $\pi^m(p, \cdot) \geq 0$ by Lemma 4.10.5 (d), and since $\omega \geq 0$ and $s \geq 0$) and it follows that the set $\{x \in M_+(G) : p \cdot x \leq p \cdot \omega + \pi^m(p, \cdot) \cdot s\}$ is non-empty and compact. Hence, by the assumptions made about preferences:

$$\text{For each } p \in C_{++}(G) \text{ and each } c \in C_\Gamma \text{ there exists an } x \in M_+(G) \text{ such that } (c, x) \in H^m(p). \tag{4.27}$$

Now note that $M_+(G)$ in the induced topology and, by construction, C_Γ are completely metrizable and separable. (See Fact 4.1.12 and Definition 3.4.12.) Thus, for any $p \in C_{++}(G)$, it follows from (4.26) that $H^m(p)$ is closed in $C_\Gamma \times M_+(G)$ (and hence closed in $C_\Gamma \times M(G)$) and then, from Lemma 4.10.1 and (4.27), that a probability measure τ on $C_\Gamma \times M(G)$ with $\tau_C = \nu$ and $\operatorname{supp} \tau \subset H^m(p)$ exists. Applying Lemma 4.10.2, we may then conclude from (4.25) and (4.26) that (a) and (c) of this lemma are true. (The convexity part in (a) follows from the definition of the Gelfand integral.) Finally, using the separation theorem (and the definition of the Gelfand integral again) we see that (b) is true. It remains to show that (d) holds. To this end, fix any $p \in C_{++}(G)$. From the definition of an economy, $\varrho = \int i_S \, d\nu_S$ and thus, since

$\pi^m(p,\cdot) \in C(\mathcal{Y}_\Gamma)$ by Lemma 4.10.5 (d), $\pi^m(p,\cdot)\cdot\varrho = \int \pi^m(p,\cdot)\cdot s\, d\nu_S(s)$. On the other hand, if τ is any probability measure on $C_\Gamma \times M(G)$ chosen according to the definition of $\varphi^m(p)$ then, by Lemma 4.10.8 (a) below, $p\cdot x = p\cdot\omega + \pi^m(p,\cdot)\cdot s$ for τ–almost all $(\succ,\omega,s,x) \in C_\Gamma \times M(G)$, whence $\int p\cdot x\, d\tau_M(x) = \int p\cdot\omega\, d\nu_\Omega(\omega) + \int \pi^m(p,\cdot)\cdot s\, d\nu_S(s)$. Thus (d) follows. ■

The final lemma is a consequence of the uniform properness condition embodied in the definition of an economy. Its proof uses in a decisive way the lattice structure of $C(G)$.

4.10.8. Lemma.

(a) *For all $m \in \mathbb{N}$ and all $p \in C_{++}(G)$, if $(\succ,\omega,s,x) \in H^m(p)$ then $p\cdot x = p\cdot\omega + \pi^m(p,\cdot)\cdot s$.*

(b) *There is a non-empty convex and compact subset B of $C_+(G)$ with $0 \notin B$ and with the following property: For each $m \in \mathbb{N}$ and each $p \in C_{++}(G)$ there is $q \in B$ and $\lambda > 0$ such that $\lambda q \le p$, $(\lambda q)\cdot x = p\cdot x$ for all $(c,x) \in H^m(p)$, and $(\lambda q)\cdot y \le p\cdot y$ for all $(Y,y) \in T^m(p)$.*

Proof. Let Γ be the properness cone from the economy under consideration and $\Gamma^0 = \{q \in C(G) :\ q\cdot z \ge 0,\ \forall z \in \Gamma\}$. Thus Γ is convex and open, $\Gamma \cap M_+(G) \ne \emptyset$, and (by the irreflexivity of preferences) $0 \notin \Gamma$. Fix any $v \in \Gamma \cap M_+(G)$ and then let $D = \{q \in \Gamma^0 :\ q\cdot v = 1\}$. Since the topology considered on $M(G)$ is the bw^*–topology it follows (by the bipolar theorem) that D is (norm) compact. Let $E = \{\lambda q :\ q \in D,\ 0 \le \lambda \le 1\}$ and $E^+ = \{q^+ : q \in E\}$. Then E is compact and by the continuity of the lattice operations in $C(G)$, E^+ is compact as well. According to Ascoli's theorem, a subset of $C(G)$ is relatively compact if and only if it is bounded and equicontinuous. Using this fact, it is straightforward to verify that $\sup A$ (the supremum of A) exists in $C(G)$ for every non-empty subset A of E^+ and, moreover, that there is a (norm) compact subset S of $C_+(G)$ such that for every such A one has $\sup A \in S$. (Cf. Theorem 12.29 in Aliprantis and Burkinshaw [2, p. 195].) Let $B' = \{q \in S :\ q\cdot v \ge 1\}$ and let B be the closed convex hull of B'. Then B' is compact and hence so is B. Moreover, $0 \notin B$.

Now fix any $m \in \mathbb{N}$ and any $p \in C_{++}(G)$. Assume (for the moment) that

for each $(c,x) \in H^m(p)$ there is $q_{c,x} \in \Gamma^0$ with $q_{c,x} \le p$ and $q_{c,x}\cdot x = p\cdot x$; moreover, $q_{c,x} \ne 0$ for some $(c,x) \in H^m(p)$; (4.28)

for each $(Y,y) \in T^m(p)$ there is $q_{Y,y} \in \Gamma^0$ with $q_{Y,y} \le p$ and $q_{Y,y}\cdot y^- = p\cdot y^-$. (4.29)

Let $A = H^m(p) \cup T^m(p)$ and then, for every $\alpha \in A$, write q_α instead of $q_{c,x}$ or $q_{Y,y}$, respectively. Thus $q_\alpha \cdot v \leq p \cdot v$ for all $\alpha \in A$, and $q_\alpha \cdot v > 0$ for some $\alpha \in A$. Hence, setting $\lambda = \sup\{q_\alpha \cdot v : \alpha \in A\}$ we have $0 < \lambda < \infty$. Let $\gamma = 1/\lambda$. Then $(\gamma q_\alpha)^+ \in E^+$ for all $\alpha \in A$. Hence $\sup\{(\gamma q_\alpha)^+ : \alpha \in A\}$ exists and it is easy to see that it belongs to B'. Clearly $\lambda \sup\{(\gamma q_\alpha)^+ : \alpha \in A\} = \sup\{q_\alpha^+ : \alpha \in A\}$. Moreover, since $q_\alpha^+ \leq p$ for all $\alpha \in A$, we have $\sup\{q_\alpha^+ : \alpha \in A\} \leq p$. Consider some $\alpha \in A$ and some $z \in M_+(G)$ and suppose $q_\alpha \cdot z = p \cdot z$. Then $q_\alpha \cdot z \leq q_\alpha^+ \cdot z \leq \sup\{q_\alpha^+ : \alpha \in A\} \cdot z \leq p \cdot z$ and hence $\sup\{q_\alpha^+ : \alpha \in A\} \cdot z = p \cdot z$. That is, for each $(c, x) \in H^m(p)$ we have $\sup\{q_\alpha^+ : \alpha \in A\} \cdot x = p \cdot x$, and for each $(Y, y) \in T^m(p)$ we have $\sup\{q_\alpha^+ : \alpha \in A\} \cdot y^- = p \cdot y^-$ whence $\sup\{q_\alpha^+ : \alpha \in A\} \cdot y \leq p \cdot y$. It thus remains to show that (4.28) and (4.29) hold and, besides, that (a) of the lemma is valid. To do so, we shall use the following fact (see [24, Lemma 2]):

> *Let E be a (real) vector space endowed with a Hausdorff, locally convex topology τ. Let A and B be convex subsets of E with A τ-open and such that $A \cap B \neq \emptyset$. Let $z \in B \cap c\ell\, A$ ($c\ell\, A$ denotes the τ-closure of A), let f be a linear functional on E and suppose $fz \leq fz'$, $\forall z' \in A \cap B$.* (4.30)
> *Then there exist linear functionals f_1 and f_2 on E such that f_1 is τ-continuous, $f_1 z \leq f_1 a$, $\forall a \in A$, $f_2 z \leq f_2 b$, $\forall b \in B$, and $f_1 + f_2 = f$.*

Let $c = (\succ, \omega, s, x) \in H^m(p)$. In particular, then, $x \in M_+(G)$. Since Γ is a properness cone for $\succ$, we have $(\{x\} + \Gamma) \cap M_+(G) \subset \{x' \in M_+(G) : x' \succ x\}$. Hence, by the definition of $H^m(p)$, $p \cdot x \leq p \cdot z$ for all $z \in (\{x\} + \Gamma) \cap M_+(G)$. Moreover, since $\Gamma \cap M_+(G) \neq \emptyset$ we have $(\{x\} + \Gamma) \cap M_+(G) \neq \emptyset$. Two points can be deduced. First, since Γ is a cone we have $p \cdot x = p \cdot \omega + \pi^m(p, \cdot) \cdot s$. Thus (a) of the lemma holds. Second, since $\{x\} + \Gamma$ is convex and open, and since x belongs to the closure of $\{x\} + \Gamma$, it follows from (4.30) that we can write $p = q_{c,x} + t_{c,x}$ with $t_{c,x}$ and $p_{c,x}$ elements of $C(G)$ satisfying $q_{c,x} \cdot x \leq q_{c,x} \cdot z$ for all $z \in \{x\} + \Gamma$ and $t_{c,x} \cdot x \leq t_{c,x} \cdot z$ for all $z \in M_+(G)$, respectively. The first inequality means $q_{c,x} \in \Gamma^0$, and from the latter we have $t_{c,x} \geq 0$ and $t_{c,x} \cdot x = 0$, whence $q_{c,x} \leq p$ and $q_{c,x} \cdot x = p \cdot x$. In particular, if $p \cdot x > 0$ then $q_{c,x} \neq 0$. By virtue of the definition of C_Γ, (4.27) in the proof of Lemma 4.10.7 guarantees that there are elements $(\succ, \omega, s, x)$ of $H^m(p)$ with $\omega \neq 0$. But for such elements one has $p \cdot \omega + \pi^m(p, \cdot) \cdot s > 0$ (note that $\pi^m(p, \cdot) \cdot s \geq 0$) and hence, by part (a) of this lemma, $p \cdot x > 0$. Thus (4.28) follows.

Next, consider any $(Y, y) \in T^m(p)$. Note that $y \in M^m(G)$. (Recall that $M^m(G) = \{z \in M(G) : z^+(G) \leq m\}$). Let $Z_y = \{z \in M(G) : z \leq y^+\}$. Since Γ is a properness cone for Y we have $(\{y\} - \Gamma) \cap Z_y \subset Y$. Since

$Z_y \subset M^m(G)$ we have, indeed, $(\{y\} - \Gamma) \cap Z_y \subset Y \cap M^m(G)$. Hence $p \cdot y \geq p \cdot z$ for all $z \in (\{y\} - \Gamma) \cap Z_y$ by the definition of $T^m(p)$. Moreover, since $\Gamma \cap M_+(G) \neq \emptyset$ we have $(\{y\} - \Gamma) \cap Z_y \neq \emptyset$. Note also that if t is any linear functional on $M(G)$ with $t \cdot y \geq t \cdot z$ for all $z \in Z_y$ then $t \geq 0$ (because $-M_+(G) \subset Z_y$) and, therefore, also $t \cdot y^- = 0$ (because $y^+ \in Z_y$). Again using (4.30), it follows that there exists a $q_{Y,y} \in \Gamma^0$ such that $q_{Y,y} \leq p$ and $q_{Y,y} \cdot y^- = p \cdot y^-$. Thus (4.29) holds. ∎

Completion of the proof of Theorem 3.5.6.

To simplify the notation, we write $\bar{\omega}$ for $\int i_\Omega \, d\nu_\Omega$ in the following.

For every $m \in \mathbb{N}$ and every $p \in C_{++}(G)$ let

$$Z^m(p) = \varphi^m(p) - \phi^m(p) - \{\bar{\omega}\}.$$

As noticed in Lemma 4.10.7 (b),

$$\varphi^m(p) \subset M_+(G) \quad \text{for each } p \in C_{++}(G) \text{ and all } m, \tag{4.31}$$

and as noticed in Lemma 4.10.5 (b), $\phi^m(p) \subset M^m(G)$ for each $p \in C_{++}(G)$ and all m (recall: $M^m(G) = \{z \in M(G): \ z^+(G) \leq m\}$). Combining these two statements shows:

For every m there is a real number b_m such that $z^-(G) \leq b_m$ for all $z \in Z^m(p)$ and all $p \in C_{++}(G)$. (4.32)

Furthermore, combining Lemma 4.10.5 (f) and Lemma 4.10.7 (d), we see that

$$p \cdot z = 0 \quad \text{for all } z \in Z^m(p), \text{ each } p \in C_{++}(G), \text{ and each } m. \tag{4.33}$$

Next, we claim:

For every $m \in \mathbb{N}$, given any non-empty compact and convex subset A of $C_{++}(G)$, there is $\bar{p} \in A$ and $\bar{z} \in Z^m(\bar{p})$ such that $p \cdot \bar{z} \leq 0$ for all $p \in A$. (4.34)

To prove this claim, fix any m and let $A \subset C_{++}(G)$ be as hypothesized. From Lemma 4.10.5 (c) and Lemma 4.10.7 (c), if $p_i \to p$ in A and $z_i \in Z^m(p_i)$ for each i then a subsequence of $\{z_i\}$ converges to some $z \in Z^m(p)$. Since compactness in $C(G)$ implies sequential compactness, and since a converging sequence in $M(G)$ must be $\|\cdot\|_V$-bounded, it follows that there is a compact and convex subset Z of $M(G)$ such that $Z^m(p) \subset Z$ for all $p \in A$. Moreover, noting that Z (in the induced topology) is metrizable, and noting that the mapping $(p, z) \mapsto p \cdot z$ is continuous on $A \times Z$, it follows that the correspondence $\Psi : A \times Z \to 2^{A \times Z}$ given by

$$\Psi(p, z) = \{p' \in A : \quad p' \cdot z \geq p'' \cdot z, \ \forall p'' \in A\} \times Z^m(p)$$

has a closed graph. Consider any $(p, z) \in A \times Z$. Since A is compact and convex, the set $\{p' \in A : p' \cdot z \geq p'' \cdot z, \forall p'' \in A\}$ is non-empty and convex, and from Lemma 4.10.5 (a) and Lemma 4.10.7 (a), so is $Z^m(p)$. By the Fan–Glicksberg fixed point theorem, we conclude that Ψ has a fixed point, say $(\bar{p}, \bar{z})$. In view of (4.33), it is evident that $(\bar{p}, \bar{z})$ satisfies the requirements in (4.34).

Now consider $Y^a(\varrho)$, the aggregate production set of the economy under consideration. From its definition (and the definition of the Gelfand integral) it follows that it is convex, and according to assumption (A3) it is closed. Since $-M_+(G) \subset Y^a(\varrho)$ according to assumption (A2), using the separation theorem we deduce that

$$\text{if} \quad y \in Y^a(\varrho) \quad \text{and} \quad z \in M_+(G) \quad \text{then} \quad y - z \in Y^a(\varrho), \tag{4.35}$$

and since $Y^a(\varrho) \cap M_+(G) = \{0\}$ according to assumption (A1), we also get the following fact (using the compactness of $\{x \in M_+(G) : \|x\|_V = 1\}$):

Let $\{x_i\}$ be a sequence in $M_+(G)$ and $\{y_i\}$ a sequence in $Y^a(\varrho)$. For each i let $z_i = x_i - y_i - \bar{\omega}$. If the sequence $\{z_i\}$ is $\|\cdot\|_V$-bounded, in particular if $\{z_i\}$ is convergent, then $\{x_i\}$ and $\{y_i\}$ are $\|\cdot\|_V$-bounded (and hence have convergent subsequences according to Fact 4.1.5). (4.36)

Besides, directly from the definition of $Y^a(\varrho)$ and $\phi^m(p)$, we have

$$\phi^m(p) \subset Y^a(\varrho) \quad \text{for each } m \text{ and each } p \in C_{++}(G); \tag{4.37}$$

Now let us fix a subset B of $C_+(G)$, constructed according to Lemma 4.10.8 (b). In particular, B is compact and convex and we have $0 \notin B$. Moreover, given any m and any $p \in C_{++}(G)$, there is $q \in B$ and $\lambda > 0$ such that (a) $(\lambda q) \leq p$, (b) if η is as in the definition of $\phi^m(p)$ then $(\lambda q) \cdot y \leq p \cdot y$ for η_M-almost all $y \in M(G)$, (c) if τ is as in the definition of $\varphi^m(p)$ then $(\lambda q) \cdot x = p \cdot x$ for τ_M-almost all $x \in M(G)$. That is, if $x' \in \varphi^m(p)$ then $(\lambda q) \cdot x' = p \cdot x'$ and if $y' \in \phi^m(p)$ then $(\lambda q)y' \leq p \cdot y'$. Besides, $(\lambda q) \cdot \bar{\omega} \leq p \cdot \bar{\omega}$ since $(\lambda q) \leq p$. Thus, if $z \in Z^m(p)$ then $(\lambda q) \cdot z \geq p \cdot z$ whence, from (4.33), $q \cdot z \geq 0$. We thus have:

For each m, each $p \in C_{++}(G)$, and each $z \in Z^m(p)$ there is $q \in B$ with $q \cdot z \geq 0$. (4.38)

Next, let us choose a sequence $\{d_i\}$ in $C_+(G)$, converging to 0 and such that, given any $z \in M(G)$, $d_i \cdot z \leq 0$ for every i implies $z \leq 0$. Since $C(G)$ (and hence $C_+(G)$) is separable, such a sequence certainly exists. Clearly, we can assume

$d_1 = 1_G$. Let D be the closed convex hull of the set $\{d_i : i \in \mathbb{N}\}$. Then D is compact. Let $\Delta = B + D$, and for each $n \in \mathbb{N}$ let $\Delta^n = \Delta + \{(1/n)1_G\}$. Then Δ and Δ^n for each n are compact and convex. Moreover, since $B \subset C_+(G)$, we have $\Delta \subset C_+(G)$ and $\Delta^n \subset C_{++}(G)$. Besides, since $0 \notin B$, we have $0 \notin \Delta$.

Temporarily fix any $m \in \mathbb{N}$. According to (4.34), for each n we can choose $p_n^m \in \Delta^n$ and $z_n^m \in Z^m(p_n^m)$ such that $p \cdot z_n^m \leq 0$ for all $p \in \Delta^n$. By (4.38) we can then choose $q_n^m \in B$ with $q_n^m \cdot z_n^m \geq 0$. By construction, we have $q_n^m + (1 + 1/n)1_G \in \Delta^n$, whence $z_n^m(G) \leq 0$, and thus (4.32) implies that the sequence $n \mapsto z_n^m$ is $\|\cdot\|_V$-bounded. For each n choose an $x_n^m \in \varphi^m(p_n^m)$ and a $y_n^m \in \phi^m(p_n^m)$ such that $x_n^m - y_n^m - \bar{\omega} = z_n^m$. Letting $n \to \infty$, glancing at (4.31) and (4.37), applying (4.36), and passing to subsequences if necessary, we can assume $x_n^m \to x^m$ for some $x^m \in M_+(G)$ and $y_n^m \to y^m$ for some $y^m \in Y^a(\varrho)$ (recalling that $Y^a(\varrho)$ is closed). Let $z^m = x^m - y^m - \bar{\omega}$. Thus $z_n^m \to z^m$ as $n \to \infty$. Besides, since B is compact, we can assume $q_n^m \to q^m$ for some $q^m \in B$ as $n \to \infty$. In particular, $q_n^m \cdot z_n^m \to q^m \cdot z^m$ as $n \to \infty$, whence $q^m \cdot z^m \geq 0$. Consider any $p' \in D$. Then $(q^m + p' + (1/n)1_G) \cdot z_n^m \leq 0$ for each n, hence, letting $n \to \infty$, we have $(q^m + p')z^m \leq 0$, whence $p' \cdot z^m \leq 0$. Thus, by the properties of D, $z^m \leq 0$, i.e., $x^m - y^m - \bar{\omega} \leq 0$, hence $x^m \in Y^a(\varrho) + \{\bar{\omega}\}$ by virtue of (4.35). Letting now $m \to \infty$, another application of (4.36) shows that the sequence $\{x^m\}$ has a subsequence, say $\{x^{m_k}\}$, which converges to some $\bar{x} \in M_+(G)$. In particular, then, $\bar{x} \in Y^a(\varrho) + \{\bar{\omega}\}$ since $Y^a(\varrho)$ is closed.

According to Fact 4.1.12, $M_+(G)$ is metrizable in the induced topology. Let d_{M_+} be a corresponding metric. By the preceding constructions, for each $k \in \mathbb{N}$ we can choose a pair (p^k, x^k) where $p^k \in \Delta^n$ for some $n \in \mathbb{N}$ and $x^k \in \varphi^{m_k}(p^k)$ and so that $d_{M_+}(x^k, x^{m_k}) \leq 1/k$ (where $m_k \in \mathbb{N}$ is as above, i.e. so that $x^{m_k} \to \bar{x}$ as $k \to \infty$). Then $x^k \to \bar{x}$ as $k \to \infty$. Moreover, for each k we have $p^k = \hat{p}^k + (1/n)1_G$ for some $\hat{p}^k \in \Delta$ and some $n \in \mathbb{N}$. Thus, since Δ is compact, we can assume $p^k \to \bar{p}$ for some $\bar{p} \in C(G)$ and since $\Delta \subset C_+(G)$ and, as noted above, $0 \notin \Delta$, we must have $\bar{p} \in C_+(G)$ and $\bar{p} \neq 0$.

By construction, for each k we have $x^k = \int i_M \, d\tau_M^k$ for some probability measure τ^k on $C_\Gamma \times M(G)$ with $\operatorname{supp} \tau^k \subset H^{m_k}(p^k)$ and $\tau_C^k = \nu$. Hence, by using Lemma 4.10.3, we can find a probability measure τ on $C_\Gamma \times M(G)$ such that (i) $\tau_C = \nu$; (ii) $\int i_M \, d\tau_M$ exists and $\int i_M \, d\tau_M \leq \bar{x}$; and (iii) (passing to a subsequence of $\{(p^k, x^k, \tau^k)\}$ and appropriately relabeling terms if necessary):

for each $(c, x) \in \operatorname{supp} \tau$ *there is a sequence* $\{(c^k, z^k)\}$ *in* $C_\Gamma \times M(G)$ *with* $(c^k, z^k) \to (c, x)$ *and* $(c^k, z^k) \in H^{m_k}(p^k)$ *for each* k. (4.39)

From (4.39) we obtain, using Lemma 4.10.6 (d) and Lemma 4.10.8 (a):

$$\text{if} \quad (\succ, \omega, s, x) \in \operatorname{supp} \tau \quad \text{then} \quad x \in M_+(G) \\ \text{and} \quad \bar{p} \cdot x \geq \bar{p} \cdot \omega + \pi(\bar{p}, \cdot) \cdot s, \tag{4.40}$$

and, since the space of preferences is endowed with the topology of closed convergence:

$$\text{if} \quad (\succ, \omega, s, x) \in \operatorname{supp} \tau \quad \text{then} \quad x' \succ x \quad \text{implies} \\ \bar{p} \cdot x' \geq \bar{p} \cdot \omega + \pi(\bar{p}, \cdot) \cdot s. \tag{4.41}$$

(According to Lemma 4.10.6 (a), $\pi(\bar{p}, \cdot)$ is a lower semi-continuous, nonnegative extended real-valued function, so $\pi(\bar{p}, \cdot) \cdot s$ for each $s \in S$ and, below, $\pi(\bar{p}, \cdot) \cdot \varrho$ are well-defined.)

Now let $\bar{y} = \int i_M \, d\tau_M - \bar{\omega}$. Since $\int i_M \, d\tau_M \leq \bar{x}$ and $\bar{x} \in Y^a(\varrho) + \{\bar{\omega}\}$ it follows from (4.35) that $\bar{y} \in Y^a(\varrho)$. By the definition of $Y^a(\varrho)$, we can choose a probability measure η on $\mathcal{Y}_\Gamma \times M(G)$ such that $\bar{y} = \int i_M \, d\eta_M$, $\eta_Y = \varrho$, and such that $y \in Y$ for η–almost all $(Y, y) \in \mathcal{Y}_\Gamma \times M(G)$. In particular, $\bar{p} \cdot y \leq \pi(\bar{p}, Y)$ for η–almost all $(Y, y) \in \mathcal{Y}_\Gamma \times M(G)$, so that

$$\bar{p} \cdot \bar{y} = \int \bar{p} \cdot y \, d\eta_M(y) \leq \int \pi(\bar{p}, Y) \, d\varrho(Y) \equiv \pi(\bar{p}, \cdot) \cdot \varrho. \tag{4.42}$$

From Lemma 4.10.6 (c), $\pi(\bar{p}, \cdot) \cdot \varrho = \int \pi(\bar{p}, \cdot) \cdot s \, d\nu_S(s)$. Since $\int i_M \, d\tau_M = \bar{y} + \bar{\omega}$ we can conclude with the aid of (4.42) (recalling that $\bar{\omega}$ stands for $\int i_\Omega \, d\nu_\Omega$):

$$\begin{aligned} \int \bar{p} \cdot x \, d\tau_M(x) &= \bar{p} \cdot \bar{\omega} + \bar{p} \cdot \bar{y} \\ &\leq \bar{p} \cdot \bar{\omega} + \pi(\bar{p}, \cdot) \cdot \varrho \\ &= \int \bar{p} \cdot \omega \, d\nu_\Omega(\omega) + \int \pi(\bar{p}, \cdot) \cdot s \, d\nu_S(s). \end{aligned}$$

Hence, by (4.40), $\bar{p} \cdot x = \bar{p} \cdot \omega + \pi(\bar{p}, \cdot) \cdot s$ for τ–almost all $(\succ, \omega, s, x) \in C_\Gamma \times M(G)$. Thus $\int \bar{p} \cdot x \, d\tau_M(x) = \int \bar{p} \cdot \omega \, d\nu_\Omega(\omega) + \int \pi(\bar{p}, \cdot) \cdot s \, d\nu_S(s)$. Consequently, by the preceding calculation, $\pi(\bar{p}, \cdot) \cdot \varrho = \bar{p} \cdot \bar{y}$ whence $\bar{p} \cdot y = \pi(\bar{p}, Y)$ for η–almost all $(Y, y) \in \mathcal{Y}_\Gamma \times M(G)$. Thus, in view of (4.41), $(\bar{p}, \eta, \tau)$ constitutes a quasi-equilibrium for the economy under consideration, and the proof of Theorem 3.5.6 is finished. ■

4.11. Proof of Theorem 3.5.7

Let (p, η, τ) be a quasi-equilibrium for the economy (ϱ, ν). In particular, then, $p \in C(G)$ and $p \neq 0$. If $p(g) < 0$ for some $g \in G$, it follows immediately that

(p, η, τ) is an equilibrium since all consumption sets are equal to $M_+(G)$ and since preferences are continuous. Thus suppose $p \in C_+(G)$. Let

$$E = \{(\succ, \omega, s) \in C_\Gamma : \quad p \cdot \omega + \pi(p, \cdot) \cdot s > 0\}.$$

From the definition of a quasi-equilibrium it follows:

For τ-almost all $(\succ, \omega, s, x) \in C_\Gamma \times M(G)$ we have:
$$p \cdot x \leq p \cdot \omega + \pi(p, \cdot) \cdot s; \tag{4.43}$$
if $(\succ, \omega, s) \in E$ then $x' \succ x$ implies $p \cdot x' > p \cdot \omega + \pi(p, \cdot) \cdot s$.

(The latter again from the hypothesis that consumption sets are equal to $M_+(G)$ and that preferences are continuous.) Also from the definition of a quasi-equilibrium,

$$\nu(E) = \tau(E \times M(G)). \tag{4.44}$$

We thus have to show that $\nu(E) = 1$. (Using Lemma 4.10.6 (b), one easily verifies that E is a Borel subset of C_Γ so $\nu(E)$ is well-defined.)

Let $y = \int i_M \, d\eta_M$ and consider any $y' \in Y^a(\varrho)$. From the definition of $Y^a(\varrho)$ we have $p \cdot y' \leq \pi(p, \cdot) \cdot \varrho$. On the other hand, from the definition of a quasi-equilibrium again, $p \cdot y = \pi(p, \cdot) \cdot \varrho$. Thus

$$p \cdot y \geq p \cdot y' \qquad \text{for all} \quad y' \in Y^a(\varrho). \tag{4.45}$$

It follows, in particular, that $\nu(E) > 0$. Indeed, since $p \in C_+(G)$ and $p \neq 0$, (4.45) and (A4) imply $p \cdot y + p \cdot \int i_\Omega \, d\nu_\Omega > 0$. Hence, by Lemma 4.10.6 (c), we have $\int p \cdot \omega \, d\nu_\Omega(\omega) + \int \pi(p, \cdot) \cdot s \, d\nu_S(s) > 0$ whence $\nu(E) > 0$.

Now suppose that (A5) holds and let G_D and G_P be as hypothesized in this assumption. Since $\nu(E) > 0$ it follows from (a) of (A5), (4.43), and (4.44) that $p(g) > 0$ for each $g \in G_D$. But therefore, from (b) of (A5) and (4.45), for each $g \in G_P$ we have $p(g) > 0$ as well and it follows from (c) of (A5), then, that $\nu(E) = 1$ because $p \geq 0$ and $\pi(p, \cdot) \geq 0$ (the latter since $0 \in Y$ for each $Y \in \mathcal{Y}_\Gamma$).

Finally, assume that (A6) holds. Let $A = C_\Gamma \setminus E$ and suppose $\nu(A) > 0$. Let $G_A = \bigcup_{(\succ, \omega, s) \in A} \operatorname{supp} \omega$. Then, since $p \in C_+(G)$ and $\pi(p, \cdot) \geq 0$, the definition of A implies

$$p(g) = 0 \qquad \text{for all} \quad g \in G_A. \tag{4.46}$$

Choose elements z and u of $M_+(G)$ so that u satisfies (a) of (A6), i.e. so that $\operatorname{supp} u \subset G_A$, and so that (b) and (c) of (A6) are fulfilled with respect to y and τ. Since $p \geq 0$ and $\nu(E) > 0$, it follows from (c) of (A6), (4.43), and (4.44) that $p \cdot z > 0$. But then, from (b) of (A6) and (4.45), $p \cdot u > 0$. Hence, since $\operatorname{supp} u \subset G_A$, we must have $p(g) > 0$ for some $g \in G_A$. However, this contradicts (4.46) and we conclude that $\nu(E) = 1$, as desired. The proof of Theorem 3.5.7 is complete. ■

4.12. Proof of Theorem 3.5.9

If $\int i_\Omega \, d\nu_\Omega + \bar{y} \in \Gamma \cap M_+(G)$ for some $\bar{y} \in Y^a(\varrho)$, we can construct the set B of Lemma 4.10.8 (b) by choosing $\int i_\Omega \, d\nu_\Omega + \bar{y}$ to play the role of v in the proof of that lemma. Then $p \cdot (\int i_\Omega \, d\nu_\Omega + \bar{y}) \geq 1$ for all $p \in B$ and so the proof of Theorem 3.5.6 yields a quasi-equilibrium (p, η, τ) with $p \cdot \int i_\Omega \, d\nu_\Omega + p \cdot \bar{y} \geq 1$. By the argument which has led to (4.45) above, $p \cdot \int i_M \, d\eta_M \geq p \cdot \bar{y}$ whence, from Lemma 4.10.6 (c), $\int p \cdot \omega \, d\nu_\Omega(\omega) + \int \pi(p, \cdot) \cdot s \, d\nu_S(s) > 0$. As in the proof of Theorem 3.5.7, we can then conclude from (A5) or (A6) that (p, η, τ) is, in fact, an equilibrium. ■

References

1. C. D. Aliprantis and D. J. Brown, Equilibria in markets with a Riesz space of commodities, *Journal of Mathematical Economics* **11** (1983), 189–207.

2. C. D. Aliprantis and O. Burkinshaw, *Positive operators*, Pure an Applied Mathematics Series No. 119, Academic Press, London Orlando, 1985.

3. K. J. Arrow and G. Debreu, Existence of an equilibrium for a competitive economy, *Econometrica* 22 (1954), 265–290.

4. J. Diestel and J. J. Uhl, *Vector measures*, Mathematical Surveys No. 15, American Mathematical Society, Rhode Island, 1977.

5. I. Fradera, Perfect competition with product differentiation, *International Economic Review* 27 (1986), 529–538.

6. O. Hart, Monopolistic competition in a large economy with commodity differentiation, *Review of Economic Studies* **46** (1979), 1–30.

7. O. Hart, Monopolistic competition in a large economy with commodity differentiation: a correction, *Review of Economic Studies* **49** (1982), 313–314.

8. O. Hart, Monopolistic competition in the spirit of Chamberlin: a general model, *Review of Economic Studies* **52** (1985), 529–546.

9. O. Hart, Monopolistic competition in the spirit of Chamberlin: special results, *Economic Journal* **95** (1985), 889–908.

10. W. Hildenbrand, *Core and equilibria of a large economy*, Princeton University Press, Princeton, 1974.

11. R. Holmes, *Geometrical functional analysis and its applications*, Springer, New York Heidelberg Berlin, 1975.

12. L. Jones, Existence of equilibria with infinitely many consumers and infinitely many commodities: a theorem based on models of commodity differentiation, *Journal of Mathematical Economies* 12 (1983), 119–138.

13. L. Jones, A competitive model of commodity differentiation, *Econometrica* 52 (1984), 507–530.

14. L. Jones, The efficiency of monopolistically competitive equilibria in large economies: commodity differentiation with gross substitutes, *Journal of Economic Theory* 41 (1987), 356–391.

15. A. Mas-Colell, A model of equilibrium with differentiated commodities, *Journal of Mathematical Economics* 2 (1975), 263–295.

16. A. Mas-Colell, The price equilibrium problem in topological vector lattices, *Econometrica* 54 (1986), 1039–1053.

17. A. Mas-Colell, Valuation equilibrium and Pareto optimum revisited, in: *Contributions to mathematical economics* (eds. W. Hildenbrand and A. Mas-Colell), North–Holland, New York, 1986.

18. J. Ostroy, The existence of Walrasian equilibrium in large–square economies, *Journal of Mathematical Economics* 13 (1984), 143–163.

19. J. Ostroy and W. R. Zame, Non-atomic economies and the boundaries of perfect competition, *Econometrica* 62 (1994), 593–633.

20. M. Pascoa, Noncooperative equilibrium and Chamberlinian monopolistic competition, *Journal of Economic Theory* 60 (1993), 335–353.

21. M. Pascoa, Monopolistic competition and non-neighboring–goods, *Economic Theory* 9 (1997), 129–142.

22. K. Podczeck, Walrasian equilibria in large production economies with differentiated commodities, University of Vienna, Working Paper, 1985.

23. K. Podczeck, General equilibrium with differentiated commodities: the linear activity model without joint production, *Economic Theory* 2 (1992), 247–263.

24. K. Podczeck, Equilibria in vector lattices without ordered preferences or uniform properness, *Journal of Mathematical Economics* 25 (1996), 465–485.

25. S. Richard, Competitive equilibria in Riesz spaces, Carnegie–Mellon University, Pittsburgh, Working Paper, 1986.

26. S. Richard, A new approach to production equilibria in vector lattices, *Journal of Mathematical Economics* 18 (1989), 41–56.

27. L. Schwartz, *Radon measures on arbitrary topological spaces and cylindrical measures*, Oxford University Press, London, 1973.

28. V. S. Varadarajan, Weak convergence of measures on separable metric spaces, *Sankhya* 19 (1958), 15–22.

29. N. C. Yannelis, On a market equilibrium theorem with an infinite number of commodities, *Journal of Mathematical Analysis and its Applications* 108 (1985), 595–599.

30. N. C. Yannelis and W. R. Zame, Equilibria in Banach lattices without ordered preferences, Preprint No. 71, Institute for Mathematics and its Applications, University of Minnesota (1984). A shortened version has appeared in *Journal of Mathematical Economics* 15, 85–110.

31. W. R. Zame, Markets with a continuum of traders and infinitely many commodities, SUNY at Buffalo, Working Paper, 1986.

32. W. R. Zame, Equilibria in production economies with an infinite dimensional commodity space, *Econometrica* 55 (1987), 1075–1108.

Konrad Podczeck
Institut für Wirtschaftswissenschaften
Universität Wien
Hohenstaufengasse 9
A-1010 Wien
Austria
email: konrad.podczeck@univie.ac.at

AN EXACT IMPLEMENTATION OF THE NASH BARGAINING SOLUTION IN DOMINANT STRATEGIES

WALTER TROCKEL*

Abstract. For any abstract bargaining problem a non-cooperative one stage strategic game is constructed whose unique dominant strategies Nash equilibrium implements the Nash solution of the bargaining problem.

1. Introduction

After his seminal paper [8] where he introduced his bargaining solution Nash [9] also suggested to base cooperative axiomatic solutions on equilibria of non-cooperative strategic games. This idea is known as the "Nash program" in the game theoretic literature. Thorough discussions of the Nash program are contained in [1] and [10]. Extensive treatments of axiomatic bargaining theory are [12] and [17].

Nash [9] himself started with a contribution to the Nash program by introducing what is called today the *simple Nash demand* game. The continuum of equilibria of this game, which coincides with the efficient boundary of the bargaining problem, is not suited to single out the Nash solution. A modification, also due to Nash, with informational trembling which anticipated features of Selten's [15] perfectness resulted in an approximate implementation of the Nash solution. The most prominent later contribution to the Nash program is Rubinstein's [13] alternate bargaining model in which the Nash solution turns out to be implementable as a limit of perfect equilibria in a sequence of certain multi–stage games. As [1] remarks, "none of the non-cooperative bargaining models who have been studied implement the Nash bargaining solution *exactly*. In each case the implementation is *approximate* (or exact only in the limit)."

An *exact* implementation of the Nash solution by a subgame perfect equilibrium of an extensive game has been provided recently by Howard [6]. His game has been reproduced in [11].

In the present paper we suggest a different non-cooperative approach to the bargaining problem, which is based on Shapley's [16] idea of endogenously determined transfer rates between players' utilities. Here a bargaining problem is viewed as a specific Arrow–Debreu economy with production and private ownership (cf. [18]). It turns out that the unique (stable) Walrasian equilibrium

*Helpful discussions with Bernd Korthues, Jörg Naeve and particularly with Till Requate are gratefully acknowledged.

of this economy coincides with the Nash solution of the bargaining problem. It is this competitive approach to the bargaining problem which then suggests a modification of Nash's simple demand game where the payoff function is derived from demand and supply in the economy. The unique Nash equilibrium, which is even an equilibrium in dominant strategies, implements *exactly* the Nash bargaining solution.

2. Motivation and description of the game

To introduce and discuss our game we first recall how Nash described a bargaining situation.

A two-person bargaining problem is a pair (S, d) where the *feasible set* S is a convex subset of $\mathbb{R}^2$ and d is an element of S. The point d is interpreted as the *status quo* or *disagreement point*, while each point in S is interpreted as a feasible pair of utility levels of the two players. Implicitly this interpretation is based on the idea of some underlying unspecified economic or social model whose different states evaluated by the two players' cardinal utility functions result in the elements of S. In the most frequent interpretation one thinks of von Neumann–Morgenstern utility functions in a lottery setup.

It was Nash [9] himself who tried to support his axiomatic bargaining solution by a strategic non-cooperative model. In his static model both players announce "demands" on utility levels for themselves. If these are compatible, i.e., if the pair of announced demands builds an element of S, each player receives the amount of utility he demanded. Otherwise the disagreement point results. The continuum of Nash equilibria of this *simple demand game* contains all efficient elements of S as well as the disagreement point.

Nash extended his analysis to a more elaborate version where he considered *perturbed demand games.* Anticipating features of Selten's [15] perfectness Nash singled out those equilibria which are robust with respect to specific perturbations reflecting some uncertainty about the outcome. The set of those equilibria of the perturbed games generating agreement with positive probability is the smaller the smaller the perturbation is and converges to the Nash solution of (S, d).

In the present paper we consider quite a different modification of Nash's simple demand game. We think of S as of a technology set representing all feasible production plans where pairs of utility levels for the two players are the two commodities which may be produced jointly without any input. (This assumption might appear confusing but it can be easily removed by introducing a third input commodity without direct utility to the players). Imagine now that all points y in the efficient boundary ∂S are evaluated by their associated efficiency prices $p(y)$, which are assumed to be normalized by $p(y) \cdot y = 1$. For any efficient joint production plan $y \in \partial S$ half of its value $\frac{1}{2}p(y) \cdot y$ is made available to each player. He may choose now his *demand*, i.e., the optimal bundle in his budget set. Considered as an agent of an economy, each player is interested only in one of the two commodities, namely his own utility. Accordingly, both players' preferences are described by indifference lines parallel to their axes. Their optima are the corner solutions $\left(\frac{1}{2p_1(y)}, 0\right)$ for player 1 and $\left(0, \frac{1}{2p_2(y)}\right)$ for player 2. The possible conflict between the proposed supply vector $y \in \partial S$ and the resulting aggregate demand $\left(\frac{1}{2p_1(y)}, \frac{1}{2p_2(y)}\right)$ determines our equilibrium approach and is the key to the subsequent strategic approach.

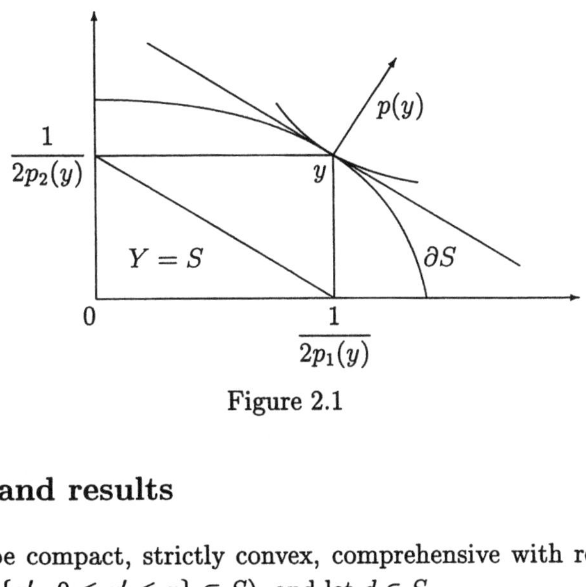

Figure 2.1

3. Model and results

Let $S \subset \mathbb{R}^2$ be compact, strictly convex, comprehensive with respect to $\mathbb{R}^2_+$ (i.e. $x \in S \Rightarrow \{x' : 0 \leq x' \leq x\} \subset S$), and let $d \in S$.

The pair (S, d) is called a *two–person bargaining problem* if there is some $x \in S$ such that $x \gg d$. ($x \gg x' \Leftrightarrow \forall i = 1,2 \quad x_i > x'_i; \quad x \geq x' \Leftrightarrow \forall i = 1,2 \quad x_i \geq x'_i; \quad x > x' \Leftrightarrow (x \geq x'$ and $x \neq x'))$.

Having cardinality in mind we may and do assume $d = 0$ without loss of generality. Let $\mathcal{B}$ denote the set of all two–person bargaining problems $(S, 0)$. A *bargaining solution* is a map

$$\varphi : \mathcal{B} \to \mathbb{R}^2 \qquad \text{such that} \qquad (S, 0) \mapsto \varphi(S, 0) \in S.$$

The *Nash bargaining solution* φ_N, originally defined by Nash [8] through some axioms, has been characterized (also by Nash) by

$$(S,0) \overset{\varphi_N}{\longmapsto} \arg\max_{x\in S} x_1 \cdot x_2,$$

i.e., by associating with each $(S,0)$ the point where the so called "Nash product" becomes maximal.

Next we consider any bargaining problem $(S,0)$ as a specific *Arrow–Debreu economy* with production and private ownership.

Given $(S,0)$, consider an economy $\mathcal{E}_S$ defined as follows: $\mathcal{E}_S = (\succsim_1, \succsim_2;\ e_1, e_2;\ \vartheta_1, \vartheta_2;\ Y)$ with production possibility set $Y \subset \mathbb{R}^2$, preferences $\succsim_i$, endowment vectors e_i, and ownership shares in production ϑ_i, $i = 1, 2$. Specifically we define $Y := S$, $e_1 = e_2 = (0,0)$, $\vartheta_1 = \vartheta_2 = \frac{1}{2}$, and $\succsim_i$ by $x \succ_i x' \Leftrightarrow x_i > x'_i$, $i = 1, 2$. The interpretation of this economy is as follows.

Two agents jointly own a firm which can produce joint utilities for the two agents. Each agent is interested only in one of the two commodities, namely his utility. The only income results from shares in profit from production. The confusing property of production without inputs could be easily removed by introduction of a third (input) commodity, a fixed amount of which is required for production, consumption of which, however, does not give utilities to the players.

An *equilibrium* of the economy is characterized as follows: Take a point y in the efficient boundary ∂Y and evaluate it by its *efficiency price* (a normal to ∂Y at y). Make half of the resulting value available as *income* to each player. Determine both players' *individual demands* and add them up to get the *aggregate demand.* This aggregate demand has to coincide with y.

Without loss of generality we assume that all efficiency prices $p(y)$ for $y \in \partial Y$ are normalized by $p(y) \cdot y = 1$. This normalization, which has been motivated in [2], will be notationally convenient.

Note that for any efficient production plan $y \in \partial Y$ both players have the budget set $B(y) := \left\{ x \in \mathbb{R}^2_+ :\quad p(y)\cdot x \le \frac{1}{2} p(y)\cdot y = \frac{1}{2} \right\}$. Given their specific preferences, the individual demand vectors are $\left(\frac{1}{2p_1(y)}, 0\right)$ and $\left(0, \frac{1}{2p_2(y)}\right)$ for player 1 and 2, respectively. The resulting aggregate demand $\left(\frac{1}{2p_1(y)}, \frac{1}{2p_2(y)}\right)$ can be interpreted as the demand of a hypothetical *representative consumer* whose preference is represented by the Cobb–Douglas utility function $(x_1, x_2) \mapsto x_1^{1/2} x_2^{1/2}$.

Therefore the equilibria of $\mathcal{E}_S$ coincide with the optimal choices of this representative agent on Y. Due to convexity of Y and strict quasi-concavity of the Cobb–Douglas utility function the representative consumer has a unique optimum. Hence the economy $\mathcal{E}_S$ has a unique competitive equilibrium. But this unique optimum is just the point which maximizes the Nash product $x_1 \cdot x_2$ on $S = Y$.

So we have established the following result which for n–person bargaining games has been stated in [18].

3.1. Proposition. *Given a two–person bargaining problem $(S, 0)$, the economy $\mathcal{E}_S$ has a unique Walrasian equilibrium. The equilibrium production plan and the aggregate equilibrium consumption coincide with the Nash solution $N = \varphi_N(S, 0)$ of $(S, 0)$.*

Next we define a one–stage non-cooperative game in strategic form whose payoff functions are derived from the individual demand functions in $\mathcal{E}_S$. To simplify the arguments we assume in addition that ∂S is smooth, i.e., at each point $y \in \partial S$ there is a unique (normalized) $p(y)$.

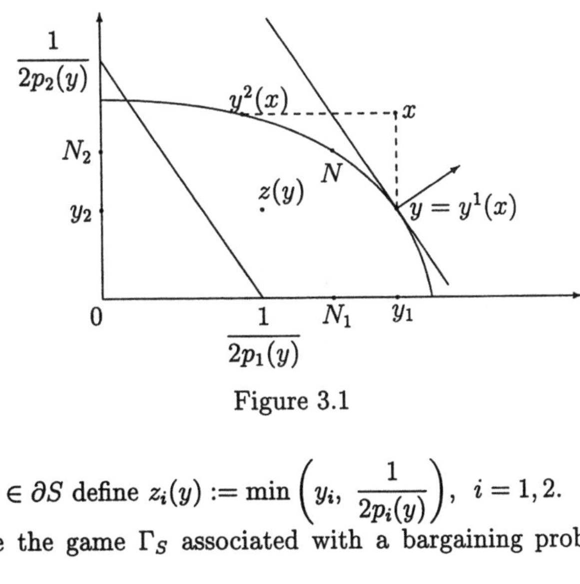

Figure 3.1

For any $y \in \partial S$ define $z_i(y) := \min\left(y_i, \dfrac{1}{2p_i(y)}\right), \quad i = 1, 2.$

We define the game Γ_S associated with a bargaining problem $(S, 0)$ as follows.

The strategy sets S_i are the projections of S to the i–th axis, i.e. $S_i = \text{proj}_i S, \quad i = 1, 2.$

Next observe that any $x = (x_1, x_2) \in S_1 \times S_2$ defines two points $y^1(x)$, $y^2(x) \in \partial S$ by $y^1_1(x) = x_1$ and $y^2_2(x) = x_2$, which coincide if and only if $x \in \partial S$.

The payoff functions ζ_i for the game Γ_S are defined by

$$\zeta_i(x) := z_i(y^i(x)).$$

So we get $\Gamma_S = (S_1, S_2;\ \zeta_1, \zeta_2)$.

Now we can state our main result.

3.2. Proposition. *Given a two-person bargaining problem $(S, 0)$, the game Γ_S has a unique Nash equilibrium. This is even an equilibrium in dominant strategies, and it coincides with the Nash solution $N = \varphi_N(S, 0)$.*

Proof. First observe that for $N \in \partial S$ we have $y^i(N) = N$, $i = 1, 2$, and therefore $\zeta_i(N) = z_i(N) = N_i = \dfrac{1}{2p_i(N)}$ by Proposition 3.1.

Next observe that ∂S is the graph of two strictly decreasing, strictly concave functions f_1 and f_2 of y_1 and of y_2, respectively. Accordingly, the functions $y_i \mapsto \dfrac{1}{2p_i(y_i, f_i(y_i))}$, $i = 1, 2$, are strictly decreasing. Therefore we get for any $y \in \partial S$ one of the following three cases:

1. $y_i = N_i = \dfrac{1}{2p_i(N)}, \quad i = 1, 2;$

2. $y_1 > N_1 = z_1(N) > \dfrac{1}{2p_1(y)} \quad$ and $\quad y_2 < N_2 = z_2(N) < \dfrac{1}{2p_2(y)};$

3. $y_1 < N_1 = z_1(N) < \dfrac{1}{2p_1(y)} \quad$ and $\quad y_2 > N_2 = z_2(N) > \dfrac{1}{2p_2(y)}.$

This implies that for any $x \in S$, for both $i = 1, 2$, either $x_i = N_i$ or $\zeta_i(x) = z_i(y^i(x)) < N_i$.

This shows that any $x \in S_1 \times S_2$ results in a feasible payoff vector $(\zeta_1(x), \zeta_2(x)) \leq N$ and that N_i is the unique optimal choice for player i independently of the other player's choice. Therefore $N = (N_1, N_2)$ is the unique Nash equilibrium of Γ_S with dominant strategies N_i, $i = 1, 2$, for the two players. ■

4. Concluding remarks

What we have achieved in this paper is a direct one–stage implementation of the Nash bargaining solution in dominant strategies. Rather than giving a prescription for the organization of real bargaining, our results shed some light on how Nash's solution is to be interpreted. As required in the Nash program, it gives one specific non-cooperative foundation which appears quite different from non-cooperative approaches in the literature. Of particular interest is the fact that our Nash equilibrium is unique and in dominant strategies.

The equal shares assumption in our economy $\mathcal{E}_S$, which has its impact also on the payoff functions in Γ_S, reflects Shapley's [16] concept of "equity." Evaluating two players' utility levels by suitable prices defines endogenously transfer rates. In the equilibrium (i.e. the Nash solution) the rates of the two utilities used in the transition from 0 to N coincide with the transfer rates at N represented by $p(N)$.

The equilibrium "equilibrates" two opposite interests of the agents: On the one hand either of the agents wants "his own" commodity to be expensive thereby inducing the firm to produce a large quantity of it and making a high profit. On the other hand each agent would like it to be inexpensive thereby enabling him to buy a large amount of it. If any proposed efficient bundle y could be sold at prices $p(y)$, then each agent could receive half of the resulting revenue and trade at prices $p(y)$ on the hypothetical markets. In all but one cases this would make exactly one of the two agents worse off because he would be unable to buy back, with his available money, the quantity of "his" commodity he sold before. Only the Nash solution gives to each agent his demand at $p(N)$ without having to make use of any hypothetical market transaction. Only N leaves no hypothetical arbitrage possibilities. This is quite the same as with Shapley's λ-transfer value which is the unique one in a family of Shapley values of TU games associated with a given NTU game, which can be realized without making use of transfer possibilities.

The results seem to suggest an interpretation of the Nash solution as an agreement which is forced by some kind of competitive pressure. It would be interesting to see our "hypothetical market opportunities" replaced by a real market structure in an explicit dynamical competitive model like, for instance, [3, 4, 14, 7].

An alternative way of looking at $\mathcal{E}_S$ is as a coalition production economy with a unit interval of players of two types (cf. [5]), where both types are represented by half of the interval. Any change of the weights of the two types would lead to an economy with different shares represented by α, $1-\alpha \in (0,1)$. These weights could be interpreted as the bargaining power. The unique equilibrium of a representing economy $\mathcal{E}_S^\alpha$ would be an asymmetric Nash solution with weights α, $1-\alpha$.

Also, Γ_S could be modified to Γ_S^α in a straightforward way. Again the asymmetric Nash solution would turn out as the unique Nash equilibrium (in dominant strategies) of Γ_S^α. Different bargaining powers would be represented by different budgets resulting in different payoff functions in Γ_S^α.

An extension of these results to $n > 2$ appears to be possible. The dominant strategy equilibrium underlines the competitive aspect by the fact that every player can influence only his own payoff, without any influence on the other players' payoffs.

The game proposed in the present paper might appear at first sight similar to a game where each player i proposes an amount of his utility and receives N_i if he proposes N_i, and 0 otherwise. This is a way in which the implementation via equilibrium in dominant strategies could always be established. But here the player i has to know N_i.

In the game Γ_S, however, the players need not know the concept of the Nash solution nor the specific point $\varphi_N(S,0) = N$. They only have to be able to compare any efficient point with the resulting demand. This requires knowledge of $(S,0)$ but not of the point $N = \varphi_N(S,0)$, nor even of the concept of the Nash solution.

References

1. K. Binmore, Nash bargaining theory I, in: *The Economics of Bargaining* (eds. Binmore and Dasgupta), Basil Blackwell, Cambridge, 1987.
2. G. Debreu, The coefficient of resource utilization, *Econometrica* **19** (1951), 273–291.
3. D. Gale, Bargaining and competition, part I: characterization, *Econometrica* **54** (1986), 785–806.
4. D. Gale, Bargaining and competition, part II: existence, *Econometrica* **54** (1986), 807–818.
5. W. Hildenbrand, *Core and Equilibria of a Large Economy*, Princeton University Press, Princeton, 1974.
6. J. V. Howard, A social choice rule and its implementation in perfect equilibrium, *Journal of Economic Theory* **56** (1992), 142–159.
7. A. McLennan and H. Sonnenschein, Sequential bargaining as a noncooperative foundation for Walrasian equilibrium, *Econometrica* **59** (1991), 1395–1424.
8. J. F. Nash, The bargaining problem, *Econometrica* **18** (1950), 155–162.
9. J. F. Nash, Two–person cooperative games, *Econometrica* **21** (1953), 128–140.
10. M. J. Osborne and A. Rubinstein, *Bargaining and Markets*, Academic Press, San Diego, 1990.
11. M. J. Osborne and A. Rubinstein, *A Course in Game Theory*, MIT Press, Cambridge, 1994.
12. A. E. Roth, *Axiomatic Models of Bargaining*, Springer, Berlin, 1979.

13. A. Rubinstein, Perfect equilibrium in a bargaining model, *Econometrica* 50 (1982), 207–211.

14. A. Rubinstein and A. Wolinsky, Equilibrium in a market with sequential bargaining, *Econometrica* 53 (1985), 1133–1150.

15. R. Selten, Reexamination of the perfectness concept for equilibrium points in extensive games, *International Journal of Game Theory* 4 (1975), 25–55.

16. L. S. Shapley, Utility comparison and the theory of games, in: *La Decision: Agregation et Dynamique des Ordres de Preference*, Paris, 1969, 251–263.

17. W. Thomson and T. Lensberg, *Axiomatic Theory of Bargaining with a Variable Number of Agents*, Cambridge University Press, Cambridge, 1989.

18. W. Trockel, A Walrasian approach to bargaining games, *Economics Letters* 51 (1996), 295–301.

Walter Trockel
Institute of Mathematical Economics
University of Bielefeld
Postfach 10 01 31
D-33501 Bielefeld
Germany
email: wtrockel@wiwi.uni-bielefeld.de

ON THE EXISTENCE OF A TEMPORARY UNEMPLOYMENT EQUILIBRIUM

DEMETRIUS YANNELIS*

Abstract. For a two-period model, in which firms set the wages of workers they employ, it is shown that an unemployment equilibrium may exist while the goods market clears. The unemployment that occurs in period one is due to pessimistic expectations of the firms concerning the level of their sales in period two. Firms do not have any incentive to lower the wage rate since this may decrease rather than increase the level of effective demand. Therefore, the rigidity of the wage rate is not the cause of unemployment which is due to the deficient effective demand.

1. Introduction

Since the pathbreaking article by Clower [6], there has been generated an extensive literature on the so-called disequilibrium models.[1] Most of the writers following Hick's notion of the fixed price method have shown the existence of equilibrium in an economy where quantities (instead of prices) move to equate demand and supply, since prices are by assumption fixed. Such models are capable of examining disequilibrium situations such as unemployment, inflation, etc, but their main shortcoming is the assumption of fixed prices.

In most of these models, the writers have stressed the possibility of the existence of an unemployment equilibrium, but none of them have shown the existence of such an equilibrium. An exception is Glustoff [8], who showed the existence of such an equilibrium by keeping the wage rate fixed at an arbitrary level.

Some writers, following the suggestion of Arrow [2] that in disequilibrium situations agents behave in some respects as monopolists, have allowed for price setting behavior either by the firms or households or both (see [5, 7, 9, 11]). Futia [7] shows the existence of an unemployment equilibrium, which is the result of a rigid wage rate. But unlike Glustoff, he gives a reason as to why the wage rate does not fall. Our aim is to answer the same question as Futia did, but from a different point of view. The crucial question is: "Under what conditions can we have unemployment while the wage rate does not show any tendency to fall?"

We shall allow for price setting behavior by the firms, where they set the wages of the workers that they employ. The prices of goods are determined in the goods market under the forces of perfect competition. This implies that

*I would like to thank Rick Harris for helpful comments and suggestions.

[1] For full references see Grandmont [10].

firms behave as price takers in the goods market, and workers behave as perfect competitors under all circumstances. Our model is a two–period model and is similar to the one presented by Grandmont and Laroque [9]. Firms control the wages, which they announce at the outset of every period. The prices of the goods are also announced in the same period by some "auctioneer."

In a given period (say, period 1), firms try to forecast the effective demand for their product in period 2, which comes from the wages and money balances of consumers at date 2. So, production is not instantaneous and goods which are produced during period 1 are available from the outset of period 2. Hence, in any given period the firms' demand for labor depends greatly on the expected proceeds from the sales of their product in the next period, something that was noted by Keynes [14] and demonstrated later by Patinkin [16], and Barro and Grossman [4]. We will find that, in such a model, if firms have pessimistic expectations about the sales of their product in period 2, then involuntary unemployment may result in period 1. In such a case the wage rate does not show any tendency to fall, since firms do not see any advantage in lowering it.

2. Description of the economy

The structure of the economy at date 1 is as follows. There are $i = 1, 2, \ldots, n$ households, $j = 1, 2, \ldots, m$ firms, and $h = 1, 2, \ldots, N$ goods. Good 1 is the only input in every production process and we call it labor. We denote by the scalar $m \in \mathbb{R}_+$ the money balances of the agents in the economy and we assume that the total stock of money is fixed. There is only one future market for one physical commodity, which is money.

The money prices of all goods are represented by a vector $p \in \mathbb{R}^N$. The price of money is by definition equal to one. Since we want to abstain from monetary considerations, the criticism of Hahn [13] applies to this model, where we assume a priori that the accounting price of money is greater than zero.

3. Producer behavior

The typical producer in period 1 must combine goods and labor services represented by a vector $y_j \in \mathbb{R}^N$ in order to get outputs of goods, which will be available from the outset of period 2. The production possibilities set of the j–th producer is a subset $Y_j \subset \mathbb{R}^N \times \mathbb{R}^N$ which is assumed to be closed and strictly convex, since we want to deal with functions rather than correspondences. The producer at date 1 has to decide what his production vector will be and what wage rate he will quote at the outset of period 2. At date 1 he has an endowment of goods y^0_{j+}, which is the result of the production process of the previous period 0. In other words, we assume that storage activities can be done by firms.

The producer must forecast what the effective demand for his output will be at period 2. He has some expectations about the prices that will prevail at date 2, and given any signals he may receive at date 1 (such as current prices), he must predict what wages to quote in order to sell a quantity $y \in \mathbb{R}^N_+$ at period 2.

The whole procedure results in a function $R_j(y_{j+},\ p^2_1,\ y^2_{j_1})$ (where y_{j+} is a vector with nonnegative components), which is the maximum revenue expected by the entrepreneur at date 1 from the sale of his output at date 2. The parameters in this function are the wage rate p^2_1 at date 2 and the amount of unemployment $y^2_{j_1}$ at date 2. Assuming that R_j is differentiable, we clearly have that $\partial R/\partial p^2_1 > 0$, which means that, whenever the wages quoted by the producer are high enough, his revenue will be high as well. If we take the amount of unemployment as variable, then this function says how much labor the producer would employ given his expectations about the proceeds that he will receive. In other words, this function may very well represent Keynes' "supply price."[2]

Then, for a given production vector and the current level of prices, the producer's maximum profit is $\Pi = \max R_j(y_{j+},\ p^2_1,\ y^2_{j_1}) - p_1 y^1_{j_1}$ with respect to vector y_{j+}, subject to $y_{j+} \leq y^1_{j+}$ and $y_j \in Y_j$. The condition $y_{j+} \leq y^1_{j+}$ means that the output that the entrepreneur expects to sell at period 2 cannot exceed that of period 1. The maximum expected profit is now a function of y^1_j, which the producer maximizes, i.e.,

$$\max \Pi(y^1_j, p) \qquad \text{s.t.} \quad p(y^1_j - y^0_j) + m^1_j - m^0_j = 0, \\ m \geq 0, \quad y_j \in Y_j.$$

This maximization procedure will give the producer his current production vector and, implicitly, a choice of wages which will be announced from the outset of period 2. By this type of procedure, the producer is able to control the main purchasing power of the consumer and hence his effective demand. So he always tries to select the proper wage rate in order to keep the effective demand as high as possible, so as to maximize his profits.

[2]See [14, pp. 24–25].

4. Consumer behavior

A representative consumer chooses at period 1 a consumption plan $(x_i^1, x_i^2) \in X_i \times X_i$. The current consumption vector x_i^1 of the i–th consumer is the result of the following procedure. First we assume that:

(a) The consumption possibility set X_i is closed, strictly convex and bounded from below.

(b) The utility function $u_i(x^1, x^2)$ is continuous and strictly concave.

The consumption plan (x^1, x^2) of the consumer depends on his current consumption vector and on his expectations about the prices that will prevail at date 2. So he has to forecast the prices for goods and labor that will prevail at date 2, as well as the effective demand for his labor. We assume that there are no storage activities on the consumers' part and hence no intertemporal transfer of goods. Furthermore, there is no distribution of profits by the firms, so the sole income of the consumer in any given period is his labor income and the money balances from the previous period.

Let $R_i(x_{i1})$ represent the maximum income expected by consumer i at date 1 from the sale of his labor at date 2. If the consumer is always able to sell as much labor as he wants, then involuntary unemployment cannot exist.

For any given current consumption vector ${x_i}^1 \in X_i$, the maximum expected utility of ${x_i}^1$ is, by definition,

$$U \equiv \max_{x^2} u_i(x^1, x^2) \quad \text{such that} \quad p^2 x_{i_t}^2 + m_i^2 = R_i(x_{i1}) + m_i^1,$$
$$(x_i^1, x_i^2) \in X_i \times X_i, \quad m^1, m^2 \geq 0.$$

The maximum of the derived utility is now a function of the current consumption plan x_i^1, which the consumer maximizes subject to the current constraint $p^1 x_i^1 + m_i^1 - m_i^0 = 0$; $x_i^1 \in X_i$, $m^1, m^0 \geq 0$. This maximization procedure gives the consumption vector x_i of the i–th consumer.

5. Existence of unemployment

How can unemployment result in such an economy? In order for the producer to maximize his profits at his desired wages, there may be people unemployed. This occurs, for example, whenever producers have pessimistic expectations about the sale of their product and this is reflected in a movement to the left of their demand curve for labor. This happens since the demand for labor includes as a parameter the expected sales (proceeds) of the entrepreneurs. This movement to the left is going to press down the wage rate. Here the crucial question arises whether or not the wage rate is going to fall. This depends heavily on the elasticity of the demand curve for labor. If it is inelastic, then a fall in the wage rate will increase employment, but will result in an aggregate fall of the total wage bill. This, of course, will lower the effective demand for the product of the entrepreneurs and hence their aggregate profits. Hence we postulate the following.

5.1. Assumption. *The demand for labor is inelastic for every firm.*

This assumption ensures that each producer in an unemployment equilibrium has no reason to decrease the wage rate, since this will eventually result in a fall in his expected profits. From this point of view, all firms can be regarded as identical, all facing an inelastic demand curve for labor and acting in the same way.

This of course does not mean that wage rigidity is the cause of unemployment. The real cause is the deficient demand which induces the entrepreneurs to demand less labor because of their pessimistic expectations for the sale of their product.

Since in such a case consumers are constrained in the labor market, we have to distinguish, according to Clower [6], between the notional and effective or realized demand. We shall be interested in the case where firms can realize their plans whereas consumers cannot because of lack of effective demand. The following two assumptions formalize the above concepts.

5.2. Assumption. $y^{1e} \neq y^1$ *only if* $x_1^1 - y_1^1 < 0$, *where the superscript e denotes the effective or realized production vector.*

So Assumption 5.2 states that a firms' notional and effective plans will differ only if they are unable to buy the labor they want. In every other case they may realize their plans. The next assumption has to do with the realization of plans by the part of consumers.

5.3. Assumption. $x_+^1 \neq x_+^{1e}$ *implies* $y^1 = y^{1e}$, *where the subscript* $+$ *denotes a vector with nonnegative components.*

This assumption states that if households do not realize their plans, then they have not sold the labor they wanted. This in turn implies that firms have bought all the labor they wanted. Hence, given Assumptions 5.2 and 5.3, and especially that $x_+^1 \neq x_+^{1e}$ in period 1, Walras' law in this case cannot contain the demand for labor as noted by Clower [6]. Walras' law in this case takes the form

$$\sum_{h=2}^{N} p_h z_h + z_m = 0,$$

where

$$z_h = \sum_i \sum_j (y^1_{jh+} + x^1_{ih+} - y^0_{jh+})$$

and z_m is the excess demand for money. This can be seen if we write the budget constraint of the consumers as $px^{1e}_+ + m^1 = m^0$, under the assumption $x^1_+ \neq x^{1e}_+$. By Assumptions 5.2 and 5.3, firms can realize their plans, and so we have

$$px^{1e}_+ + m^1 - m^0 = p(y^0_+ - y^1_+) + m^0 - m^1,$$

or

$$p(x^{1e}_+ + y^1_+ - y^0_+) + m^1 - m^0 = 0,$$

which gives us Walras' law, that is,

$$\sum_{h=2}^{N} p_h z_h + z_m = 0.$$

In other words, when the notional excess demand for labor is negative, Walras' law does not contain the excess demand for labor. Hence we have proved the following result.

5.4. Lemma. *Under Assumptions* 5.2 *and* 5.3 *Walras' law is*

$$\sum_{h=2}^{N} p_h z_h + z_m = 0.$$

We can now define the concept of an unemployment equilibrium for a given price system that prevails in period 1. But first we want to be sure that in such an equilibrium the wage rate cannot be zero. Actually, the wage rate can fall to zero whenever the expectations of the entrepreneurs are very pessimistic. But this is an unusual case that we can exclude by postulating the following.

5.5. Assumption. $p_1 = 0$ *implies* $x_1 = 0$.

In words this means that nobody would be willing to work at a wage rate equal to zero. Hence the involuntary unemployment would not exist in such a case. This in turn implies that there will be no output and hence no profits. So, firms would never choose a zero wage rate.

5.6. Definition. *A price vector p is an equilibrium if the following conditions are true:*

(a) $z_h \leq 0,\ z_1 = 0,$ *and* $z_h < 0$ *if* $p_h = 0$ *for* $h = 2, 3, \ldots, N,$

(b) $z_h \leq 0,\ z_1 < 0,\ \hat{p}_1 > 0,$ *and* $z_h < 0$ *if* $p_h = 0$ *for* $h = 2, 3, \ldots, N.$

The first notion of equilibrium is that of full employment where excess demand for labor is zero. The second notion is that of an unemployment equilibrium with a given positive wage rate $\hat{p}_1 > 0$ and a zero or negative excess demand in the other markets.

Before proving the existence of such an equilibrium, we must note that since we are concerned with a monetary economy and the price of money is equal to one, there can be no tatonnement on its price. So we can regard the excess demand for money as irrelevant, or we may for simplicity set its value equal to zero (but not identically equal to zero). Any tatonnement that would have taken place on its price is reflected by a change in the money prices of all other commodities. Hence this does not harm the validity of the next theorem.

5.7. Theorem. *Let $z_h(p)$ be the excess demand functions that satisfy the following conditions:*

(a) $z_h(p) > -K$ *for all* $h = 1, 2, \ldots, N,$ *where* K *is a positive scalar,*

(b) $z_h(p)$ *is continuous for* $h = 1, 2, \ldots, N,$

(c) $z_h(\lambda p) = z_h(p), \quad \lambda > 0, \quad h = 1, 2, \ldots, N,$

(d) *For all* $p,$ $\sum_{h=2}^{N} p_h z_h + z_m = 0.$

Then there exists a price vector p^* which is an equilibrium.

Proof. Let $S = \left\{\bar{p} : \ \bar{p} \geq 0, \ \sum_{h=1}^{N} \bar{p}_h = 1\right\}$. For each $K > 0$ define the mappings

$$T_1(\bar{p}) = \frac{\max[\hat{p}_1,\ \bar{p}_1 + K z_1(\bar{p})]}{\lambda}, \tag{5.1}$$

$$T_h(\bar{p}) = \frac{\max[0,\ \bar{p}_h + K z_h(\bar{p})]}{\lambda}, \quad h = 2, 3, \ldots, N, \tag{5.2}$$

and let

$$\lambda = \sum_{h=2}^{N} \max[0,\ \bar{p}_h + K z_h(\bar{p})] + \max[\hat{p}_1,\ \bar{p}_1 + K z_1(\bar{p})].$$

Assuming that the mapping $T = (T_1, \ldots, T_N)$ is continuous, it can be seen that T maps the unit N-1 simplex into itself. Hence by Brower's fixed point theorem, there exists a fixed point such that $T(\bar{p}^*) = \bar{p}^*$. We want to show that this fixed point satisfies the definition of our equilibrium. First note that

$\lambda > 0$. Assume the contrary, that is, that $\lambda = 0$. Then in view of (5.2) we have $\max[0, \bar{p}_h^* + K z_h] = 0$.

If $\bar{p}_h^* + K z_h > 0$, then $0 = \bar{p}_h^* + K z_h(\bar{p}) > 0$, a contradiction. If $\bar{p}_h^* + K z_h(\bar{p}) < 0$, then multiplying by $\bar{p}_h^*$, we obtain $\bar{p}_h^{*2} + \bar{p}_h^* K z_h(\bar{p}) < 0$. Therefore, summing up over all h we obtain $\sum_h \bar{p}_h^{*2} + K \sum_h p_h{}^* z_h = 0$. Since $\sum_h \bar{p}_h^{*2} > 0$ and $\sum_h \bar{p}_h^* z_h = 0$ by Walras' law, we have $0 < \sum_h \bar{p}_h^{*2} < 0$, a contradiction. Hence, $\lambda > 0$.

Since at the fixed point $\bar{p}_h^* = \dfrac{\max[0,\ \bar{p}_h^* + K z_h(\bar{p})]}{\lambda}$, we obtain

$$z_h(\bar{p}) = \frac{(\lambda - 1)\bar{p}_h^*}{K}.$$

Multiplying by $\bar{p}_h^*$ and summing up over all h yields

$$\sum_{h=2}^{N} \bar{p}_h^* z_h = \frac{\sum(\lambda - 1)\bar{p}_h^{*2}}{K} = 0,$$

whence $\lambda = 1$. Therefore, from (5.1) and (5.2) we have that

$$\bar{p}_1^* = \max[\hat{p}_1,\ \bar{p}_1^* + K z_1(\bar{p})], \quad \text{and} \quad \bar{p}_h^* = \max[0,\ \bar{p}_h^* + K z_h(\bar{p})].$$

If $\hat{p}_1 < \bar{p}_1^* + K z_1(\bar{p})$, then $\bar{p}_1^* < \bar{p}_1^* + K z_1(\bar{p})$ and hence $z_1(\bar{p}) = 0$. On the other hand, from (5.2) we see that if $\bar{p}_h^* = 0$, then $z_h \leq 0$. If $\bar{p}_h^* > 0$, then $z_h = \dfrac{(\lambda - 1)\bar{p}_h^*}{K} = 0$, since $\lambda = 1$. So the first notion of equilibrium (the full employment) is satisfied.

Now suppose that $\hat{p}_1 > \bar{p}_1^* + K z_1(\bar{p})$. Then $\bar{p}_1^* = \hat{p}_1 > \bar{p}_1^* + K z_1(\bar{p})$ and therefore $z_1(\bar{p}) < 0$. By considering (5.2), we again have

$$\begin{aligned} z_h &= 0 \quad \text{if} \quad \bar{p}_h^* > 0, \text{ and} \\ z_h &< 0 \quad \text{if} \quad \bar{p}_h^* = 0, \quad \text{for } h = 2, 3, \ldots, N. \end{aligned}$$

So, the second notion of equilibrium is satisfied as well, which means that an unemployment equilibrium may exist. ■

6. Conclusions

We have shown that in an economy where producers have pessimistic expectations for the sale of their product, unemployment is a possible outcome. In such an unemployment equilibrium, producers do not find it profitable to decrease the established wage rate. We must stress here that this is different from saying that the wage rate is rigid and does not fall. The wage rate is not rigid. It does not fall, since such a fall would worsen, rather than improve, the level of employment. This is the case Keynes[3] was concerned with. The notion of unemployment equilibrium is different from the one to be found in the modern literature for the fixed price models, where there is an excess supply in both goods and labor markets. It might be thought that the notion of unemployment equilibrium given here is similar to the classical unemployment as defined by Malinvand [15], where there is an excess supply of labor, but firms can sell all of their output. But although firms can sell their products, households cannot buy the amount of goods they want. So, there is an excess demand for goods and the goods market does not clear. Hence the notion of classical equilibrium is different from the one presented in this paper.

To conclude, the main result of this paper asserts that the cause of unemployment is due to the deficiency of effective demand, and the wage cut does not restore the full unemployment equilibrium, as the classics used to believe. A possible extension would be to generalize the above result to infinite-dimensional spaces using the techniques developed in [1].

References

1. C. D. Aliprantis, D. J. Brown, and O. Burkinshaw, *Existence and Optimality of Competitive Equilibria*, Springer, Berlin, 1989.
2. K. Arrow, Towards a theory of price adjustment, in: *The allocation of economic resources* (ed. Abramowitz), Stanford University Press, 1959.
3. K. Arrow and F. Hahn, *General Competitive Analysis*, Holden–Day, San Francisco, 1971.
4. R. Barro and H. Grossman, A general disequilibrium model of income and employment, *American Economic Review* **61** (1971), 82–93.
5. J. P. Benassy, The disequilibrium approach to monopolistic price setting and general monopolistic competition, *Rev. Econom. Stud.* **43** (1976), 69–81.
6. R. Clower, The Keynesian counterevolution: a theoretical appraisal, in: *The Theory of Interest Rates* (eds. F. Hahn and F. P. R. Brechling), Macmillan, London, 1965.
7. C. Futia, Excess supply equilibria, *J. Econom. Theory* **14** (1977), 200–220.
8. E. Glustoff, On the existence of a keynesian equilibrium, *Rev. Econom. Stud.* **35** (1968), 327–334.

[3]See [14, Ch. 19, p. 261]. Cf. Hahn's discussion [12, p. 34], where Hahn points out that: "The wage is neither fixed nor flexible. It is what it is because no agent finds it advantageous to change it."

9. J. M. Grandmont and Laroque, On temporary keynesian equilibria, *Rev. Econom. Stud.* **43** (1976), 53–67.

10. J. M. Grandmont, Temporary general equilibrium theory, *Econometrica* **45** (1977), 535–572.

11. F. Hahn, On non-Walrasian equilibria, *Rev. Econom. Stud.* **45** (1978), 1–17.

12. F. Hahn, Keynesian economics and general equilibrium theory: Reflections of some current debates, in: *The Microfoundations of Macroeconomics* (ed. G. Harcourt), Macmillan, London, 1978.

13. F. Hahn, On some problems of proving the existence of an equilibrium in a monetary economy, in: *The Theory of Interest Rates* (eds. F. Hahn and F. P. R. Brechling), Macmillan, London, 1965.

14. J. M. Keynes, *General Theory of Employment Interest and Money*, Macmillan, London, 1936.

15. E. Malinvand, *The Theory of Unemployment Reconsidered*, Basil Blackwell, Oxford, 1977.

16. Don Patinkin, *Money Interest and Prices*, 2nd edition, Harper and Row, New York, 1965.

17. H. Varian, Non-Walrasian equilibria, *Econometrica* **45** (1977), 573–590.

Demetrius Yannelis
Department of Economics
University of Piraeus
Piraeus 18534
Greece
email: yannelis@unipi.gr

ON THE EXISTENCE OF A BAYESIAN NASH EQUILIBRIUM

NICHOLAS C. YANNELIS

Abstract. We provide a new proof (based on a Caratheodory–type selection theorem) of the existence of a Bayesian Nash equilibrium.

1. Introduction

The purpose of this note is to present an alternative proof of an existence result for Bayesian Games (or games with differential information) given in [8].[1] The result we provide is identical to Theorem 4.1 in [8]. However, not only the present argument is different, but it also has the advantage that it follows the footsteps of the argument given in [13] and therefore it can be used to generalize the Kim–Yannelis theorem to abstract Bayesian economies *a là* [3]. Our argument combines several measure theoretic and functional analytic results. In particular, we employ a Caratheodory–type selection theorem, a result on weak compactness (known as Diestel's theorem), the Fatou Lemma in infinite–dimensional spaces, and the Fan–Glicksberg fixed point theorem.

2. Notation and definitions

Let 2^X denote the set of all non-empty subsets of the set X. If X and Y are sets, the *graph* of the set–valued function (or correspondence) $\phi : X \to 2^Y$ is $G_\phi = \{(x, y) \in X \times Y : y \in \phi(x)\}$. Let $(\Omega, \mathcal{F}, \mu)$ be a complete, finite measure space, and Y be a separable Banach space. The correspondence $\phi : \Omega \to 2^Y$ is said to have a *measurable graph* if $G_\phi \in \mathcal{F} \otimes B(Y)$, where $B(Y)$ denotes the Borel σ–algebra on Y and $\otimes$ denotes the product σ–algebra. The correspondence $\phi : \Omega \to 2^Y$ is said to be *lower measurable* if $\{\omega \in \Omega : \phi(\omega) \cap V \neq \phi\} \in \mathcal{F}$ for every V open subset of Y. We denote by $L_1(\mu, Y)$ the space of equivalence classes of Y–valued Bochner integrable functions. Recall that $L_1(\mu, Y)$ is a separable Banach space provided that $(\Omega, \mathcal{F}, \mu)$ is separable.

3. The game with differential information

Let $(\Omega, \mathcal{F}, \mu)$ be a complete, finite, separable measure space, where Ω denotes the set of states of the world and the σ–algebra $\mathcal{F}$ denote the set of events. Let Y be a separable Banach space and T be a set of agents (either finite or infinite).

[1]The reader is referred to [1] or [8] for a discussion of the related literature.

A *Bayesian game* (or a *game with differential information*) is a set $G = \{(X_t, u_t, \mathcal{F}_t, g_t) : t \in T\}$, where

1. $X_t : \Omega \to 2^Y$ is the *action set-valued function* of agent t, where $X_t(\omega)$ is the set of actions available to t when the state is ω,
2. for each $\omega \in \Omega$, $u_t(\omega, \cdot) : \prod_{s \in T} X_s(\omega) \to \mathbb{R}$ is the *utility function* of agent t, which can depend on the states,
3. $\mathcal{F}_t$ is a sub σ–algebra of $\mathcal{F}$ which denotes the *private information* of agent t,
4. $q_t : \Omega \to \mathbb{R}_{++}$ is the *prior* of agent t, which is a Radon–Nikodym derivative such that $\int q_t(\omega)\, d\mu(\omega) = 1$.

Let L_{X_t} denote the set of all Bochner integrable and $\mathcal{F}_t$–measurable selections from the action set–valued function $X_t : \Omega \to 2^Y$ of agent t, i.e.,

$$L_{X_t} = \{\tilde{x}_t \in L_1(\mu, Y) : \quad \tilde{x}_t \text{ is } \mathcal{F}_t\text{-measurable and } \tilde{x}_t(\omega) \in X_t(\omega) \ \mu\text{–a.e.}\}.$$

The typical element of L_{X_t} is denoted as $\tilde{x}_t$, while that of $X_t(\omega)$ as $x_t(\omega)$ (or x_t). Let $L_X = \prod_{t \in T} L_{X_t}$ and $L_{X_{-t}} = \prod_{s \neq t} L_{X_s}$. Given a Bayesian game G, a *strategy* for agent t is an element $\tilde{x}_t$ in L_{X_t}.

Throughout the paper, we assume that for each $t \in T$ there exists a finite or countable partition π_t of Ω. Moreover, the σ–algebra $\mathcal{F}_t$ is generated by π_t. For each $\omega \in \Omega$, let $E_t(\omega) \in \pi_t$ denote the smallest set in $\mathcal{F}_t$ containing ω, and we assume that for all t

$$\int_{\omega' \in E_t(\omega)} q_t(\omega')\, d\mu(\omega') > 0.$$

For each $\omega \in \Omega$, the *conditional (interim) expected utility function* of agent t, $\nu_t(\omega, \cdot, \cdot) : L_{X_{-t}} \times X_t(\omega) \to \mathbb{R}$, is defined as:

$$\nu_t(\omega, \tilde{x}_{-t}, x_t) = \int_{\omega' \in E_t(\omega)} u_t(\omega', \tilde{x}_{-t}(\omega'), x_t) q_t(\omega' | E_t(\omega))\, d\mu(\omega'),$$

where

$$q_t(\omega' | E_t(\omega)) = \begin{cases} 0 & \text{if } \omega' \notin E_t(\omega) \\ \dfrac{q_t(\omega')}{\int_{\tilde{\omega} \in E_t(\omega)} q_t(\tilde{\omega})\, d\mu(\tilde{\omega})} & \text{if } \omega' \in E_t(\omega). \end{cases}$$

The function $\nu_t(\omega, \tilde{x}_{-t}, x_t)$ is interpreted as the conditional expected utility of agent t using the action x_t when the state is ω and the other agents employ the strategy profile $\tilde{x}_{-t}$, where $\tilde{x}_{-t}$ is an element of $L_{X_{-t}}$.

A *Bayesian Nash equilibrium* for G is a strategy profile $\tilde{x}^* \in L_x$ such that for all $t \in T$

$$\nu_t(\omega, \tilde{x}^*_{-t}, \tilde{x}^*_t(\omega)) = \max_{y_t \in X_t(\omega)} \nu_t(\omega, \tilde{x}_{-t}, y_t) \quad \mu\text{–a.e.}$$

4. Statement of the theorem and proof

We state the assumptions needed for our theorem.

(A.1) $X_t : \Omega \to 2^Y$ is a non-empty, convex, weakly compact–valued and integrably bounded correspondence having a $\mathcal{F}_t$–measurable graph, i.e., $G_{X_t} \in \mathcal{F}_t \otimes B(Y)$.

(A.2) 1. For each $\omega \in \Omega$, $u_t(\omega, \cdot, \cdot) : \prod_{s \neq t} X_s(\omega) \times X_t(\omega) \to \mathbb{R}$ is continuous, where $X_s(\omega)$ $(s \neq t)$ is endowed with the weak topology and $X_t(\omega)$ with the norm topology.

2. For each $x \in \prod_{x \in T} Y_x$ with $Y_s = Y$, $u_t(\cdot, x) : \Omega \to \mathbb{R}$ is $\mathcal{F}$–measurable.

3. For each $\omega \in \Omega$ and $x_{-t} \in \prod_{s \neq t} X_s(\omega)$, $u_t(\omega, x_{-t}, \cdot) : X_t(\omega) \to \mathbb{R}$ is concave.

4. u_t is integrably bounded.

4.1. Theorem. *Assume that T is a countable set. Let $G = \{(X_t, u_t, \mathcal{F}_t, q_t) : t \in T\}$ be a Bayesian game satisfying* (A.1)–(A.2). *Then there exists a Bayesian Nash equilibrium for G.*

Proof. It follows from a standard argument (see, for example, [14, p. 40] or [2] or [8, Lemma A1]) that for each $\omega \in \Omega$, $\nu_t(\omega, \cdot, \cdot) : L_{X_{-t}} \times X_t(\omega) \to \mathbb{R}$ is continuous, where L_{X_s} $(s \neq t)$ is endowed with the weak topology and $X_t(\omega)$ with the norm topology. It is easy to see that for each $(\tilde{x}_{-t}, x_t) \in L_{X_{-t}} \times X(t)$, $\nu_t(\cdot, \tilde{x}_{-t}, x_t) : \Omega \to \mathbb{R}$ is $\mathcal{F}_t$–measurable. Moreover, it follows from the concavity of u_t in x_t that so is ν_t. For each $t \in T$, define $P_t : \Omega \times L_{X_{-t}} \to 2^Y$ by

$$P_t(\omega, \tilde{x}_{-t}) = \{y_t \in X_t(\omega) : \quad \nu_t(\omega, \tilde{x}_{-t}, y_t) > \nu_t(\omega, \tilde{x}_{-t}, x_t) \text{ for all } x_t \in X_t(\omega)\}.$$

By Lemma 2.13 in [14], for each fixed $\tilde{x}_{-t} \in L_{X_{-t}}$, $P_t(\cdot, \tilde{x}_{-t})$ has a measurable graph. Since ν_t is concave in x_t, we can conclude that for each $(\omega, \tilde{x}_{-t}) \in \Omega \times L_{X_{-t}}$, $P_t(\omega, \tilde{x}_{-t})$ is convex–valued. It follows from the continuity of ν_t that for each $\omega \in \Omega$, $P_t(\omega, \cdot)$ has an open graph in $L_{X_{-t}} \times X_t(\omega)$, where L_{X_s} $(s \neq t)$ is endowed with the weak topology and $X_t(\omega)$ with the norm topology. Therefore, we can conclude that for each $(\omega, \tilde{x}_{-t})$, $P_t(\omega, \tilde{x}_{-t})$ is open in $X_t(\omega)$ and that for each $(\omega, y_t) \in \Omega \times X_t(\omega)$ the set $P_t^{-1}(\omega, y_t) = \{(\tilde{x}_{-t}) \in L_{X_{-t}} : y_t \in P_t(\omega, \tilde{x}_{-t})\}$ is open in $L_{X_{-t}}$, i.e., for each $\omega \in \Omega$, $P_t(\omega, \cdot)$ has open lower sections. The latter together with the fact that for each $\tilde{x}_{-t} \in L_{X_{-t}}$, $P_t(\cdot, \tilde{x}_t)$ has a measurable graph, enables us to conclude that $P_t(\cdot, \cdot)$ has a measurable graph. For each $t \in T$, let

$$U_t = \{(\omega, \tilde{x}_{-t}) \in \Omega \times L_{X_{-t}} : \quad P_t(\omega, \tilde{x}_{-t}) \neq \phi\}.$$

Restrict P_t to U_t and notice that it satisfies all conditions of the Caratheodory Selection Theorem in [11, Theorem 4.1]. Hence, there exists a function $f_t : U_t \to Y$ such that $f_t(\omega, \tilde{x}_{-t}) \in P_t(\omega, \tilde{x}_{-t})$ for all $(\omega, \tilde{x}_{-t}) \in U_t$. Moreover, for each $\omega \in \Omega$, $f_t(\omega, \cdot)$ is continuous on the set $U_t^\omega = \{\tilde{x}_{-t} \in L_{X_{-t}} : P_t(\omega, \tilde{x}_{-t}) \neq \phi\}$ and for each $\tilde{x}_{-t} \in L_{X_{-t}}$, $f_t(\cdot, \tilde{x}_{-t})$ is measurable on the set $U_t^{\tilde{x}_{-t}} = \{\omega \in \Omega : P_t(\omega, \tilde{x}_{-t}) \neq \phi\}$. Also, by Proposition 3.1 in [11], $f_t(\cdot, \cdot)$ is jointly measurable.

For each $t \in T$, define $F_t : \Omega \times L_{X_{-t}} \to 2^Y$ by

$$F_t(\omega, \tilde{x}_{-t}) = \begin{cases} \{f_t(\omega, \tilde{x}_{-t})\} & \text{if } (\omega, \tilde{x}_{-t}) \in U_t \\ Y & \text{otherwise.} \end{cases}$$

By Lemma 2.12 in [14], for each $\tilde{x}_{-t} \in L_{X_{-t}}$, $F_t(\cdot, \tilde{x}_{-t})$ is lower measurable. Since $F_t(\cdot, \cdot)$ is closed–valued, we can conclude that for each $\tilde{x}_{-t} \in L_{X_{-t}}$, $F_t(\cdot, \tilde{x}_{-t})$ has a measurable graph. Clearly for each $(\omega, \tilde{x}_{-t}) \in \Omega \times L_{X_{-t}}$, $F_t(\omega, \tilde{x}_{-t})$ is non-empty.

By Lemma 6.1 in [13], for each $\omega \in \Omega$, $F_t(\omega, \cdot)$ is (weakly) u.s.c.[2] For each $t \in T$ define $F_t : L_{X_{-t}} \to 2^{L_X}$ by

$$F_t(\tilde{x}_{-t}) = \{y \in L_X : \quad y_t(\omega) \in F_t(\omega, \tilde{x}_{-t}) \ \ \mu\text{–a.e.}\}.$$

Since for each $\tilde{x}_{-t} \in L_{X_{-t}}$, $F_t(\cdot, \tilde{x}_{-t})$ has a measurable graph by virtue of the Aumann measurable selection theorem, there exists an $\mathcal{F}_t$–measurable function $g_t : \Omega \to Y$ such that $g_t(\omega) \in F_t(\omega, \tilde{x}_{-t})$ μ–a.e. Since for each $(\omega, \tilde{x}_{-t}) \in \omega \times L_{X_{-t}}$, $F_t(\omega, \tilde{x}_{-t}) \subset X_t(\omega)$ and $X_t(\cdot)$ is integrably bounded, it follows that $g_t \in L_{X_t}$. Hence, $g_t \in F_t(\tilde{x}_{-t})$, i.e., F_t is non-empty–valued. By Diestel's theorem[3] L_{X_t} is a weakly compact subset of $L_1(\mu, Y)$.

Since the weak topology for a weakly compact subset of a separable Banach space is metrizable [4, p. 434], we can conclude that L_{X_t} is metrizable, and since T is countable, so is L_X. It follows from the Fatou Lemma in infinite–dimensional spaces (see, for example, [12]) that for each $t \in T$, $F_t(\cdot)$ is (weakly) upper semicontinuous, and it is obviously convex and closed–valued. Define $\phi : L_X \to 2^{L_X}$ by

$$\Phi(x) = \prod_{t \in T} F_t(\tilde{x}_{-t}).$$

Clearly, L_X is compact, convex, non-empty, and Φ is a (weakly) u.s.c. convex, closed, non-empty–valued correspondence from L_X to 2^{L_X}. By the Fan–Glicksberg fixed point theorem there exists $\tilde{x} \in L_X$ such that $x^* \in \Phi(x^*)$. One

---

[2] i.e., for each $\omega \in \Omega$, the set $\{\tilde{x}_{-t} \in L_{X_{-t}} : F_t(\omega, \tilde{x}_{-t}) \subset V\}$ is open in $L_{X_{-t}}$ (where each L_{X_s} $(s \neq t)$ is endowed with the weak topology) for every (norm) open subset V of Y.

[3] For a proof using James' theorem on weak compactness, see [12, Theorem 3.1].

can easily check that x^* is by construction a Bayesian Nash equilibrium for G. ■

Remark. One can introduce a constraint correspondence for each agent (*a là* [3] and [10]) and define an abstract Bayesian economy. The argument adopted in the above proof can be used to cover abstract Bayesian economies as well.

References

1. E. J. Balder and A. Rustichini, An equilibrium result for games with private information and infinitely many players, *J. Econ. Theory* 62 (1994), 385–393.
2. E. J. Balder and N. C. Yannelis, On the continuity of expected utility, *Economic Theory* 3 (1993), 625–643.
3. G. Debreu, A social equilibrium existence theorem, *Proc. Natl. Acad. Sci. USA* 38 (1952), 886–893.
4. N. Dunford and J. T. Schwartz, *Linear Operators, Part* I, Interscience, New York, 1958.
5. K. Fan, Fixed point and minimax theorems in locally convex topological linear spaces, *Proc. Natl. Acad. Sci. USA* 38 (1952), 131–136.
6. A. Fryszkowski, Caratheodory type selectors of set–valued maps of two variables, *Bull. Acad. Polon. Sci.* 25 (1977), 41–46.
7. I. L. Glicksberg, A further generalization of the Kakutani fixed point theorem, with applications to Nash equilibrium points, *Proc. Amer. Math. Soc.* 3 (1952), 170–174.
8. T. Kim and N. C. Yannelis, Existence of equilibrium in Bayesian games with infinitely many players, *J. Econ. Theory*, to appear.
9. J. F. Nash, Noncooperative games, *Ann. Math.* 54 (1951), 386–295.
10. W. Schafer and H. Sonnenschein, Equilibrium in abstract economies without ordered preferences, *J. Math. Econ.* 2 (1975), 345–348.
11. N. C. Yannelis, Set–valued function of two variables in economic theory, in: *Equilibrium Theory with Infinitely Many Commodities* (eds. M. A. Khan and N. C. Yannelis), Studies in Economic Theory, Springer, 1991, 36–72.
12. N. C. Yannelis, Integration of Banach–valued correspondences, in: *Equilibrium Theory with Infinitely Many Commodities* (eds. M. A. Khan and N. C. Yannelis), Studies in Economic Theory, Springer, 1991, 2–35.
13. N. C. Yannelis and N. D. Prabhakar, Existence of maximal elements and equilibria in linear topological spaces, *J. Math. Econ.* 12 (1983), 233–245.

14. N. C. Yannelis and A. Rustichini, Equilibrium points of non-cooperative random Bayesian games, in: *Positive operators, Riesz spaces and Economics* (eds. C. D. Aliprantis, K. Border, and W. A. J. Luxemburg), Springer, 1991, 23–48.

Nicholas C. Yannelis
Department of Economics
University of Illinois
Champaign, IL 61801
USA
email: nyanneli@uiuc.edu

MIX
Papier aus verantwortungsvollen Quellen
Paper from responsible sources
FSC® C105338

If you have any concerns about our products,
you can contact us on
ProductSafety@springernature.com

In case Publisher is established outside the EU,
the EU authorized representative is:
Springer Nature Customer Service Center GmbH
Europaplatz 3, 69115 Heidelberg, Germany

Printed by Libri Plureos GmbH
in Hamburg, Germany